'The authors have provided a remarkably comprehensive discussion of many topics relating to planning for climate change. No matter your background and prior knowledge, you will be rewarded with new insights from this book.'

John Ottensmann, *Professor Emeritus, Indiana University–Purdue University Indianapolis.*

'This is an excellent and timely overview of the methods and tools for assessing climate-related vulnerabilities and identifying solutions through the use of GIS. The authors describe traditional GIS functionalities as well as emerging geospatial technologies that have great potential to assist in research and operations and to inform climate adaptation policy. Educators, students, and practitioners will find this book a useful resource, as it presents a balanced combination of theory, methods, and practical GIS examples.'

Olga Wilhelmi, *National Center for Atmospheric Research.*

Geospatial Applications for Climate Adaptation Planning

Climate adaptation is a timely yet complex topic that does not fit squarely into any one disciplinary realm.

Geospatial Applications for Climate Adaptation Planning presents an overview of the range of strategies, tools, and techniques that must be used to assess myriad overlapping vulnerabilities and to formulate appropriate climate-relevant solutions at multiple scales and in varying contexts. Organized into four sections, the book includes 15 chapters. Each chapter is grounded in the literature and presents case studies designed by the authors, as well as many examples from a diverse international group of scholars and entities in the public and private sectors. Areas covered include:

- Climate Change and Climate Adaptation Planning: Context and Concepts
- Geospatial Technologies: Fundamentals and Terminology
- GIS and Climate Vulnerability Assessments
- Technical Approaches to Formulating Mitigation and Adaptation Strategies

Geospatial Applications for Climate Adaptation Planning is aimed at advanced students, researchers, and entities in the public and private sectors. It also provides supplementary reading for courses in planning, public administration, policy studies, and disaster management.

Diana Mitsova is Associate Professor in the School of Urban and Regional Planning and Director of the Visual Planning Technology Lab at Florida Atlantic University. Her research focuses on the use of geographic information systems in disaster planning, critical infrastructure protection, coastal resilience, and climate adaptation of urban systems. Her research has been funded by the National Science Foundation, USGS, the National Park Service, The Nature Conservancy, and the Florida Sea Grant. She holds a Master's of Public Affairs from the School of Public and Environmental Affairs at Indiana University–Purdue University, Indianapolis, and a Ph.D. in Regional Development Planning from the University of Cincinnati. She is also a recipient of a NATO Democratic Institutions Research Fellowship.

Ann-Margaret Esnard is a Distinguished University Professor in the Department of Public Management and Policy at Georgia State University. Her expertise encompasses urban planning, disaster planning, and hazard and vulnerability assessment. Esnard has been involved in a number of research initiatives, including NSF-funded projects on the topics of population displacement from catastrophic disasters, school recovery after disasters, long-term recovery, and community resilience. She holds degrees in Agricultural Engineering (B.Sc., University of the West Indies–Trinidad), Agronomy and Soils (M.S., University of Puerto Rico–Mayaguez), and Regional Planning (Ph.D., UMASS–Amherst). She also completed a two-year post-doc at UNC–Chapel Hill.

Geospatial Applications for Climate Adaptation Planning

Diana Mitsova and
Ann-Margaret Esnard

Routledge
Taylor & Francis Group
NEW YORK AND LONDON

First published 2019
by Routledge
711 Third Avenue, New York, NY 10017

and by Routledge
2 Park Square, Milton Park, Abingdon, Oxon, OX14 4RN

Routledge is an imprint of the Taylor & Francis Group, an informa business

Library of Congress Cataloging-in-Publication Data
A catalog record for this book has been requested

ISBN: 978-1-4987-5548-1 (hbk)
ISBN: 978-1-351-11395-3 (ebk)

Typeset in Times New Roman
by Apex CoVantage, LLC

Contents

Figures

Tables

Acknowledgments

We thank CRC Press for entrusting us with this book project, which addresses an important topic. Natalja Mortensen supported this book project and guided us through the process. We are also grateful to our funding agencies (the National Science Foundation [Grants CMMI-1541089, CMMI-1634234, and CMMI-0726808], the United States Geological Survey (USGS), the National Park Service, The Nature Conservancy, and the Florida Sea Grant) for their support over the years.

We are also deeply grateful to our academic institutions and students for their support and assistance at various stages of the book project. We would like to thank Ryan Alhawiti, Daniel Mantell, Estefania Mayorga, Connor Bailey, and Catherine Velarde-Perez, graduate research assistants in the School of Urban and Regional Planning, and Meagan Weisner, a Ph.D. student in the Department of Geosciences at Florida Atlantic University, for their untiring support with collecting data and completing research tasks related to the case studies. Three graduate research assistants at GSU—Erik Brownsword, Christopher Wyczalkowski, and Natasha Malmin—were also instrumental in helping with literature reviews, summaries of geospatial applications, and formatting of citations and references. You all helped tremendously as we developed and refined the book chapters. Our many research collaborators over the decades and years deserve special mention as well. You helped us expand our knowledge base in a wide range of topics, as well as our methodological approaches to a wide range of complex climate hazard and disaster topics.

Last, but most important, our immediate families deserve special mention given the innumerable hours that we spent working on this book project. Diana thanks her daughter Ivet Boneva for her understanding and patience over the years, as well as her family—Stanka, Nikolai, Alex, Peter, and Sophia. Ann-Margaret thanks her husband, Joseph Esnard, and sons (Joshua Esnard and Kriston Esnard) for their unwavering love, support, and patience throughout the years.

Acronyms

Chapter 1

AR5	Fifth Assessment Report
CCA	climate change adaptation
COP21	21st Conference of the Parties
FEMA	Federal Emergency Management Agency
GHG	greenhouse gas
GIS	geographic information system
GPS	global positioning system
IPCC	Intergovernmental Panel on Climate Change
NASA	National Aeronautical and Space Agency
PPGIS	public participation geographic information system
SIDS	Small Island Developing States
UNEP	United Nations Environment Programme

Chapter 2

AIP	American Institute of Physics
CARB	California Air Resources Board
CFC	chlorofluorocarbons
COP21	Conference of the Parties
EIA	Energy Information Administration
GDP	gross domestic product
GHG	greenhouse gas
INDC	intended nationally determined contribution
IPCC	Intergovernmental Panel on Climate Change
LGM	Last Glacial Maximum
LIA	Little Ice Age
LPAA	Lima to Paris Action Agenda

MIT	Massachusetts Institute of Technology
NASA	National Aeronautics and Space Administration
NAZCA	Non-State Actor Zone for Climate Action
NRC	National Research Council
NSF	National Science Foundation
SB	Senate Bill
UN	United Nations
UNCED	United Nations Conference on Environment and Development (also known as the Earth Summit)
UNEP	United Nations Environment Programme
UNFCCC	United Nations Framework Convention on Climate Change
WIM	Warsaw International Mechanism

Chapter 3

CCA	climate change adaptation
COP21	Conference of the Parties
DRR	disaster risk reduction
GEF	Global Environment Facility
ICLEI	International Council for Local Environmental Initiatives
IEA	International Energy Agency
IPCC	Intergovernmental Panel on Climate Change
LDCF	Least Developed Countries Fund
NRC	National Research Council
OBM	Office of Management and Budget
OECD	Organisation for Economic Co-operation and Development
SADC	Southern African Development Community
SCCF	Spatial Climate Change Fund
SIDS	Small Island Developing States
UN	United Nations
UN DESA	United Nations Department of Economic and Social Affairs
UNDP	United Nations Development Programme
UNEP	United Nations Environment Programme
UNFCCC	United Nations Framework Convention on Climate Change
UNISDR	United Nations Office for Disaster Risk Reduction
WWF	World Wildlife Fund

Chapter 4

CANARI	Caribbean Natural Resources Institute
CAPF	Climate Action Planning Framework
CAS	Climate Adaptation Strategy
CBA	community-based adaptation
CBDRM	community-based disaster risk management
CCA	climate change adaptation
CCI	Climate Change Initiative
CEOS	Committee on Earth Observation Satellites

CNRA	California Natural Resources Agency
CRISP	Community Resilience to Climate and Disaster Risk Project
DRR	disaster risk reduction
ESA	European Space Agency
FAO	Food and Agriculture Organization
FEWS NET	Famine Early Warning Systems Network
GDP	gross domestic product
GFCS	Global Framework for Climate Services
GFDRR	Global Facility for Disaster Reduction and Recovery
GHG	greenhouse gas
GIS	geographic information system
GLA	Greater London Authority
GMFS	Global Monitoring for Food Security
GPS	global positioning system
HFA	Hyogo Framework of Action
HSDRSS	Hurricane and Storm Damage Risk Reduction System
ICAM	Integrated Climate Adaptation Model
ICLEI	International Council for Local Environmental Initiatives
IHNC	Inner Harbor Navigation Canal
KAP	Kiribati Adaptation Program
LDCF	Least Development Country Fund
LEGGI	London Energy and Greenhouse Gas Inventory
LKCCAP	Local Knowledge and Climate Change Adaptation Project
MCA	Multi-Criteria Analysis
NCICD	National Capital Integrated Coastal Development
NGO	nongovernmental organization
P3DM	participatory 3D modeling
PACE	Property Assessed Clean Energy
PLA	participatory learning and action
RCOF	Regional Climate Outlook Forum
SIDS	Small Island Developing States
SFRCCC	Southeast Florida Regional Climate Change Compact
SSA	sub-Saharan Africa
ULI	Urban Land Institute
UNISDR	United Nations Office for Disaster Risk Reduction
USAID	United States Agency for International Development
WGClimate	Working Group Climate
WMO	World Meteorological Organization

Chapter 5

CANARI	Caribbean Natural Resources Institute
CBDRM	community-based disaster risk management
CCKP	Climate Change Knowledge Portal
CMSP	coastal and marine spatial planning
FEMA	Federal Emergency Management Agency

GIS	geographic information system
GPS	global positioning system
GRaBS	Green and Blue Space Adaptation for Urban Areas and Eco-Towns
HMS	Hazard Mapping System
IHRC	International Hurricane Research Center
LiDAR	Light Detection and Ranging
MODIS	Moderate Resolution Imaging Spectroradiometer
MRLC	Multi-Resolution Land Characteristics
MSS	Multispectral Scanner
NASA	National Aeronautics and Space Administration
NCEI	National Centers for Environmental Information
NDMC-UNL	National Drought Mitigation Center at the University of Nebraska–Lincoln
NOAA	National Oceanic and Atmospheric Administration
NWS	National Weather Service
P3DM	participatory 3D modeling
PPGIS	public participation geographic information system
SLOSH	Sea, Lake, and Overland Surge from Hurricanes
SSHS	Saffir-Simpson Hurricane Scale
SSHWS	Saffir-Simpson Hurricane Wind Scale
STAR	surface temperature and runoff
TM	Thematic Mapper
TNC	The Nature Conservancy
USGS	United States Geological Survey

Chapter 6

ABM	agent-based model
ASF	Atlanta Satellite Facility
ASTER	Advanced Spaceborne Thermal Emission and Reflection Radiometer
BAEM	Built-up Area Extraction Method
CA	cellular automata
CCAM	Climate Change Adaptation Modeler
DEM	digital elevation model
DSM	digital surface model
DSS	decision support system
DTM	digital terrain model
EBK	Empirical Bayesian Kriging
ESRI	Environmental Systems Research Institute
ETM	Enhanced Thematic Mapper
FEMA	Federal Emergency Management Agency
GHG	greenhouse gas
GIS	geographic information system
GPS	global positioning system
IDW	Inverse Distance Weighting
InSAR	interferometric synthetic aperture radar

LCM	Land Change Modeler
LEAM	Land-Use Evolution and Impact Assessment Model
LiDAR	Light Detection and Ranging
LISA	local indicators of spatial autocorrelation
LP DAAC	Land Processes Distributed Active Archive Center
LST	land surface temperature
MNDWI	Modified Normalized Difference Water Index
MODIS	Moderate Resolution Imaging Spectroradiometer
MSS	Multispectral Scanner
MWK	Moving Window Kriging
NDBI	Normalized Difference Built-up Index
NDVI	Normalized Difference Vegetation Index
NDWI	Normalized Difference Water Index
NOAA	National Oceanic and Atmospheric Administration
NOS	National Ocean Service
OLI	Operational Land Imager
PSS	planning support system
RBF	radial basis function
RS	remote sensing
SAC	spatial autocorrelation coefficient
SDSS	spatial decision support system
SLEUTH	Slope, Land use, Exclusion, Urban extent, Transportation, Hillshade
SRS	satellite remote sensing
TIN	triangulated irregular network
TIRS	Thermal Infrared Sensor
TM	Thematic Mapper
TSA	Trend Surface Analysis
UAV	unmanned aircraft vehicle
UGM	Urban Growth Model
USGS	United States Geological Survey

Chapter 7

BRIC	baseline resilience indicators for communities
CCAST	climate change adaptation strategy tool
CCFVI	Coastal City Flood Vulnerability Index
CDRI	Community Disaster Resilience Index
CEVI	Coastal Economic Vulnerability Index
DHS	Department of Homeland Security
DROP	Disaster Resilience of Place
EPA	Environmental Protection Agency
ESI	Environmental Sustainability Index
GIS	geographic information system
HVRI	Hazards Vulnerability Research Institute
ND-GAIN	University of Notre Dame Global Adaptation Index
NOAA	National Oceanic and Atmospheric Agency

OECD	Organisation for Economic Co-operation and Development
SoVI	Social Vulnerability Index
USAID	United States Agency for International Development
USGCRP	United States Global Change Research Program
WEF	World Economic Forum

Chapter 8

AIRS	Atmospheric Infrared Sounder
AMSR-E	Advanced Microwave Scanning Radiometer
AMSU-A	Advanced Microwave Sounding Unit
AOGCMs	Atmosphere-Ocean General Circulation Models
AR5	Fifth Assessment Report
ASTER	Advanced Spaceborne Thermal Emission and Reflection Radiometer
ATMS	Advanced Technology Microwave Sounder
AVHRR	Advanced Very High Resolution Radiometer
CCI SM	Climate Change Initiative Soil Moisture
CERES	Clouds and the Earth's Radiant Energy System
CMIP5	Coupled Model Intercomparison Project phase 5
CrIS	Cross-track Infrared Sounder
CTP	convective triggering potential
DSI	Drought Severity Index
ECMWF	European Centre for Medium-Range Weather Forecasts
ENSO	El Niño Southern Oscillation
EOS	Earth Observing System
ET	evapotranspiration
FEMA	Federal Emergency Management Agency
FVC	Fractional Vegetation Cover
GHRSST	Group for High-Resolution Sea Surface Temp
GIMMS	Global Inventory Monitoring and Modeling Systems
GLACE	Global Land–Atmosphere Coupling Experiment
GRACE	Gravity Recovery and Climate Experiment
HI	humidity index
HVI	Heat Vulnerability Index
ICESat	Ice, Cloud, and land Elevation Satellite
InSAR	Interferometric Synthetic Aperture Radar
IPCC	Intergovernmental Panel on Climate Change
IRS	Indian Remote Sensing
JPSS	Joint Polar Satellite System
JRA-25	Japanese 25-year Reanalysis
LiDAR	Light Detection and Ranging
LISS	Linear Imaging Self Scanning images
LST	land surface temperatures
MERRA	Modern Era Retrospective-analysis for Research and Applications
MODIS	Moderate Resolution Imaging Spectroradiometer
NASA	National Aeronautics and Space Administration

NCAR	National Center for Atmospheric Research
NCEP	National Centers for Environmental Prediction
NDVI	Normalized Difference Vegetation Index
NDWI	Normalized Difference Water Index
NESDIS	National Environmental Satellite, Data, and Information Service
NOAA	National Oceanographic and Atmospheric Administration
NPOESS	National Polar-orbiting Operational Environmental Satellite System
NPP	net primary productivity
OLI	Operational Land Imager
OMPS	Ozone Mapping Profiler Suite
ORAP5	Ocean Reanalysis Pilot 5
ORAS4	Ocean Reanalysis System 4
ORTA4	Ocean Real-Time Analysis System 4
PCMDI	Program for Climate Model Diagnosis and Intercomparison
PDO	Pacific Decadal Oscillation
RAOB	Radiosonde Observation
SAR	synthetic aperture radar
SDCI	Scaled Drought Condition Index
S-NPP	Suomi National Preparatory Project
SRS	satellite remote sensing
SST	sea surface temperatures
SUHI	surface urban heat island
TCI	Temperature Condition Index
TIRS	Thermal Infrared Sensor
TRMM	Tropical Rainfall Measuring Mission
TWSA	Total Water Storage Anomalies
USDM	United States Drought Monitor
USGS	United States Geological Survey
VCI	Vegetation Condition Index
VHI	Vegetation Health Index
VIIRS	Visible Infrared Imaging Radiometer Suite

Chapter 9

AAA	Adaptation Action Area
AHP	Analytic Hierarchy Process
ASTER	Advanced Spaceborne Thermal Emission and Reflection Radiometer
CCSP	Climate Change Science Program
CEMP	Comprehesive Emergency Management Plan
CEVI	Coastal Economic Vulnerability Index
CHHA	Coastal High Hazard Area
DTM	digital terrain model
EPA	Environmental Protection Agency
FCPA	Florida Community Planning Act
FEMA	Federal Emergency Management Agency
GDEM	Global Digital Elevation Model

GIS	geographic information system
IPCC	Intergovernmental Panel on Climate Change
LE	linear error
LiDAR	Light Detection and Ranging
OECD	Organisation for Economic Co-operation and Development
MHHW	Mean Higher High Water
MnSA	minor statistical area
MPA	Marine Protected Area
NLCD	National Land Cover Dataset
NOAA	National Oceanic and Atmospheric Administration
RMSE	root mean square error
SCAPE	Soft Cliff and Platform Erosion
SFRCCC	Southeast Florida Regional Climate Change Compact
SFRPC	South Florida Regional Planning Council
SLR	sea level rise
SRES	Special Report on Emissions Scenarios
SRTM	Shuttle Radar Topography Mission
UNDP	United Nations Development Programme
USACE	United States Army Corps of Engineers
USGS	United States Geological Survey

Chapter 10

BCPA	Broward County Property Appraiser
CCSP	Climate Change Science Program
DEM	digital elevation model
EWRI	extreme weather risk indicator
FDEM	Florida Division of Emergency Management
FOIR	Florida Office of Insurance Regulation
GEJE	Great East Japan Earthquake
GIS	geographic information system
HAT	Highest Astronomical Tide
LiDAR	Light Detection and Ranging
LOS	level of service
RMSE	root mean square error
SLOSH	Sea, Lake, and Overland Surge from Hurricanes
SLR	sea level rise
TIMM	Transit Inundation Modeling Method
TIN	Triangulated Irregular Networks

Chapter 11

ATLAS	Advanced Thermal and Land Applications Sensor
CalEPA	California EPA
CESM	Community Earth System Model
DH	degree hours
EPA	United States Environmental Protection Agency

EROS	Earth Resources Observation and Science
ESRI	Environmental Systems Research Institute
GIS	geographic information system
InVEST	Integrated Valuation of Ecosystem Services and Trade-offs
IPCC	Intergovernmental Panel on Climate Change
LEED	Leadership in Energy and Environmental Design
LSOA	Lower Super Output Area
LST	land surface temperature
LU	land use
LULC	land use/land cover
MGET	Marine Geospace Ecology Tools
MODIS	Moderate Resolution Imaging Spectroradiometer
MSA	metropolitan statistical area
NASA	National Aeronautics and Space Administration
NCEI	National Centers for Environmental Information
NDBI	Normalized Difference Built-up Index
NDVI	Normalized Difference Vegetation Index
NLCD	National Land Cover Database
NOAA	National Oceanic and Atmospheric Administration
OLI	Operational Land Imager
PRISM	Panchromatic Remote-sensing Instrument for Stereo Mapping
SFWMD	South Florida Water Management District
SRI	solar reflectance index
TIRS	Thermal Infrared Sensor
TOA	top-of-atmosphere
UHI	urban heat island
UHII	Urban Heat Island Index
USGBC	United States Green Building Council
USGS	United States Geological Survey
UTM	Universal Transverse Mercator
WGS	World Geodetic System

Chapter 12

CASPER	Community Assessment for Public Health Emergency Response
CDC	Centers for Disease Control and Prevention
ED	emergency department
GIS	geographic information system
HMS	Hazard Mapping System
HTCI	Human Thermal Comfort Index
IPCC	Intergovernmental Panel on Climate Change
LHA	local health area
NDVI	Normalized Difference Vegetative Index
OLS	ordinary least squares
PTSD	post-traumatic stress disorder
SAVI	Soil-Adjusted Vegetation Index
WHO	World Health Organization

Chapter 13

ABM	agent-based model
ACIIS	Abeokuta Critical Infrastructure Information System
API	application program interface
AR5	Fifth Assessment Report
ASCE	American Society of Civil Engineers
ASU	Arizona State University
BBC	British Broadcasting Corporation
CIMS	Critical Infrastructure Modeling System
DHS	Department of Homeland Security
DoD	Department of Defense
EEA	European Environment Agency
ESRI	Environmental Systems Research Institute
ESTDM	Event-based Spatio-Temporal Data Model
FAIT	Fast Analysis Infrastructure Tool
FEMA	Federal Emergency Management Agency
GAO	Government Accountability Office
GIS	geographic information system
HHM	Hierarchical Holographic Modeling
IEISS	Interdependent Energy Infrastructure Simulation System
INL	Idaho National Laboratory
IPCC	Intergovernmental Panel on Climate Change
ISR	inventory-to-sales ratio
JESS	Java Expert System Schell
LANL	Los Alamos National Laboratory
LBCII	Location-Based Critical Infrastructure Interdependency
NCA	National Climate Assessment
NIAC	National Infrastructure Advisory Council
NIPP	National Infrastructure Protection Plan
NISAC	National Infrastructure Simulation and Analysis Center
PPD	Presidential Policy Directive
RDBMS	relational database management system
RFRM	Risk Filtering, Ranking, and Management
RRAP	Regional Resiliency Assessment Program
SLF	spatially localized failure
SNL	Sandia National Laboratory
TISN	Trusted Information Sharing Network
TRAGIS	Transportation Routing Analysis Geographic Information System
TRANSIMS	Transportation Analysis Simulation System
UIS	Urban Infrastructure Suite

Chapter 14

ABM	agent-based model
AR4	Fourth Assessment Report
AST	Adaptation Support Tool

CA	chain analysis
CAT	BASINS Climate Assessment Tool
CPZ	community protection zone
GCM	general circulation model
GENUS	GENerator of interactive Urban blockS
GI	green infrastructure
GIS	geographic information system
IPCC	Intergovernmental Panel on Climate Change
LandCaRe 2020	Land, Climate and Resources
LCM	Land Change Modeler
LEAM	Land-Use Evolution and Impact Assessment Model
LSM	Land Use Scanner Model
MCE	multi-criteria evaluation
NDVI	Normalized Difference Vegetation Index
NEDUM	Non-Equilibrium Dynamical Urban Model
PSS	planning support system
SELES	Spatially Explicit Landscape Event Simulator
SLEUTH	Slope, Land use, Exclusion, Urban extent, Transportation, Hillshade
SRES	Special Report on Emissions Scenarios
SWMM	Storm Water Management Model
TAR	Third Assessment Report
TM	Thematic Mapper
UGM	Urban Growth Model
UNEP	United Nations Environment Programme
WHAMED	Wildfire Hazard Mitigation and Exurban Development
WUI	wildland-urban interface

Chapter 15

AR5	Fifth Assessment Report
CAIT	Climate Analysis Indicators Tool
CCAM	Climate Change Adaptation Modeler
CDS	Climate Data Services
CIESIN	Center for International Earth Science Information Network
CMIP5	Coupled Model Intercomparison Project Phase 5
CPC	Climate Prediction Center
DST	Decision Support Tool
EBM	Ecosystem-Based Management
ESRI	Environmental Systems Research Institute
GHG	greenhouse gas
GIS	geographic information system
GraBS	Green and Blue Space Adaptation for Urban Areas and Eco-Towns
GRUMP	Global Rural-Urban Mapping Project
ICLEI	International Council for Local Environmental Initiatives

InVEST	Integrated Valuation of Ecosystem Services and Trade-offs
IPCC	Intergovernmental Panel on Climate Change
LEED	Leadership in Energy and Environmental Design
LiDAR	Light Detection and Ranging
LST	land surface temperature
NASA	National Aeronautics and Space Administration
NCAR	National Center for Atmospheric Research
NCCV	National Climate Change Viewer
NCEI	National Centers for Environmental Information
NEX	NASA Earth Exchange
NEXDCP	NASA Earth Exchange Downscaled Climate Projections
NEX-GDDP	NASA Earth Exchange Global Daily Downscaled Projections
NOAA	National Oceanic and Atmospheric Administration
NWS	National Weather Service
OCM	Office of Coastal Management
RCP	Representative Concentration Pathway
SEDAC	Socioeconomic Data and Applications Center
SLAMM	Sea Level Affecting Marshes Model
SLR	sea level rise
SoVI	Social Vulnerability Index
SPARC	Scholarly Publishing and Academic Resources Coalition
TNC	The Nature Conservancy
TPL	Trust for Public Land
UNISDR	United Nations International Strategy for Disaster Risk Reduction
USGS	United States Geological Survey
WCT	Weather and Climate Toolkit
WGS	World Geodetic System
WRI	World Resources Institute

1 Introduction

Climate adaptation is a timely yet complex topic. It does not fit squarely into any one disciplinary realm, especially given the range of strategies, tools, and techniques that must be used to further our understanding of weather-related hazards and the impacts on people and places. Assessing vulnerability and risk, developing strategies to mitigate the potential impacts of natural hazards, and related policy and governance frameworks are examples that demonstrate the complex, interdisciplinary nature of the challenges that planning practitioners and policy-makers commonly face. By examining specific themes and problems, this book provides users with an understanding of the meaningful cross-cutting themes and connections and how the combination of geospatial tools and techniques can be used to plan for mitigation of greenhouse gas emissions and adaptation to climate impacts. However, a discussion of tools and techniques devoid of context and concepts is problematic. As such, the book sets out to provide relevant context and concepts for an intended broad audience of scholars, practitioners, and students.

Global Warming and Climate Change

Global warming is the term used to describe an increase in average global temperatures (National Aeronautical and Space Agency [NASA], 2016). The term is also commonly used to indicate a relationship between the rise in global temperatures and the release of excessive amounts of carbon dioxide, methane, and other greenhouse gases. The temperature increase documented over the past century is largely attributed to the buildup of carbon dioxide from anthropogenic emissions and land clearing and is projected to continue affecting the global climate (Figure 1.1). Scientists prefer the term *climate change* as a more accurate representation of the multiple processes associated with the increases in global temperatures.

As a scientific concept, climate change was first introduced in 1975 in a paper by the U.S. geologist Wallace Broecker and later adopted by the National Academy of

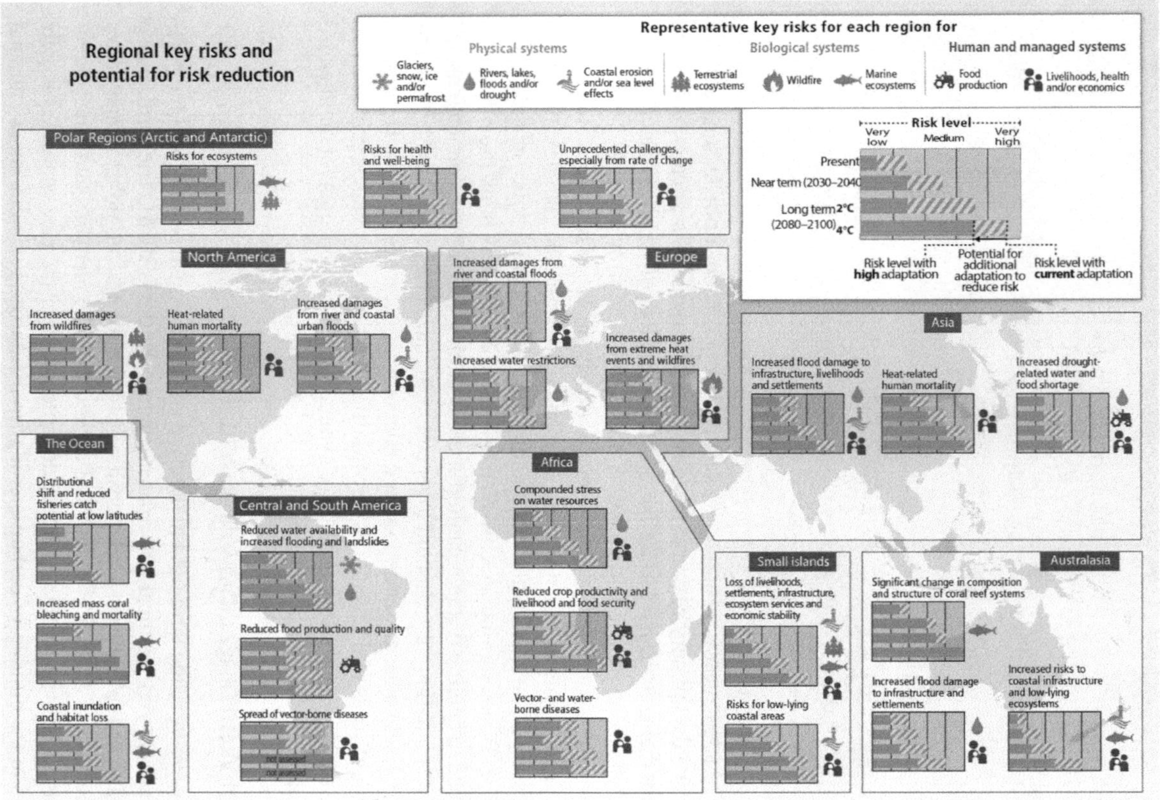

Figure 1.1 Risks by region

Source: Intergovernmental Panel on Climate Change (IPCC, 2014), Figure 2.4

Sciences in the first major study on carbon dioxide and climate published in 1979 (NASA, 2016). Established in 1989, the United States Global Change Research Program adopted the term *global change*, with climate research being one of its research pillars. *Global change* and *global environmental change* are other terms used in the scientific literature, but lesser known to the general public.

Overcoming the challenges posed by climate change will require understanding and action on a personal level and engagement of multiple sectors and governments across the globe. Successful mitigation of greenhouse gas emissions and effective climate adaptation planning depend on a multitude of actions and initiatives passed at global, national, regional, state, and local levels. In December 2015, 195 nations, Parties to the Paris Agreement, pledged their support for measures that can cap greenhouse gas (GHG) emissions and, more importantly, set the agenda for large-scale strategic planning in climate mitigation and adaptation. The Paris Climate Conference is officially known as the 21st Conference of the Parties (COP21). The proposed strategies garnered the attention of world leaders, policy-makers at the highest level of authority, regional and city governments, and businesses and citizens alike. Kim, Smith, Mack, Cook, Furlow, and Njinga Cote (2016) describe a paradigm shift in national climate adaptation planning from "short-term, project-level interventions" into strategic considerations related to longer-term development goals.

In addition to actions at the global, national, regional, and local scale, climate adaptation is contingent upon myriad actions taken by individuals in their daily lives, including mode of transportation choices, home improvements, relocation decisions, and personal involvement in local climate initiatives. The ability to visualize the spatial distribution of hazards, and analyze interrelatedness between physical, social, economic, and environmental factors, is also fundamental to planners, policy-makers, and other professionals who work on climate mitigation and adaptation issues.

Climate Change Mitigation and Adaptation

Climate change commitments fall into two fundamental response strategies: climate mitigation and adaptation. *Climate mitigation* is defined in the Intergovernmental Panel on Climate Change (IPCC) Fourth Assessment Report as "anthropogenic intervention to reduce the sources or enhance the sinks of greenhouse gases" (IPCC, 2007). *Climate mitigation* refers to actions that reduce the human contribution to the planetary greenhouse effect (Melillo, Richmond & Yohe, 2014). According to the United Nations Environment Programme (UNEP), mitigation actions can range from complex and high-tech projects (such as subway systems) to other simpler initiatives such as walkways (UNEP, 2017).

According to Climate-Eval Community of Practice (2015, p. 3), *climate change adaptation* (CCA) focuses on anticipating the risks and adverse impacts of a changing climate, taking appropriate action to prevent or minimize the damage, and seizing on potential opportunities that may arise. Climate adaptation also promotes the ability to cope with short-term climate impacts from extreme events and strategies for long-term response, including minimizing the threat of potential population displacement.

Climate change adaptation pathways may include fashioning programs, projects, and policies that try to minimize the effects of climate change. These pathways and interventions can take the form of disaster risk reduction and promotion of community resilience and capacities to respond to, cope with, and prepare communities for climate variability (Climate-Eval Community of Practice, 2015, p. 3). As such, it is important that this book provides in-depth coverage of rapid-onset and slow-onset natural hazards. Rapid-onset hazards develop with little warning and strike rapidly (e.g., volcanic eruptions, flash floods, landslides, severe thunderstorms, lightning, and wildfires) while slow-onset hazards take years to develop (e.g., droughts and sea level rise).

Climate change is intensifying the impacts of natural disasters given forecasts for more frequent hot days and nights over most land areas, more heat waves and droughts, heavier precipitation and rainfall events, and increases in intense tropical cyclones. Global warming is also expected to have an effect on cities along the banks of rivers. River flooding risk is projected to increase due to a higher prevalence and intensity of storms, and the economic value of urban infrastructure is expected to increase over time (Hirabayashi et al., 2013; Winsemius et al., 2016). Hirabayashi et al. (2013) reported notable increases in flood frequency in **Southeast Asia**, **Peninsular India**, **eastern Africa**, and the **northern half of the Andes**. From 1971 to 2010 ocean temperatures increased by 0.11°C per decade (IPCC, 2014), and the mean sea level rose by 0.19 meter from 1901 to 2010 (IPCC, 2014). In addition, the acidity of oceans (measured by pH) increased by 26% during the same period (IPCC, 2014). Coastal zones are particularly susceptible to erosion, accretion of sediments, saltwater intrusion, and sea level rise. The increased frequency and severity of storms is expected to exacerbate these processes (Masria, Iskander & Negm, 2015).

The Role of Spatial Planning and Geospatial Technologies

Climate mitigation and adaptation planning is as much policy as it is politics and science. Its success greatly depends on adequate data support. Planning for climate adaptation is unlike any other planning process given the long-term horizons, projections, uncertainty, and coupling of natural systems with human interactions. Activities, events, and phenomena all occur in a geographic context; therefore, all data are explicitly or implicitly spatial data. Spatial data are embodied in digital files, web portals, locational services, spatial data infrastructures, and real-time data capture and management. Spatial data play a significant role in storing geographic knowledge and analyzing trends and patterns of human activities. Spatial data applications range from a simple display of geographic features on a map to complex multilayered/multidimensional analyses of health, crime, hazards, real estate, transportation, business operations, social media, and more. Furthermore, many decisions are driven by their geographical contexts, with climate adaptation as a prime example.

Successful implementation of climate mitigation and adaptation policies will depend upon a robust understanding of spatially explicit information, rigorous vulnerability analysis, competent scenario planning, and visioning of alternatives and solutions. Virtually any adaptation decision begins with available scientific data, identification of knowledge gaps, and a quest for data resources sharing and capacity building. Geospatial technologies that combine the tools of a geographic information system (GIS), remote sensing, and global positioning systems (GPS) provide some of the widely available tools to build a strategic knowledge that reflects the latest advances in the scientific understanding of climate and its impacts and has the analytical leverage to inform policy decisions. Geospatial technologies are well suited to vulnerability assessment and, at the most fundamental level, to urban systems spatial modeling.

Increasingly, integrated spatial assessments of physical and social vulnerability to climate-related hazards use GIS as a visualization and analytical platform to link meteorological information with other environmental and socioeconomic data (Wilhelmi & Morss, 2013). Demographic data are an essential source of information used by planners and policy-makers to assess social vulnerability. Socioeconomic variables (e.g., population density, housing and homeownership, income, educational attainment, gender, age, and race) are often overlaid with other datasets for all phases of the disaster management cycle. Assessing vulnerability and risk, developing strategies to mitigate the potential impacts of natural hazards, and formulating policy and governance frameworks are examples that demonstrate the complex, interdisciplinary nature of the challenges that planning practitioners and policy-makers commonly face.

Engaging the "Whole Community"

Adaptation at the community level is largely dependent on buy-in and adoption of strategies and tools. Climate adaptation by nature connotes a proactive approach and inclusion of a broad range of stakeholders. Successful climate adaptation, therefore, requires the engagement of what the United States Federal Emergency Management Agency (FEMA) describes as the "whole community" approach. In the context of hazard management, a broad understanding of community risks, needs, resources, and capabilities must be captured. That approach "attempts to engage the full capacity of the private and nonprofit sectors including businesses, faith-based and disability organizations, and the general public, in conjunction with the participation of local, tribal, state, territorial, and Federal governmental partners" (FEMA, 2011).

Who should be involved in climate adaptation practice and research from various sectors remains a work in progress, but we can borrow from scholars such as Buanes, Jentoft, Maurstad, Soreng, and Karlsen (2005), Schwab (2010), FEMA (2011), and Smith (2012) with regard to some non-traditional stakeholders that need to be engaged in climate adaptation planning. These stakeholders include advocacy groups, airports, colleges, universities, chambers of commerce, financial institutions, insurance, media, public transportation systems, and utility providers. Several factors need to be considered, including engagement of community residents and planning and policy professionals for tasks such as vulnerability and assets assessment and scenario-building. As noted by Esnard (2012, p. 306), planners have increasingly filled their visualization toolkit to:

> **(8)** enhance and inform multiple planner roles and tasks (e.g., plan maker, regulator, mediator, advocate, technical analyst); (ii) facilitate participatory and communicative planning processes; (iii) integrate and interpret data from disparate sources and in a variety of formats; and (iv) facilitate iterative processes in the analysis and recursive exploration of data.

Beyond planners, a broader range of private stakeholders and local interests must be involved, although participating groups are sure to vary by country and across economic sectors. In **Norway**, for example, the list of stakeholder participation in coastal zone planning is categorized as: (i) private stakeholders including fishermen and tourist operators; (ii) public stakeholders including government officials; and (iii) local interests such as recreation organizations, cottage owners, and developers and landowners (Buanes, Jentoft, Maurstad, Soreng & Karlsen, 2005).

Overview of Chapters

By examining specific themes and problems, this book provides users with an understanding of the meaningful thematic and conceptual linkages across disciplines and how the combination of geospatial tools and techniques can be used to mitigate climate impacts or inform climate adaptation strategies. In addition to the introduction (Chapter 1), the remaining 14 chapters are organized into four sections. Each chapter is grounded in the literature and uses examples and case studies from a diverse international group of scholars and entities in the public and private sectors. The focus is on reinforcement of geospatial applications and tools that can be used to explore problems and design climate-relevant solutions at multiple scales.

Part I*, Climate Change and Climate Adaptation Planning: Context and Concepts**, includes Chapters 2, 3, and 4. ***Chapter 2 briefly discusses how climate change issues have gained political saliency and highlights some of the efforts of the international community toward legally binding commitments to reduce greenhouse gases. The chapter also features global, regional, and local initiatives and commitments toward "low-carbon cities." ***Chapter 3*** examines various conceptualizations of climate adaptation and places climate action in the context of the related concepts of risk management, hazard mitigation planning, risk reduction, resilience, and vulnerability. ***Chapter 4*** provides an overview of climate adaptation initiatives and coping strategies from across the globe. It examines how major cities have addressed both greenhouse gas emission reductions and climate adaptation challenges, what role participatory and consensus-building approaches to climate adaptation play in the Small Island Developing States (SIDS), and how smallholder farming communities in the Global South are adapting to changing environmental conditions. The chapter highlights the linkages between climate adaptation, food security, and development and concludes with a brief discussion of how climate services and geospatial technologies are facilitating climate adaptation planning and decision-making.

Part II*, Geospatial Technologies: Fundamentals and Terminology**, includes Chapters 5, 6, 7, and 8. ***Chapter 5 provides a summary of weather-related hazards and the wide array of datasets that facilitate vulnerability assessments instrumental to disaster management, adaptation planning, and increasingly population relocation planning. The chapter also discusses participatory mapping and public participation GIS (PPGIS), which allow citizens to merge indigenous knowledge with available science and technology. ***Chapter 6*** provides a brief overview of spatial analytical tools and techniques including spatial queries, geoprocessing, surface analysis, spatial statistics, and spatial interpolation methods as they apply to vulnerability assessments and climate adaptation planning. Additionally, it highlights several applications of remote sensing, geocomputation, decision support systems, and web-based GIS. ***Chapter 7*** begins with defining indicators and indices as context for understanding numerous country-specific and global index-building initiatives. Despite the growing utility and applicability of indicators and indices, there are a variety of methodological challenges and limitations. Such challenges are discussed in the final section of the chapter. ***Chapter 8*** examines the role of satellite remote sensing and GIS in the assessment of global environmental change including changes in global average air and ocean temperatures; dynamics of ice sheets, glaciers, and ice caps; and rising global average sea level. It also reviews the applicatons of satellite remote sensing and GIS in monitoring and evaluating ecological responses to climate change and the effects of rising temperatures on urban microclimate.

Part III, **GIS and Climate Vulnerability Assessments**, includes Chapters 9, 10, 11, and 12. In each of these chapters, case studies designed by the authors are presented to highlight the use of geospatial technologies to assess vulnerabilities and risks of the phenomena being studied. The chapters also discuss data and methodological issues that limit such assessments, which are collectively used to inform adaptation and risk reduction strategies and policy-making. ***Chapter 9*** summarizes a broad range of data collection efforts, and spatial analyses used by various scholars, to assess the impact of sea level rise on the built environment, environmentally sensitive areas, and economic assets. ***Chapter 10*** highlights infrastructure interdependencies by presenting examples from various cities and places across the globe. A section is dedicated to summarizing various combinations of geospatial technologies, data, and models used to assess exposure, vulnerability, and risk of critical infrastructure to climate hazards. Numerous examples are used throughout ***Chapter 11*** to showcase the combination of datasets, geospatial tools, and meteorological models that are used to quantify, assess, and visualize the impact of urban development on urban heat islands, as well as to assess pre-disposition to health and environmental vulnerabilities for socially vulnerable populations, neighborhoods, cities, and regions around the globe. ***Chapter 12*** provides an overview of several categories of health impacts (e.g., heat stress, respiratory illness, injury, food and waterborne diseases, vector-borne and zoonotic diseases, and mental health sequelae) from weather changes. Also covered in this chapter are multiple examples of the use of geospatial technologies for analysis of health events and risk factors, crisis mapping, and modeling of temporal and spatial patterns of health hazards and risks at various scales: nations and large regions, cities, and neighborhoods.

Part IV, **Technical Approaches to Formulating Mitigation and Adaptation Strategies**, includes Chapters 13, 14, and 15. ***Chapter 13*** builds on Chapter 10 by covering more advanced spatial analysis methodologies, analyses, and tools that are used to characterize and model infrastructure interdependencies, capture the complex behavior of interdependent infrastructures, and improve knowledge of the location, timing, and spatial extent of service disruptions. The chapter covers approaches for modeling spatial interdependencies through object-relational databases, the potential utility of various econometric, and agent-based models of infrastructure interdependencies. ***Chapter 14*** provides a brief overview of the evolution of the urban systems spatial modeling approaches and their various applications. The emphasis is on cellular automata and agent-based models and their integration with GIS. These models have demonstrated immense potential in forecasting spatial patterns of development, conducting long-term environmental assessments, and evaluating alternatives that can minimize adverse and unanticipated impacts. Research initiatives have also indicated that coupled urban growth–climate modeling systems can inform policy scenarios and urban growth strategies to promote long-term climate adaptation strategies. The approach adopted in ***Chapter 15*** is twofold. First, in order to be consistent with the organization of the book, we categorize the web-based applications according to their topical area (e.g., climate data portals, sea level rise visualization, flood analysis and mapping, heat effects mapping, climate change vulnerability and sensitivity mapping, ecosystem-based approaches, and GHG emissions estimation). Next, we identify tools that offer different capabilities in each topical area. In doing so, we categorize the available tools in each topical area as viewers, digital data portals, software platforms with modeling capabilities, and software with decision support tools.

Overall, this book was written with a multidisciplinary audience in mind, and it covers a vast range of topics, concepts, methods, and applications and uses examples and case studies from across the globe to demonstrate the value of geospatial and related technologies for climate adaptation planning.

References

Buanes, A., Jentoft, A., Maurstad, A., Soreng, S.U., & Karlsen, G.R. 2005. In Whose Interest? An Exploratory Analysis of Stakeholders in Norwegian Coastal Zone Planning. *Ocean and Coastal Management* 48, 658–669.

Climate-Eval Community of Practice. 2015. *Good Practice Study on Principles for Indicator Development, Selection, and Use in Climate Change Adaptation Monitoring and Evaluation, Climate Eval.* Retrieved from www.climate-eval.org/study/good-practice-study-principles-indicator-development-selection-and-use-climate-change.

Esnard, A-M. 2012. Visualizing Information. In: Weber, R., & Crane, R. (Eds.), *Handbook of Urban Planning*, Chapter 16. New York: Oxford University Press.

Federal Emergency Management Agency (FEMA). 2011. *A Whole Community Approach to Emergency Management: Principles, Themes, and Pathways for Action.* FEMA. Retrieved from www.fema.gov/media-library/assets/documents/23781.

Hirabayashi, Y., Mahendran, R., Koirala, S., Konoshima, L., Yamazaki, D., Watanabe, S., Kim, H., & Kanae, S. 2013. Global Flood Risk under Climate Change. *Nature Climate Change* 3(9), 816–821.

Intergovernmental Panel on Climate Change. 2007. *AR4 WGII: Chapter 18, Section 18.1.2—Differences, Similarities and Complementarities Between Adaptation and Mitigation*. Retrieved from www.ipcc.ch/publications_and_data/ar4/wg2/en/ch18s18-1-2.html (accessed 09/2017).

Intergovernmental Panel on Climate Change (IPCC). 2014. *Climate Change 2014: Synthesis Report. Contribution of Working Groups I, II and III to the Fifth Assessment Report of the Intergovernmental Panel on Climate Change*. Core Writing Team: R.K. Pachauri and L.A. Meyer (Eds.). Geneva, Switzerland: IPCC, 151 pp.

Kim, Y., Smith, J.B., Mack, C., Cook, J., Furlow, J., Njinga, J-L., & Cote, M. 2016. A Perspective on Climate-Resilient Development and National Adaptation Planning Based on USAID's Experience. *Climate and Development*. DOI: 10.1080/17565529.2015.1124037

Masria, A., Iskander, M., & Negm, A. 2015. Coastal Protection Measures, Case Study Mediterranean Zone, Egypt. *Journal of Coastal Conservation* 19(3), 281–294.

Melillo, J.M., Richmond, T., & Yohe, G.W. (Eds.) 2014. *Climate Change Impacts in the United States: The Third National Climate Assessment*. Washington, DC: U.S. Global Change Research Program.

National Aeronautics and Space Administration (NASA). 2016. *Climate Change: How Do We Know? Global Climate Change: Vital Signs of the Planet*. Retrieved from climate.nasa.gov/evidence.

Schwab, J. 2010. *Hazard Mitigation: Integrating Best Practices into Planning*. PAS Report 560. Chicago, IL: American Planning Association.

Smith, G. 2012. *Planning for Post-Disaster Recovery: A Review of the United States*. Washington, DC: Island Press.

United Nations Environment Programme. (2017). *Climate Change: Mitigation*. Retrieved from http://web.unep.org/climatechange/mitigation (accessed June 2017).

Wilhelmi, O.V., & Morss, R.E. 2013. Integrated Analysis of Societal Vulnerability in an Extreme Precipitation Event: A Fort Collins Case Study. *Environmental Science and Policy* 26, 49–62.

Winsemius, H. C., Aerts, J. C., van Beek, L. P., Bierkens, M. F., Bouwman, A., Jongman, B., Kwadiijk, J. C., Ligtvoet, W., Lucas, P. L., van Vuuren, D. P., & Ward, P. J. 2016. Global Drivers of Future River Flood Risk. *Nature Climate Change* 6(4), 381–385.

PART 1

Climate Change and Climate Adaptation Planning

Context and Concepts

2 Climate Change

Historical Context and Global Initiatives

Chapter Objectives

This chapter examines how climate change has transcended from science to policy and what initiatives have been undertaken at a global and regional scale to counterbalance its most immediate impacts. The chapter seeks to answer the following questions:

- How did we come to realize that the planet was warming?
- How is the global community responding to the challenges of climate change?

Why Is the Climate Changing?

In 2012, a group of graduate students and researchers from the Massachusetts Institute of Technology's (MIT) Department of Earth, Atmospheric, and Planetary Sciences traveled to the Lehman Caves in Great Basin National Park in Nevada to search for answers that could explain the causes of a great ecological shift that occurred in the region several thousand years earlier. The study, funded by the National Science Foundation (NSF), provided evidence that vast areas of Nevada, Utah, eastern California, and southern Oregon were not as dry and barren as we know them today (Chu, 2015). Nearly 8,200 years ago, these areas were covered with swampy grasslands interspersed with seasonally fluctuating lakes, springs, and pockets of lush vegetation (Steponaitis et al., 2015). The dry lake beds scattered throughout the deserts of the American Southwest at present are vestiges of these ancient times when the climate was wetter and moisture levels were high. Analyzing the chemical signatures of water and magnesium (Mg) in the stalagmites[1] recovered from the Lehman Caves, the researchers were able to reconstruct and depict the past climate of the American West (Steponaitis et al., 2015). The findings, published in *Quaternary Science Reviews*, indicated that the humid conditions persisted for approximately 4,000 years after the retreat of the massive ice sheets formed during the Last Glacial Maximum (LGM) (i.e., the last Ice Age), which peaked 21,500 years ago. Around 6,000 BP, the transition toward aridity and barren desert conditions began to unfold. According to David McGee, who led the expedition in the Great Basin, the area started drying out when the remnants of the last ice sheet in what is nowadays Canada collapsed, altering the jet stream supply of vast amounts of moisture to the south (McGee, personal communication, as cited in Chu, 2015).

On the other side of the globe, not far off the same latitudes, similar changes were taking place. Evidence from the fluvial deposits found in the dry lake beds of the West African Sahel, fossilized pollen, traces of fluctuating lake levels, and the history of dune formation all indicated that the semi-arid region was once a grassy floodplain where streams, lakes, and lowlands were chock-full of abundant life (Ruddiman, 2005). Then, around 8,000 years ago, the torrential monsoon rains that once drenched the swampy grasslands, replenishing the lakes and rivers of the Sahel, started to die down. From a geologist's perspective, these changes occurred quite precipitously, only over the span of a thousand years or so. The climatic conditions stabilized around 7400–7000 BP and remained stable with only short-term variations ever since (National Research Council [NRC], 1983). At present, the sub-Saharan Sahel is a semi-arid region, located between the Senegal and Nile rivers, mostly known for its extremely dry conditions and a persistent record of cataclysmic droughts and famines (NRC, 1983).

Inevitably, we have to ask: What precipitated these changes? The dominant hypothesis is that, when the grasses were greener, mile-thick ice sheets altered the atmospheric circulation so that moist air masses extended further south to reach areas near 30 degrees latitude[2] (Steponaitis et al., 2015; Ruddiman, 2005). While this hypothesis still holds for the American Southwest and other regions, it does not fully explain what led to the desertification of sub-Saharan Africa (Kutzbach, 1981; Ruddiman, 2005). In 1997, Kutzbach and Liu (also Kutzbach, 1981) offered an alternative explanation looking for answers not beneath the ground but above it. Several decades earlier, a Serbian astrophysicist by the name of Milutin Milankovitch had mathematically proven that the amount of solar radiation received by the earth can influence the growth and retreat of ice sheets. The basic premise of his theory was that cyclical changes in the earth's orbit, tilt, and rotation (known as *eccentricity*, *obliquity*, and *precession*)[3] affect the amount of insolation that the earth receives. Physical models of the earth's atmospheric processes have already established that, at present, the summer radiation is at its minimum, that is, at least 8% lower than what it was 8,000 years ago, due to a tidbit of a *precession* wobble. As the link between the strength of the monsoon season and the amount of solar radiation was well established (Ruddiman, 2005), Kutzbach and Liu (1997) suggested that a stronger season would increase the rainfall amounts to levels that could explain the presence of wetter conditions in sub-Saharan Africa at the time when the changes occurred.

Roughly when the American Southwest started withering, other notable developments took place. Some annual plants including wild cereals thrived in the mild climatic conditions after the Ice Age (National Geographic, 2017). Probably first in the Fertile Crescent[4] in the Near East, and then along the Indus, Yangtze, and Yellow rivers in East Asia, "our evolutionary cousins" (NRC, 2010), *Homo sapiens*, who bore striking resemblance to us, began parting ways with hunting-gathering in favor of more sedentary lifestyles that involved cultivating plants and domesticating wild animals (Potts, 2012; Zhuang & Kidder, 2014). Some scientists believe that this was when humans first began altering the climate by cutting down trees, burning forests, clearing land for cultivating rice and other crops, and expanding animal husbandry. Slowly but surely, the new inventions of the Neolithic man started to change the composition of the trace gases[5] in the atmosphere by adding more methane and CO_2 (Ruddiman, 2005).[6] The amounts were minuscule and hardly noticeable, but who knew that one day this same man, for all his ingenuity, would come to rival the forces of nature in setting off climatic change (Ruddiman, 2005)?

Historical Milestones: Climate Change Issues to the Forefront

For many, predicting climate is very much like predicting the weather. We get it wrong half of the time.[7] For this and other reasons that are beyond the scope of this book, many still believe that climate change is in the eye of the beholder and "seeing is believing." Indeed, the magnitude of change and the timescales at which it occurs are inconceivable and, at times, simply mind-boggling. If we "were to turn the geological clock" (Revkin, 1992, rev. 2012, p. 7) some 8,000 years forward starting from the time when the deserts set in and agriculture was invented, we would descend into the modern world. Around 1,000 AD, many European cities flourished during a relatively warm period known as the *Medieval Optimum* (Ladurie, 1971; Mann, 2003). This was the time when the Vikings set out to colonize Greenland and the Brits reveled in the sightings of dozens of vineyards (Revkin, 1992). Then, 300 years later, for reasons not yet fully understood, frosty weather set in. The *Little Ice Age* (LIA), as it became known, resulted in worldwide glacial expansion with a climatic minimum in the early 1600s (Mann, 2003), just around the time when the pilgrims disembarked in the New World. All proxy evidence found in fossils, sediments, glacial ice, tree rings, corals, and cave formations, as well as direct observations preserved in historical and instrumental records, pointed to the fact that it was cold and getting colder. Then, something changed.

The latest wriggle of Earth's climatic history began a little over a century ago (Crutzen, 2002; Ruddiman, 2005). By the early decades of the twentieth century, there was an abundance of written and anecdotal evidence of warmer and shorter winter seasons (Weart, 2008). The frosty, cold winters depicted in the works of Charles Dickens, John Constable, and Hans Christian Andersen had long disappeared (Lamb, 1995; Weart, 2008; Brown, 2010; Brooke, 2014). The frost fairs on the ice-covered Thames—a commonplace in the early 19th century—were a distant memory (Brown, 2010). A British engineer and amateur meteorologist by the name of Guy S. Calendar took an interest in this anecdotal evidence and began collecting historical data on CO_2 concentrations in the atmosphere (Fleming, 1998; Weart, 2008; Hawkins & Jones, 2013). Calendar claimed that there was a connection between the disappearance of the cold, icy winters and the release of CO_2 during the industrial revolution (Weart, 2008; Hawkins & Jones, 2013). In 1938, he testified before a scientific panel of the Royal Society asserting that CO_2 was partially to blame for this climatic shift (Weart, 2008). Calendar was not the first to put these ideas forward. Similar to others before him, he was unsuccessful in proving his hypothesis. The CO_2 theory of climate change was dismissed once again on the grounds of faulty computations. Interest in the topic, however, did not fade away and continued to build momentum.

An amateur field in the 19th and early 20th centuries, climate research is presently an incredibly complex field (Ruddiman, 2008; Kump, Kasting & Crane, 2010). Climate researchers study geological, geochemical, and biotic data to chronicle the geologic past over millions of years and to build robust physical, atmospheric, and geochemical climate models that can reliably predict the planet's future (Rohli & Vega, 2008; Ruddiman, 2008; Stanley, 2009; Kump, Kasting & Crane, 2010). In their quest for discoveries, climate scientists seek to explain, among other things, how the earth has remained "habitable" over millions and even billions of years (Ruddiman, 2008, p. 41). They investigate the underlying causes of the earth's metamorphosis from a "greenhouse" to an "icehouse" (Ruddiman, 2008, p. 43). They also study the relationship among climate, tectonic plate movement, and orbital cycles. Changes in the composition and chemistry of the earth's atmosphere are of particular interest for the climate researcher who seeks to explain the short-term variation in temperature and precipitation patterns, weather extremes, and other meteorological phenomena (Rohli & Vega, 2008).

The idea that the atmosphere acts as a natural "thermostat" (Ruddiman, 2008, p. 43) was first put forward by the French mathematician and naturalist Jean Baptist Joseph Fourier in the early 19th century (Weart, 2008).[8] Fourier developed a theoretical model to explain why our planet is not just a bare piece of icy rock in the middle of the solar system but a place full of life (Weart, 2008). Fourier advanced the idea that the earth's atmosphere traps heat similar to a glass box (Weart, 2008). This process became known as the *greenhouse effect* (Figure 2.1 and *Text Box 2.1*). In the 1860s, John Tyndall, a professor of physics with the British Royal Society, conducted laboratory experiments to measure the absorptive heat potential of various atmospheric gases. He was particularly interested in the molecular physics of infrared radiation (Lamb, 1995; Fleming, 1998). Tyndall's experiments proved that certain gases in the atmosphere, including water vapor, carbon dioxide (CO_2), and methane absorb radiant heat and that their concentrations affect the average global temperatures (Lamb, 1995; Weart, 2008; Brooke, 2014).

Atmospheric CO_2 science was further advanced by the work of the Swedish natural philosopher Svante Arrhenius and two Americans—T. C. Chamberlain and E. O. Hulburt—who independently constructed mathematical models to prove that an increase in the level of carbon dioxide in the atmosphere can raise global surface temperatures due to the greenhouse effect (Weart, 2008; Brooke, 2014). In 1931, Hulburt predicted quite accurately that a two-fold increase in CO_2 would raise the average global temperature by 4°C (Weart, 2008). At the time, the CO_2 theory of global warming was not mainstream science, and most of the claims put forward by its pioneers were dismissed (Weart, 2008).

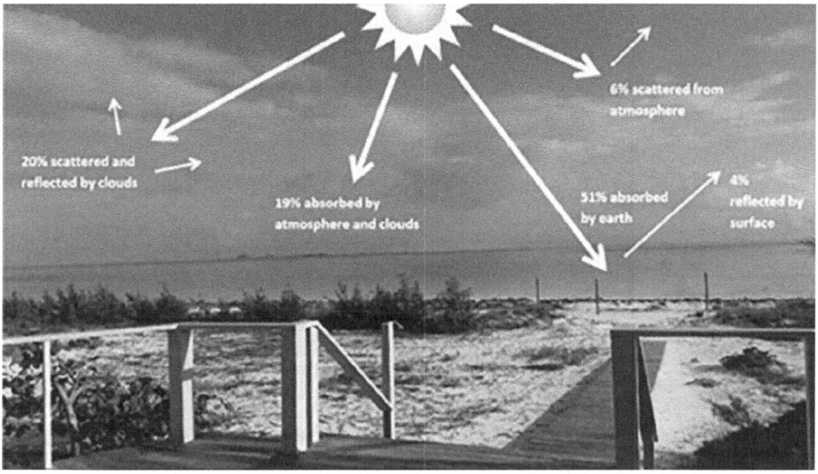

Figure 2.1 The earth receives solar radiation from the sun, one-third of which is reflected back to space; some of the thermal energy absorbed by the land and oceans is re-radiated back into the atmosphere, where it is trapped by greenhouse gases such as CO_2, methane, nitrous oxides, and water vapor

Source: Adapted from www.ucar.edu/learn/1_3_1.htm, photo credit and graphics Daniel Mantell

Text Box 2.1

Climate Change 2007: Working Group I: The Physical Science Basis

What is the greenhouse effect?

The Sun powers Earth's climate, radiating energy at very short wavelengths, predominately in the visible or near-visible (e.g., ultraviolet) part of the spectrum. Roughly one-third of the solar energy that reaches the top of Earth's atmosphere is reflected directly back to space. The remaining two-thirds is absorbed by the surface and, to a lesser extent, by the atmosphere. To balance the absorbed incoming energy, Earth must, on average, radiate the same amount of energy back to space. Because Earth is much colder than the Sun, it radiates at much longer wavelengths, primarily in the infrared part of the spectrum. Much of this thermal radiation emitted by the land and ocean is absorbed by the atmosphere, including clouds, and reradiated back to Earth. This is called the greenhouse effect.

Source: Intergovernmental Panel on Climate Change [IPCC], 2007

In the 1950s, scientists began to take a closer look at climate issues as part of a major U.S. government investment in research and development. At that time, there was still a lot of uncertainty as to what direction climate change might take. With the advent of deep-sea exploration, new evidence became available that confirmed Milankovitch's theory of orbital control of climate. Analysis of ancient corals and deep-sea cores revealed a greater number of ice ages than was previously thought, longer and shorter, appearing at regular intervals consistent with the changes in Earth's orbital parameters. At that time, some scientists proposed that we might be on the verge of a new ice age (Bryson, 1971; Sawyer, 1972). Some also believed that global cooling was "imminent" due to the aerosols emitted into the atmosphere by industrial activities (Bryson, 1971; Sawyer, 1972).

Most, however, adhered to the global warming hypothesis and the potential role of CO_2 (Broecker, 1975). Two U.S. physicists, Gilbert N. Plass and Lewis D. Kaplan, confirmed Hulburt's theory about the potential role of CO_2 in increasing global temperatures (Plass, 1958; Kaplan, 1960). In 1958, Charles D. Keeling began regular measurements of the CO_2 airborne fraction at the Mauna Loa Observatory of Scripps Institution of Oceanography. The measurements, which continue to the present day, are used to construct the *Keeling Curve* shown in Figure 2.2. As the curve indicates, there is a steady trend of rising CO_2 concentrations over the past 50 years (Scripps Institution of Oceanography, University of California at San Diego, 2017).

There is an interesting scientific debate regarding the onset of human influence on the level of CO_2 in the atmosphere.[9] Most scientists, including Paul Crutzen, a Nobel Prize laureate in chemistry, believe that this influence began with the industrial revolution and, more specifically, with the invention of the steam engine in 1784

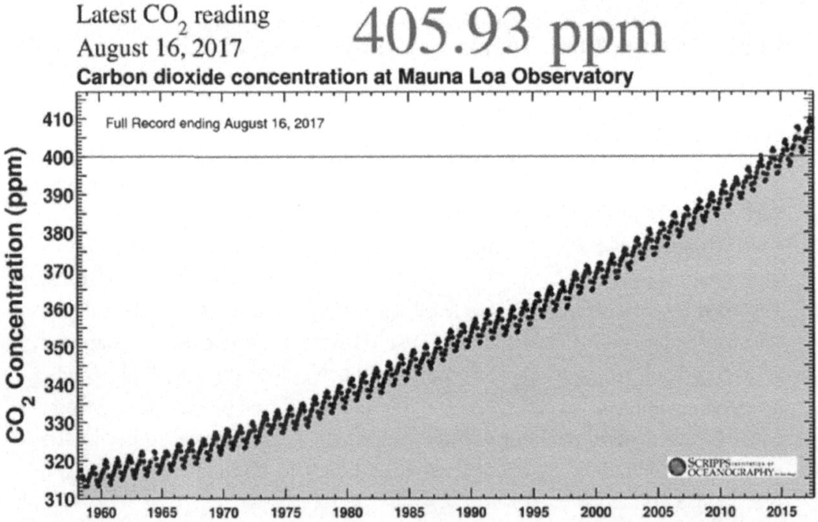

Figure 2.2 Measurements of atmospheric CO_2 at the Mauna Loa Observatory

Source: Scripps Institution of Oceanography, University of California at San Diego, https://scripps.ucsd.edu/programs/keelingcurve

(Crutzen, 2002). An interesting counterargument is proposed by William F. Ruddiman, a professor of geology, who asserts that humans began to alter the composition of the atmosphere, and hence, Earth's climate, when they first started modifying the landscape by forest clearance to expand the arable land. His book *Plows, Plagues, and Petroleum: How Humans Took Control of Climate* (2005) offers a superb analysis of complex evidence from a variety of sources supported by physical modeling searching for the causes of the anomalously high CO_2 levels of 280–285 parts per million at the dawn of the industrial era which, by all accounts, were about 40 parts per million or more above the natural background. According to Ruddiman (2005, p. 88), "Somehow, humans had seemingly added 300 billion tons or more of carbon to the atmosphere between 8,000 years ago and the start of the industrial era."

Figure 2.3 displays a curve of the CO_2 levels over the past 10,000 years. It is evident that carbon dioxide concentrations are slowly rising, reaching a point from which a steep gradient is observed coinciding roughly with the industrial activities of the past 100 years (Scripps Institution of Oceanography, University of California at San Diego, 2017). Another interesting anomaly appears at the right side of the graph, showing that the CO_2 concentrations were higher at the time of the medieval warm period followed by a sudden drop during the Little Ice Age (Figure 2.3).

In June of 1988, the hottest year on record at that time since 1870, Dr. James E. Hansen, a renowned National Aeronautics and Space Administration (NASA) scientist, testified before the U.S. Senate's Committee on Energy and Natural Resources in a hearing held by Colorado senator Timothy E. Wirth (U.S. Senate, Committee on Energy and Natural Resources, 1988, Weart, 2008; Besel, 2013). For Hansen, this

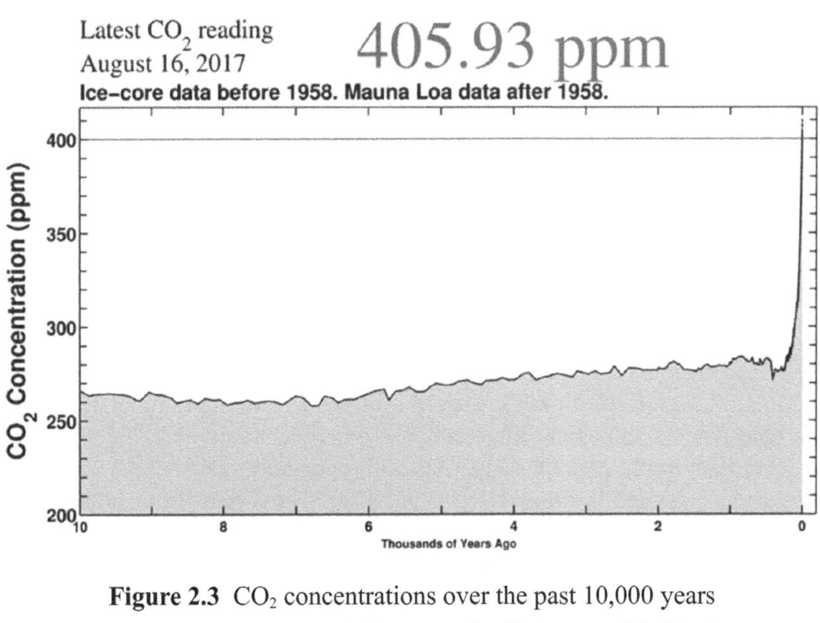

Figure 2.3 CO_2 concentrations over the past 10,000 years

Source: Scripps Institution of Oceanography, University of California at San Diego, https://scripps.ucsd.edu/programs/keelingcurve

was his third congressional hearing (Besel, 2013). The NASA scientist, and director of NASA's Goddard Institute for Space Studies, reaffirmed "with 99 percent confidence" the existence of a direct relationship between observed global warming trends and anthropogenic emissions of heat-trapping gases in the atmosphere (United States Senate, Committee on Energy and Natural Resources, 1988, p. 39). Studies in geophysics and climate modeling in the 1970s and 1980s had already revealed alarming trends, and attention was called to the dangers of global climate change. Two reports by the National Academy of Sciences were especially noteworthy. These reports predicted a rise in temperatures by 6°C by 2050 if fossil fuel combustion continued at the current pace (NRC, 1977, 1979).

At that time, global warming was largely considered a theoretical issue of no practical use and something that the future generations could attend to. (Wearth & American Institute of Physics [AIP], 2017). Despite these widely held beliefs, and quite unexpectedly, the congressional hearing of 1988 propelled global warming into the headlines of the mainstream media (Besel, 2013; Anderson, 2017). On June 23, 1988, the day of Hansen's testimony, *The New York Times* proclaimed, "Global warming has begun . . ." (Stabecoff, 1988). *Time* magazine, moving away from its time-honored tradition of designating a Person of the Year, selected the Endangered Earth as the Planet of the Year, pleading for "a universal crusade to save the planet" (*Time*, Jan. 2, 1989). "Now, more than ever, the world needs leaders who can inspire their fellow citizens," the weekly news magazine wrote, adding, "For man has reached a point in his evolution where he has the power to affect, for better or worse, the present and future state of the planet" (*Time*, Jan. 2, 1989). Michael Oppenheimer, an atmospheric physicist and chief scientist of the Environmental Defense Fund, who testified alongside James Hansen, said: "I've never seen an environmental issue mature so quickly, shifting from science to the policy realm almost overnight" (Wilford, 1988, p. 4 as cited in Besel, 2013). It might be the heightened media attention that ultimately marked the moment when climate change earned real political saliency. Whatever the reason, a Rubicon was crossed.

International Agreements and Initiatives for Climate Action

Three years after James Hansen testified to Congress, the *United Nations Conference on Environment and Development* (UNCED) (also known as the Earth Summit) brought together the heads of state and senior officials of 179 countries. One of the global conventions adopted at the summit was the 1992 *Framework Convention on Climate Change* (UNFCCC). Article 2 of the Convention articulated its ultimate objective: "stabilization of greenhouse gas concentrations in the atmosphere at a level that would prevent dangerous anthropogenic interference with the climate system" (United Nations [UN], 1992, p. 4). The *Kyoto Protocol*, adopted in 1997, was the first legally binding international treaty to implement UNFCCC. The Kyoto Protocol came on the footsteps of the successful implementation of the Montreal Protocol of 1987, which established a universally agreed-upon plan to phase out chlorofluorocarbons (CFCs) and other halogenated ozone-depleting substances (United Nations Environment Programme [UNEP], 2017). As of 2017, the Kyoto

Protocol was ratified by 192 countries with the United States and China being the most notable exceptions (CNN, 2017). The Kyoto Protocol was signed by the United States in 1998 but never ratified (Greenstone, 2015). In 2017, India, a country initially exempted from the provisions of the Kyoto Protocol, became the 80th nation to ratify the Second Commitment Period (also known as the Doha Amendment[10]), which covers national voluntary contributions for the period 2013–2020 (*Current Affairs Today*, 2017).

The Kyoto Protocol provided the legal basis for the establishment of a *Multilateral Fund* to assist developing countries in implementing stricter CO_2 control measures (UN, 2015). The Kyoto Protocol adopted some of the redistributive policy instruments that contributed to the successful elimination of CFCs by promoting burden-sharing and differentiated responsibilities in accordance with the principles instituted by the United Nations Conference on Environment and Development in 1992 (Rabe, 2007). Compared to the Montreal Protocol, the Kyoto treaty placed a stronger emphasis on market-based mechanisms and voluntary commitments (UN, 2015). The Protocol obligated the ratifying industrialized nations to achieve a 5% reduction (based on the 1990 levels) of their greenhouse gas emissions by 2012. The Protocol did not require any reduction commitments from the developing nations but encouraged them to make voluntary steps toward a low-carbon future (UN, 2015). This requirement became one of the major stumbling blocks to the universal ratification of the Kyoto Protocol as several countries including the U.S. argued against the uneven burden of responsibilities (Schiermeier, 2012).

The Kyoto Protocol has been criticized for its bureaucratic and ambiguous implementation mechanisms, as well as for dividing the world in two groups of countries—those responsible for cutting carbon loads in the atmosphere and those with no obligations (Jaimet, 2002; Harvard Magazine, 2002; Minns, 2002; Schiermeier, 2012). At the time it was written, the adopted strategy might have provided a reasonable pathway to reducing greenhouse gases in the atmosphere as most developing countries were low carbon emitters. Decades later, the largest developing countries exceeded the greenhouse gases emitted by some of their more developed counterparts (Schiermeier, 2012; Levi, 2015). Despite these limitations, progress was made. In a statement on the 10th anniversary of the Kyoto Protocol, Christiana Figueres, executive secretary of the UN Framework Convention on Climate Change, announced that global greenhouse gas emissions were reduced by 22.6% (compared to 1990 levels) at the end of the first commitment period in 2012; this exceeded by far the 5% reduction commitment targets that had been set (UNFCCC, 2015). Christiana Figueres also stressed that "without the Kyoto Protocol and its various mechanisms we would not be as far forward as we are today in respect to, for example, the growing penetration of renewable energies" (as cited in UNFCCC, 2015, online). Critics say that it remains questionable how much of this reduction can be attributed to the Kyoto Protocol and how much to other factors including the collapse of the former communist bloc, the global economic crisis, and foremost, the global shift to greener and more efficient technologies (Schiermeier, 2012). For many, including Roger Pielke, a professor in the Environmental Studies Program at the University of Colorado, the Kyoto Protocol and its architecture are "still useful," despite the flaws, drawbacks, and controversies (as cited in Schiermeier, 2012).

In December 2015, the *21st Conference of the Parties* (COP21) attended by the heads of state of 197 countries endorsed the *Paris Agreement*, which provided a comprehensive framework for further reductions in greenhouse gas (GHG) emissions. Coming into effect on November 4, 2016, the Paris Agreement is currently ratified by 159 countries (UN Climate Change Secretariat, 2017). The agreement requires transparent reporting of progress made toward meeting climate commitments. The Paris Agreement identified five critical areas of future action: mitigation, transparency, adaptation, loss and damage, and support for a clean energy future. In order to mitigate emissions, the agreement encourages all Parties to the UNFCCC to establish national climate action plans based on *intended nationally determined contributions* (INDCs) (UN, 2015). INDCs will be reviewed every 5 years at a *global stocktake* where the Parties will determine whether the net results of the collective climate initiatives are consistent with the goals established under the Paris Agreement. The *stocktakes* will not evaluate contributions by individual countries or recommend actions at a national level. Setting specific emissions targets will be "nationally determined" and will remain the prerogative of the countries party to the agreement (UN, 2015). The first *global stocktake*, which will be held in 2023, is expected to initiate the first global assessment of net changes in global GHG emissions. The agreement enacts the principle that future mitigation plans will be no less stringent than the existing ones and will contribute to steady reductions of GHG emissions to levels that will "hold the increase in the global average temperature to below 2°C" and allow for further actions that can achieve additional reductions to limit future temperature rise to below 1.5°C above pre-industrial levels (UN, 2015).

Climate Initiatives and Policies

The Paris Agreement places an unparalleled emphasis on climate adaptation, which in the past has trailed behind climate mitigation. Assessing vulnerability and risk, developing strategies to mitigate GHG emissions and strengthen adaptive capacity, and related policy and governance frameworks are examples that demonstrate the complex, interdisciplinary nature of the challenges that practitioners and policymakers commonly face. In many respects, climate mitigation and adaptation planning reflect both the science and the politics of climate change, and unlike other issues that often dominate the political agenda, they are likely to have profound intergenerational impacts (Rabe, 2007). The Paris Agreement strengthens the support for domestic policies that "incorporate climate-proofing and climate resilience measures" through international cooperation and adequate funding (UN, 2015). COP21 countries also committed to secure funding for climate initiatives estimated at US $100 billion a year by 2020. The agreement calls for expanded regional cooperation on climate adaptation including the development of regional centers and networks to strengthen ongoing efforts to reduce carbon emissions by businesses, civic groups, cities, and regions (UN, 2015).

The *Lima to Paris Action Agenda* (LPAA) and the *Non-State Actor Zone for Climate Action* (NAZCA) *Climate Action Portal*, launched in 2014 with assistance from UNFCCC, have already captured climate actions and commitments from over 2,200 cities and 150 regions with combined gross domestic product (GDP) of over

US $12.5 trillion including pledged support from over 2,600 companies and investors (NAZCA, 2015). The scope of the global climate action is expected to grow through future development assistance programs financed by United Nations agencies and major financial institutions (UN, 2015). The COP21 countries also reached a breakthrough agreement on "loss and damage" compensation measures, recognizing that not all adverse impacts associated with climate change can be successfully addressed through climate adaptation. The agreement further reinforces the *Warsaw International Mechanism* (WIM), which provides the COP21 countries with an anticipatory risk transfer mechanism "to facilitate the efforts of Parties to develop and implement comprehensive risk management strategies" (UN, 2015). Several other initiatives are also underway, many of which will be covered throughout this book. It is worth mentioning here the *C40 Cities* initiative, which includes 91 affiliated cities (as of August 2017) representing 1 in 12 people worldwide and 25% of the global GDP (C40 Cities Climate Leadership Group, Inc., 2017). By learning from each other and exploring new venues of collaboration, the C40 Cities have implemented over 10,000 innovative and effective policies and programs in building efficiency, renewable energy, ports, transportation, water conservation, greening, and waste management that can serve as models to other communities worldwide.

Under the Paris Agreement, the U.S. government has committed to decreasing total U.S. emissions below the 2005 level by 17% by 2020 and by 26%–28% by 2025 (Greenblatt,& Wei, 2016). Proposed measures to achieve these goals include an increase in energy production from renewable sources, adoption of new fuel economy standards, promotion of energy efficiency, implementation of new building codes, change in commuter and freight transportation patterns, and increase in carbon sequestration (U.S. Energy Information Administration [EIA], 2016). In a recent analysis published in *Nature Climate Change*, Greenblatt and Wei (2016) evaluate the impact of existing measures, proposed actions, and potential future policies and voluntary steps to achieve the emission reduction targets pledged as the U.S. INDCs. The authors estimate that the currently adopted policies will result in emission reduction between 9% and 24% lower than the intended 2025 contribution (Greenblatt & Wei, 2016). If all proposed policies are adopted, the United States is likely to achieve 20% reduction by 2025 (that is, 6% lower than the 26% emission target). These findings suggest that the U.S. commitments can only be met if additional actions are taken (Greenblatt & Wei, 2016).

In the United States, several states have already enacted policies for land use and transportation planning in support of regional efforts to meet the area's greenhouse gas emissions standards. *California Senate Bill (SB) 375* establishes regional targets for per-capita GHG emission reduction, introducing new measures to ensure fuel efficiency and policies to guide development that can meet the region's economic, housing, and transportation demands without compromising these goals (California Air Resources Board [CARB], 2008; Batac, Schattanek & Meyer, 2012). The state of Minnesota has passed the *Next Generation Energy Act*, which seeks to reduce the state's contribution of GHG "to at least 30 percent below 2005 levels by 2025" (State of Minnesota, 2007, p. 3), requiring adjustments of these targets based on a review of the state's climate change policies. The states of Massachusetts, New York, and Oregon have adopted policies to incentivize the implementation of regional transportation projects emphasizing measurable outcomes for reducing carbon emissions (Batac, Schattanek & Meyer, 2012). These initiatives come to suggest, despite inertia and bitter polemics at times, that state and local governments have assumed a pivotal role in enabling environmental policies, playing a key role in assuming regulatory functions and reformulating environmental governance to serve local realities. In many instances, "states are at the cutting edge of policy innovation, eager to find creative solutions to environmental problems, and 'racing to the top' with a goal of national prominence in the field" (Rabe, 2013, p. 31).

Notes

1 Stalagmites are cave formations deposited by dripping water. Mg tends to accumulate in larger quantities in stalagmites under dry conditions.

2 Most "hot" deserts are located around 30 degrees latitude in both the North and South hemispheres.

3 *Eccentricity* is the shape of the orbit of the Earth. *Obliquity* refers to the tilt of its axis. As the Earth spins around its axis, the tip of the axis slowly wobbles in a circular motion known as *precession*. These changes occur at regular intervals of 100,000, 40,000, and 22,000 years and are known as *Milankovitch cyclicity* (Kump, Kasting & Crane, 2010).

4 The Fertile Crescent includes ancient Mesopotamia and the Levant in present day's Iraq, Syria, Israel, Palestine, Jordan, and Turkey. It is also known as the "cradle of civilization" (Potts, 2012).

5 The trace gases account for only 0.1% of the atmosphere's composition but contain powerful agents such as O_3 (the ozone layer, which protects life from UV radiation) and the greenhouse gases (CO_2, methane, water vapor, and nitrous oxides, which trap the outgoing infrared radiation).

6 Some scientists contend that these early agrarians were too few in numbers and too scattered to produce a noticeable impact (Zhuang & Kidder, 2014).

7 Ruddiman (2005, p. 49) expressed this common criticism as: "How can you scientists expect to 'forecast' the weather thousands of years ago when you can't even get the weather for this coming weekend half-right?"

8 In a book and a blog entitled *The Discovery of Global Warming*, Weart (2008) and Weart and American Institute of Physics (AIP) (2017 and earlier) provide a detailed account of the scientific breakthroughs that led to the discovery and exploration of the role of greenhouse gases (GHGs) in warming up the Earth's climate system.

9 Most scientists agree that we have entered a new geological epoch, the *Anthropocene*. The concept was first suggested by Andrew Revkin in his 1992 book *Global Warming: Understanding the Forecast*. Some scientists, including Paul

Crutzen, believe it started with the industrial revolution. William Ruddiman put forward the hypothesis that it has started much earlier, probably when humans have begun changing the face of Earth to open land for agricultural activities several thousand years ago (Ruddiman, 2005).

10 The Doha Amendment to the Kyoto Protocol established a Second Commitment Period extending its climate action provisions to 2020 after the initial 2005–2012 period. Each party has to ratify the Amendment to become a party to the commitment period.

References

Anderson, A. 2017. Source Influence on Journalistic Decisions and News Coverage of Climate Change, *Climate Change Communication*, Online publication, DOI: 10.1093/acrefore/9780190228620.013.356.

Batac, T., Schattanek, G., & Meyer, M.D. 2012. *NCHRP 08–36, Task 107 Synthesis of State DOT and MPO Planning and Analysis Strategies to Reduce Greenhouse Gas Emissions*. Retrieved from http://onlinepubs.trb.org/onlinepubs/nchrp/docs/NCHRP08-36(107)_FR.pdf (accessed 03/21/2016).

Besel, R.D. 2013. Accommodating Climate Change Science: James Hansen and the Rhetorical/Political Emergence of Global Warming, *Science in Context* 26(1), 137–152.

Broecker, W. 1975. Climatic Change: Are We on the Brink of a Pronounced Global Warming? *Science* 189, 460–463.

Brooke, J.L. 2014. *Climate Change and the Course of Global History*. New York: Cambridge University Press.

Brown. A. 2010. *Has Charles Dickens Shaped Our Perception of Climate Change? The Air Vent*. https://wattsupwiththat.com/2010/12/05/has-charles-dickens-shaped-our-perception-of-climate-change/ (accessed 04/21/2016).

Bryson, R.A. 1971. A Reconciliation of Several Theories of Climate Change. In: Holdren, John P. (Ed.) *Global Ecology. Readings Toward a Rational Strategy for Man*, 78–84. New York: Harcourt Publishers Group.

C40 Cities Climate Leadership Group, Inc., 2017. *The Power of C40 Cities*. Retrieved from www.c40.org/cities (accessed 08/12/2017).

California Air Resources Board (CARB). 2008. *Climate Change Scoping Plan: A Framework for Change* (Pursuant to AB32 The California Global Warming Solutions Act of 2006). Sacramento, CA, December 2008.

Chu, J. 2015. Stalagmites pinpoint drying of American West. *MIT News*, July 27, 2015. Retrieved from http://news.mit.edu/2015/stalagmites-dry-american-west-desert-0727.

CNN. 2017. *Kyoto Protocol Fast Facts*, March 24, 2017. Retrieved from www.cnn.com/2013/07/26/world/kyoto-protocol-fast-facts/index.html (accessed 08/11/2017).

Crutzen, P. 2002. Geology of mankind. *Nature* 415, 23. DOI: 10.1038/415023a.

Current Affairs Today. 2017. India Ratifies 2nd Commitment Period of Kyoto Protocol, August 9, 2017. Retrieved from currentaffairs.gktoday.in/india-ratifies-2nd-commitment-period-kyoto-protocol-08201747178.html © GKToday (accessed 08/11/2017).

Fleming, J.R. 1998. *Historical Perspectives on Climate Change*. Oxford: Oxford University Press, 194 pp.

Greenblatt, J.B., & Wei, M. 2016. Assessment of the Climate Commitments and Additional Mitigation Policies of the United States. *Nature Climate Change* 6, 1090–1093.

Greenstone, M. 2015. Surprisingly, a Voluntary Climate Treaty Could Actually Work. *The New York Times*, February 13, 2015.

Harvard Magazine. 2002. Problems with the Protocol, November–December, 2002. Retrieved from http://harvardmagazine.com/2002/11/problems-with-the-protoc.html (accessed 08/07/2017).

Hawkins, E., & Jones, P. 2013. On Increasing Global Temperatures: 75 Years After Callendar. *Quarterly Journal of the Royal Meteorological Society*. DOI: 10.1002/qj.217.

Intergovernmental Panel on Climate Change (IPCC). 2007. Climate Change 2007: The Physical Science Basis. Contribution of Working Group I to the Fourth Assessment Report of the Intergovernmental Panel on Climate Change [Solomon, S., D. Qin, M. Manning, Z. Chen, M. Marquis, K.B. Averyt, M. Tignor, and H.L. Miller (eds.)]. Cambridge, UK, and New York, NY: Cambridge University Press. Retrieved from www.ipcc.ch/publications_and_data/ar4/wg1/en/faqs.html.

Jaimet, K. 2002. A Primer on Kyoto. *The Ottawa Citizen*, 30 August 2002: A4.

Kaplan, L.D. 1960. The Influence of Carbon Dioxide Variations on the Atmospheric Heat Balance. *Tellus* 12(2), 204–208.

Kump, L.R., Kasting, J.F., & Crane, R.G. 2010. *The Earth System* (3rd edition). Upper Saddle River, NJ: Pearson Education, Inc.

Kutzbach, J.E. 1981. Monsson Climate of the Early Holocene: Climate Experiment with the Earth's Orbital Parameters for 9000 Years Ago. *Science* 214(4516), 59–61.

Kutzbach, J.E., & Liu, Z. 1997. Response of the African Monsoon to Orbital Forcing and Ocean Feedbacks in the Middle Holocene. *Science* 278, 440–443.

Ladurie, E. le R. 1971. *Times of Feast, Times of Famine: A History of Climate Since the Year 1000. Barbara Bray*. Garden City, NY: Doubleday

Lamb, H.H. 1995. *Climate History and the Modern World*. London, UK: Routledge (reprinted 1997).

Levi, M.A. 2015. Two Cheers for the Paris Agreement on Climate Change. *Council on Foreign Relations (CFR)*. Retrieved from www.cfr.org/blog/two-cheers-paris-agreement-climate-change (accessed 08/01/2017).

Mann, M. 2003. Little Ice Age. In: MacCracken, M.C., & Perry, J.S. (Eds.) *Encyclopedia of Global Environmental Change, Volume 1, The Earth System: Physical and Chemical Dimensions of Global Environmental Change*. New York: John Wiley & Sons.

Minns, A. 2002. An Introduction to the Kyoto Protocol. *changingclimate.org*, 15 March 2002. Retrieved from www.changingclimate.org/content/articles/article/data/section_4/article_85/part_433/ (accessed 08/07/2017).

National Geographic. 2017. Genographic Project/The Development of Agriculture. Retrieved from https://genographic.nationalgeographic.com/development-of-agriculture/ (accessed 08/02/2017).

National Research Council (NRC). 1977. *Studies in Geophysics: Energy and Climate. Geophysics Study Committee, Geophysics Research Board, Assembly of Mathematics and Physical Sciences*. Washington, DC: National Academy of Sciences.

National Research Council (NRC). 1979. *Carbon Dioxide and Climate, National Academy of Sciences*. Washington, DC: National Academy of Sciences.

National Research Council (NRC). 1983. *Environmental Change in the West African Sahel*. Washington, DC: The National Academies Press. https://doi.org/10.17226/19468.

National Research Council (NRC). 2010. *Understanding Climate's Influence on Human Evolution*. Washington, DC: The National Academies Press.

Non-State Actor Zone for Climate Action (NAZCA). 2015. *About LPAA*. Retrieved from http://climateaction.unfccc.int/about-lpaa (accessed 04/15/2016).

Plass, G.N. 1958. The Carbon Dioxide Theory of Climatic Change. *Tellus* 8(2), 140–154.

Potts, D.T. (Ed.) 2012. *A Companion to the Archaeology of the Ancient Near East*. New York: Wiley-Blackwell, pp. 575–584.

Rabe, B.G. 2007. Beyond Kyoto: Climate Change Policy in Multilevel Governance Systems. *Governance* 20(3), 423–444.

Rabe, B.G. 2013. Racing to the Top, the Bottom, or the Middle of the Pack? The Evolving State Government Role in Environmental Protection. In Vig, N., & Kraft, M. (Eds.) *Environmental Policy: New Directions for the 21st Century*, 30–53. Washington, DC: Sage/CQ Press.

Revkin, A. 1992. *Global Warming: Understanding the Forecast*. New York: Abbeville Press (revised 2012).

Rohli, R.V., & Vega, A.J. 2008. *Climatology*. Sudbury, MA: Jones & Bartlett Publishers, LLC.

Ruddiman, W.F. 2005. *Plows, Plagues, and Petroleum: How Humans Took Control of Climate*. Princeton, NJ: Princeton University Press.

Ruddiman, W.F. 2008. *Earth's Climate: Past and Future* (2nd edition). New York, NY: W.H. Freeman and Company.

Sawyer, J.S. 1972. Man-Made Carbon Dioxide and the "Greenhouse" Effect. *Nature* 239, 23–26. DOI:10.1038/239023a0.

Schiermeier, Q. 2012. The Kyoto Protocol: Hot Air. *Nature* 491(7426), November 28, 2012. Retrieved from www.nature.com/news/the-kyoto-protocol-hot-air-1.11882#before (accessed 04/15/2016).

Scripps Institution of Oceanography, University of California San Diego. 2017. *Scripps CO_2 Program*. Retrieved from http://scrippsco2.ucsd.edu/ (accessed 08/17/2017).

Stabecoff, P. 1988. Global Warming Has Begun, Expert Tells Senate. *The New York Times*, June 24, 1988. Retrieved from www.nytimes.com/1988/06/24/us/global-warming-has-begun-expert-tells-senate.html?pagewanted=all (accessed 08/08/2017).

Stanley, S.M. 2009. *Earth System History*. New York, NY: W.H. Freeman & Company.

State of Minnesota. 2007. *Next Generation Energy Act*. St. Paul, MN. Retrieved from www.revisor.mn.gov/data/revisor/slaws/2007/0/136.pdf (accessed 03/25/2016).

Steponaitis, E., Andrews, A., McGee, D., Quade, J., Hsieh, Y.T., Broecker, W.S., Shuman, B.N., Burns, S.J., & Cheng, H. 2015. Mid-Holocene Drying of the US

Great Basin Recorded in Nevada Speleothems. *Quaternary Science Reviews* 127, 174–185. DOI: 10.1016/j.quascirev.2015.04.011.

Time. 1988. Planet of the Year, *Time Magazine*, January 02, 1989. Retrieved from http://content.time.com/time/magazine/article/0,9171,956627,00.html (accessed 08/07/2017).

United States Energy Information Administration (EIA). 2016. *International Energy Outlook*, Chapter 9: Energy-related CO_2 Emissions. Retrieved from www.eia.gov/outlooks/ieo/pdf/emissions.pdf (accessed 03/12/2017).

United States Senate, Committee on Energy and Natural Resources. 1988. *Greenhouse Effect and Global Climate Change: Hearing before the Committee on Energy and Natural Resources, 100th Cong., 2nd sess*. Washington DC: U.S. Government Printing Office. Retrieved from https://catalog.loc.gov/vwebv/search?searchCode=LCCN&searchArg=88601376&searchType=1&permalink=y (accessed 08/02/2017).

United Nations Framework Convention on Climate Change (UNFCCC). 2015. *Kyoto Protocol 10th Anniversary: Timely Reminder Climate Agreements Work, UNFCCC Newsroom*. Retrieved from http://newsroom.unfccc.int/unfccc-newsroom/kyoto-protocol-10th-anniversary-timely-reminder-climate-agreements-work/ (accessed 07/21/2017).

United Nations Environment Programme (UNEP). 2017. *The Montreal Protocol on Substances that Deplete the Ozone Layer*. Retrieved from http://ozone.unep.org/en/treaties-and-decisions/montreal-protocol-substances-deplete-ozone-layer (accessed 07/21/2017).

United Nations (UN). 1992. *United Nations Framework Convention on Climate Change FCCC/INFORMAL/84 GE.05–62220 (E) 200705*. UNFCCC Secretariat: Bonn, Germany. Retrieved from https://unfccc.int/resource/docs/convkp/conveng.pdf.

United Nations (UN). 2015. *United Nations Framework Convention on Climate Change FCCC/CP/2015/L.9/Rev.1*. UNFCCC Secretariat: Bonn, Germany. Retrieved from https://unfccc.int/resource/docs/2015/cop21/eng/l09.pdf (accessed 03/14/2016).

United Nations (UN) Climate Change Secretariat. 2017. *Progress Tracker1 Work Programme Resulting from the Relevant Requests Contained in Decision 1/CP.21*. Retrieved from http://unfccc.int/files/paris_agreement/application/pdf/pa_progress_tracker_200617.pdf (accessed 08/12/2017).

Weart, S. & American Institute of Physics (AIP). 2017. *The Discovery of Global Warming, The Public and Climate Change (cont.—since 1980)*. Retrieved from https://history.aip.org/climate/public2.htm (accessed 08/09/2017).

Weart, S.R. 2008. The Discovery of Global Warming: Revised and Expanded Edition (2nd edition). Copyright © 2004–2008 Spencer Weart and the American Institute of Physics, Melville, NY: AIP Publishing.

Wilford, J.N. 1988, July. His Bold Statement Transforms the Debate on Greenhouse Effect. *New York Times*. Retrieved from www.nytimes.com/1988/08/23/science/his-bold-statement-transforms-the-debate-on-greenhouse-effect.html

Zhuang, Y., & Kidder, T.R. 2014. Archeology of the Anthropocene in the Yellow River Region, China, 8000–2000 cal. BP. *The Holocene* 24(11), 1602–1623.

3 Climate Adaptation

A Nexus of Science, Policy, and Planning

Chapter Objectives

This chapter seeks to answer the following questions:

- How is climate adaptation defined?
- How do "mechanisms of development" including international policy and financial infrastructure address climate adaptation?
- What are the drivers of and impediments to climate adaptation actions at the local level?
- How does climate adaptation relate to the concepts of risk management, hazard mitigation planning, risk reduction, vulnerability, and resilience?
- How can climate adaptation measures mitigate or exacerbate issues of poverty and social vulnerability?

Impacts of Climate Change and Adaptation Needs

Climate change is expected to alter large-scale atmospheric processes known to trigger weather extremes (Knutson, Delworth, Dixon & Stouffer, 1999; Washington et al., 2000; Nozawa et al., 2000; Gordon et al., 2000; Flato & Boer, 2001; McGuffie & Henderson-Sellers, 2001). Complex interactions between climatic conditions and ecological shifts in marine and freshwater ecosystems, wetlands, forests, and grasslands carry the potential for long-term changes in water resources, agricultural productivity, and fisheries (Thomalla et al., 2006; Abeygunawardena et al., 2009). The changes are projected to take effect on global, regional, and local scales (Seneviratne et al., 2012). Although the magnitude of change in weather-related extremes attributed to warming climatic conditions is uncertain, physical reasoning and scientific evidence suggest that rising temperatures and altered precipitation patterns may increase flood risk and exacerbate drought conditions and wildfires in many regions worldwide, including densely populated areas (Seneviratne et al., 2012; Knapp et al., 2016).

The multifaceted, long-range, and uncertain nature of climate change presents major challenges to urban areas which currently host 54% of the world's population, a proportion that is projected to increase to 66% by 2050 (United Nations Department of Economic and Social Affairs [UN DESA], 2017). Rising sea levels increase the vulnerability of the coastal regions to storm surge and flooding. Although sea level rise unfolds over extended periods of time, many coastal areas are already experiencing its effects. Nicholls, Hoozemans, and Marchand (1999) reported in the late 1990s that by the year 2080, the number of people who live in areas flooded by storm surge may increase more than five times, while Neumann, Yohe, Nicholls, and Manion (2000) noted that as the sea level rises, the flooded areas are expected to expand further into the mainland, increasing the risk zone by more than 30%. Findings from a research report by Climate Central indicate that between 147 to 216 million people worldwide will be exposed to regular flooding by 2100 (Strauss & Kulp, 2014). A rise in sea level also causes erosion of beaches and bluffs, increases vulnerability to storm damage, and leads to permanent inundation of low-lying areas, saltwater intrusion into aquifers and surface waters, and elevated water table. A higher water table resulting from sea level rise can also increase the potential for inland flooding of low-lying areas, cause damage to underground infrastructures such as utility lines and pipelines, reduce the efficiency of wastewater treatment facilities and stormwater management practices, and cause health effects (Deyle, Bailey & Matheny, 2007; Bloetscher, Romah, Berry, Hammer & Cahill, 2012; Obeysekera & Park, 2013; Bloetscher & Romah, 2015; Bloetscher et al., 2016).

Climate change can also aggravate poverty and jeopardize development in many developing countries (Abeygunawardena et al., 2009). It is expected that several regions will experience reduced access to drinking water and increased risks to food security, especially in Africa, Asia, and Latin America (Abeygunawardena et al., 2009). In the most impoverished nations, climate change threatens livelihood choices by decreasing crop yields and limiting access to natural resources, further aggravating famine and compounding existing vulnerabilities (Abeygunawardena et al., 2009). Observed data already indicate measurable effects of climate change on crop yield (Lobell, Schlenker & Costa-Roberts, 2011). Exacerbated by poverty, worsening environmental conditions, lack of social safety networks, limited financial resources, and insufficient institutional capacity, these limitations can further marginalize individuals, communities, and even entire nations (Thomalla et al., 2006; Angueloski et al., 2016; Kashem, Wilson & van Zandt, 2016, Vachette, 2017).

In a critical review of how cities approach the adaptation challenges imposed by climate change, Hunt and Watkiss (2011, p. 15) identify five consensus areas of significant impacts and adaptation needs: (i) hazard reduction strategies to minimize the effects of storm surge, tsunamis, and sea level rise in coastal cities; (ii) redesigning and retrofitting existing infrastructure; (iii) expanding the renewable energy portfolio and increasing energy use efficiency; (iv) developing public health policies to address mortality and morbidity from heat and vector-borne and waterborne diseases linked to the effects of climate change; and (v) improving water resource management.

Climate Adaptation as a Process and Outcome

Over the past two decades, efforts to address climate change impacts have shown that even in the absence of enabling national policy frameworks, effective local initiatives can be successfully implemented and significant reductions in greenhouse gas emissions can be achieved (Edwards & Haines, 2007; Girardet, 2008; Davoudi, Crawford & Mehmood, 2009; International Council for Local Environmental Initiatives and World Wildlife Fund [ICLEI-WWF], 2015). Often local and fragmented, these actions have been increasingly unified by a shared understanding of the need for strategic approaches that lead to "farther reaching and systematic changes" (Bulkeley, Castan Broto, Hodson & Marvin, 2011, p. 25). As part of these wide-ranging strategies, cities have planned for their long-term protection against climate-related hazards, re-envisioned a future where self-reliance on energy resources is key, and sought opportunities for global stewardship through networking, forging alliances, and creating strategic partnerships (Bulkeley & Betsill, 2003; While, Jonas & Gibbs, 2010; Bulkeley, Castan Broto, Hodson & Marvin, 2010; ICLEI-WWF, 2015). While *climate mitigation* to reduce greenhouse gas emissions has long been a focal theme and strategic goal of the United Nations Framework Convention on Climate Change (UNFCCC), *climate adaptation* remained, to a certain extent, a secondary priority. First introduced into the negotiation process with the 2001 Marrakesh Accords to UNFCCC (Adger, Huq, Brown, Conway & Hulme, 2003; Webber, 2016), climate adaptation was officially adopted as a universal goal under the *Paris Agreement* (Streck, 2015). The agreement adopted by the Conference of the Parties (COP21) to the United Nations Framework Convention on Climate at the end of 2015 recognized that climate adaptation has not been on equal footing with climate mitigation priorities (UNFCCC, 2015; Streck, 2015) and stipulated:

> Adaptation is a global challenge faced by all with local, subnational, national, regional and international dimensions, and that it is a key component of and makes a contribution to the long-term global response to climate change to protect people, livelihoods, and ecosystems.
>
> (UNFCCC, 2015, p. 24)

The Intergovernmental Panel on Climate Change (IPCC, 2007) defines adaptation in the context of climate change as a necessary adjustment in response to actual or projected changes in natural and human systems to moderate adverse effects or exploit beneficial opportunities. Climate change adaptation (CCA) is further conceptualized as *anticipatory* vs. *reactive* to distinguish between actions taken before and after the occurrence of climate-induced impacts; *planned* vs. *autonomous* to make a distinction between deliberate and purposeful policy implementation to address climate impacts and changes that occur spontaneously in ecological and socioeconomic systems in response to climate *stimuli*; and *public* vs. *private* adaptation to recognize that the agents of change operate at all levels of human society from government and non-profit entities to private corporations, households, and individuals (IPCC, 2007). Furthermore, in the context of climate change, *adaptive capacity* is characterized by the ability to engage institutional resources, political processes, funding mechanisms, and knowledge creation to enable and implement adaptation options (Klein, Nicholls & Thomalla, 2003; Thomalla, Downing, Spanger-Siegfried, Han & Rockström, 2006; IPCC, 2007). *Adaptation assessment* is a related concept focusing on the evaluation criteria to appraise adaptation options in terms of costs and benefits, feasibility and legitimacy, and effectiveness and efficiency (IPCC, 2007). Supplementary definitions of adaptation also include a reference to "changes in processes, practices, and structures to moderate potential damages or to benefit from opportunities associated with climate change" (IPCC, 2007, online).

Analyzing the climate adaptation definitions, the Annex I Expert Group of the Organisation for Economic Co-operation and Development (OECD) and the International Energy Agency (IEA) noted that the range of keywords used to describe climate adaptation (e.g., "actions," "adjustment," "process," and "outcome") "can be interpreted differently by various stakeholders" (OECD, 2006, p. 5). The OECD experts emphasized the need for further clarifications, noting:

> Expectations from adaptation as an outcome might be much higher than expectations from it as a process. Funding aspirations and evaluation of achieved results would also vary accordingly.
>
> (OECD, 2006, p. 5)

In the context of monitoring adaptation, the United Nations Development Programme (UNDP, 2008, p. 3) identifies the following outcomes as a key part of the climate adaptation portfolio: (1) building institutional capacity to manage climate change risks; (2) mainstreaming and integrating climate change risk into effective policies at various levels; (3) implementing pilot projects and practices in key areas of climate adaptation; and (4) developing tools for climate adaptation decision support. Furthermore, the UNDP's Monitoring Framework highlights several processes that guarantee development benefits while reducing vulnerability to climate change impacts: (1) establishing relevant policies and plans; (2) raising awareness and building institutional capacity; (3) effective use of information technologies; (4) guiding future investments; and (5) ensuring sustainable livelihoods and resource management (UNDP, 2008, p. 6).

The chapter reviews the shifts in climate adaptation research and policy formulation. In addition, the chapter examines various conceptualizations of climate adaptation and places climate actions in the context of the related concepts of risk management, hazard mitigation planning, risk reduction, resilience, and vulnerability. It discusses the role of local climate adaptation through the lenses of incremental and transformative approaches. The chapter concludes with a brief discussion of the drivers of climate adaptation and its limits and barriers.

Climate Adaptation and Development

Over the past 20 years, the focus in climate adaptation research and policy formulation has gradually shifted from local initiatives toward "the existing mechanisms of development," including international policy and financial infrastructure (Webber, 2016, p. 406). Webber (2016, p. 404) identifies three mechanisms through which adaptation is "folded" into the development process, that is, "enveloped and subsumed within the institutions, policies, and practices." These mechanisms include: (1) identification of the most vulnerable nations and population groups with higher levels of exposure to extremes and less adaptive capacity; (2) deployment of targeted assistance through the multilateral financial institutions; and (3) integration or "mainstreaming" of climate projections and other relevant information into current and future investments and development plan to facilitate climate adaptation activities (Webber, 2016, pp. 404–405).

Climate adaptation funds are increasingly allocated through existing multilateral development banks and the UNFCCC climate-related funding mechanisms—primarily the Global Environment Facility (GEF) established by the World Bank in 1991 in coordination with UNDP and the United Nations Environment Programme (UNEP) (Webber, 2016, p. 405). Two additional multilateral funds—the Least Developed Countries Fund (LDCF) and the Spatial Climate Change Fund (SCCF)—were created under UNFCCC (Decisions 3/CP.11 and 5/CP.9, respectively) to promote the integration of climate adaptation in development activities and poverty reduction strategies (Biagini, Biertbaum, Stults, Dobardzic & McNeeley, 2014). Another source of funding for implementing climate adaptation measures is the Adaptation Fund established under the Kyoto Protocol in 2001 (Decision 10/CP.7). The Adaptation Fund is supported by 2% of the proceeds from the *certified emission reductions* under the Clean Development Mechanism. Over the past eight years, the Fund has committed US$477 million to support climate adaptation and resilience projects in 76 countries, including 28 Least Developed Countries and 19 Small Island Developing States (Adaptation Fund, 2018).

Biagini, Biertbaum, Stults, Dobardzic, and McNeeley (2014) developed a typology of climate adaptation activities analyzing 92 projects financed under UNFCCC in 70 countries. The findings indicate that an overwhelming majority of adaptation activities fall under the category of capacity building, followed by management and planning policy, information, and physical infrastructure (Biagini, Biertbaum, Stults, Dobardzic & McNeeley, 2014). These findings dovetail with previous scholarly work (Huntjens et al., 2012) and IPCC Fifth Assessment Report (AR5) conclusions (Mimura et al., 2014) and suggest that the implementation of adaptation activities is contingent upon "developing human resources, institutions, and communities" to create enabling institutional environments for the implementation of climate adaptation actions (Biagini, Biertbaum, Stults, Dobardzic & McNeeley, 2014, p. 104).

Millner and Dietz (2011, p. 4) explore the linkages between climate adaptation and development from two distinct perspectives. The first perspective contends that, unless a climate adaptation mechanism is in place, the potential for future growth in developing countries will be severely curtailed due to the impacts of climate change. According to Millner and Dietz (2011), this is fundamentally the view adopted by the international development agencies and multilateral banks. The second perspective affirms that the best defense for developing nations against climate change is "their own continued development" (Schelling, 1992, p. 6) given that literacy, per capita income, institutional capacity, and development resources are positively associated with increased adaptive capacity and reduced vulnerability to extreme events (Brooks, Adger & Kelly, 2005).

Addressing Risk and Vulnerability to Natural Hazards through Climate Adaptation

Scholarly research on climate adaptation has long emphasized the linkages between natural hazards, risk management, disaster risk reduction, and climate change (Klein, Nicholls & Thomalla, 2003; Thomalla, Downing, Spanger-Siegfried, Han & Rockström, 2006; Birkmann & von Teichman, 2010; Webber, 2016; Vachette, 2017; Nemakonde & van Niekerk, 2017; Adger, Brown & Surminski, 2018). Central to this perspective is the concept of resilience, which draws attention to the ability of interconnected socio-ecological systems to withstand, mitigate, absorb shocks, and recover from extreme events (Adger, 2002; Adger, Hughes, Folke, Carpenter & Rockström, 2005; Adger, 2006; Becker, 2012; Beichler, Hasibovic, Davidse & Deppisch, 2014). Researchers have also noted that while resilience and adaptive capacity are critical to forging strategies aimed at addressing the effects of climate change, they "will not necessarily translate to adaptation" due to the magnitude of the anticipated changes and the differences in wealth and institutional capacity across regions, nations, communities, and economic sectors (Berrang-Ford, Ford & Paterson, 2011, p. 25). As Thomalla, Downing, Spanger-Siegfried, Han, and Rockström (2006, p. 41) point out:

> In many of the world's developing and least developed countries extreme events occur so frequently that they tend to overwhelm their coping capacity and hamper long-term progress because attention and resources desperately needed for poverty reduction and economic development are diverted to disaster relief and reconstruction.

Thus, the unequal adaptive capacity between high-income and low-income nations and communities remains a significant challenge to future adaptation actions on local, regional, and global scales.

Earlier studies on climate adaptation have focused on identification of vulnerabilities and risks (Klein, Nicholls & Thomalla, 2003; National Research Council [NRC], 2010; Adger, Brown & Surminski, 2018), conceptualizing climate adaptation as a primarily local hazard mitigation priority (Berke, 2014; Webber, 2016). This approach aligns well with *natural hazards management* using the extant knowledge base and future scenarios to frame policies and develop implementation programs (Berke, 2014). However, damage, loss, and population displacement associated with the impacts of climate change (McNamara, Bronen, Fernando & Klepp, 2018) have brought to light deficiencies in both preparedness for weather-related extremes and relevant adaptation responses despite decades of natural hazards risk management (Berrang-Ford, Ford & Paterson, 2011; Ford, Berrang-Ford & Paterson, 2011). Two important factors contribute to these outcomes. First, traditional risk management strategies are based on "reductive" reasoning based on historical probabilities of occurrence of extreme events (Adger, Brown & Surminski, 2018, p. 3). The U.S. Office of Management and Budget (OMB) defines *risk assessment* as the assembly and synthesis of "scientific information to determine whether a potential hazard exists and/or the extent of possible risk to human health, safety or the environment" (OMB, 2006, p. 23). Policy decisions are often distilled from risk assessment methodologies to prioritize "standard rules and guidance to reduce the prospect for adverse consequences" (Adger, Brown & Surminski, 2018, p. 2). It has been argued that the effectiveness of such policies may be limited under future conditions, especially when the probable outcomes are not well known and the number of alternatives is uncertain (Adger, Brown & Surminski, 2018). Second, there is a well-established need to address uncertainty in the complex interactions between changes occurring in biophysical systems and the responses to risk in social and economic systems. As Adger, Brown, and Surminski (2018, p. 6) argue, existing risk assessment methodologies are "ill-equipped to deal with interaction effects, and with multiple time scales."

In 1999, the United Nations General Assembly established the Office for Disaster Risk Reduction (UNISDR) (Resolution 56/195) to facilitate intergovernmental and regional collaboration in disaster response and recovery. A successor of the Hyogo Framework for Action (2005–2015), the Sendai Framework for Disaster Risk Reduction (DRR) adopted in 2015 focuses on seven target areas and four priorities for action to manage disaster risk: (1) identifying the causal risks of disasters; (2) promoting disaster risk governance; (3) increasing investment in resilience; and (4) expanding disaster preparedness to improve response and recovery and "build back better" (United Nations [UN], 2015). The framework addresses three sets of factors that amplify disaster risk including (1) socioeconomic factors such as poverty, inequality, and marginalization; (2) the role of good governance in increasing the resilience of socioeconomic and environmental systems; and (3) the need to reduce exposure of people and economic assets to climate-related hazards (UN, 2015). Principle 2 of the Sendai Framework calls further attention to the pivotal role of *disaster risk governance*, stressing that "strengthening disaster risk governance for prevention, mitigation, preparedness, response, recovery, and rehabilitation is necessary and fosters collaboration and partnership across mechanisms and institutions for the implementation of instruments relevant to disaster risk reduction" (UN, 2015).

In a critical review of local and national adaptation strategies, Birkmann and von Teichman (2010) identify several gaps in knowledge and institutional challenges impeding the integration of disaster risk reduction and climate change adaptation, including: (1) incongruity of spatial, temporal, and functional scales; (2) absence of supportive regulatory environment and the need to overcome normative, cultural, and institutional barriers; and (3) inability to effectively communicate the scientific evidence to decision-makers and provide guidance and relevant data to practitioners. Birkmann and von Teichman (2010) also highlight several pathways to integrating climate adaptation into the mitigation, preparedness, response, and recovery phases of the disaster cycle. Nemakonde and van Niekerk (2017, p. 362) argue that "substantial reduction of disaster risk emanating both from natural hazards and climate risks is contingent on institutional and governance structures to implement measures." Reflecting on experiences derived from expert knowledge in the Southern African Development Community (SADC) member states, Nemakonde and van Niekerk (2017) propose a conceptual framework for integrating institutional capacities developed under existing disaster risk reduction and climate adaptation programs. The framework is structured around eight themes elicited from expert interviews that highlight existing and potential institutional arrangements to support integration of DRR and CCA, including logistical support for the coordination of activities; distinguishing short-term natural hazards risks from long-term climate risks; identifying roles, overlaps, and integrative functions; seeking formalization of responsibilities and; promoting institutional capacities that enable climate adaptation measures across organizational and jurisdictional boundaries (Nemakonde & van Niekerk, 2017).

Local Planning and Place-Based Climate Adaptation

Over the past 20 years, local governments around the world have designed and implemented a multitude of local initiatives to address observed and anticipated impacts of climate change (Koski & Siulagi, 2016; Webler, Tuler, Dow, Whitehead & Kettle, 2016; Meerow & Mitchell, 2017; Torabi, Dedekorkut-Howes & Howes, 2017; Dilling, Pizzi, Berggren, Ravikumar & Andersson, 2017; Wamsler & Brink, 2018). Klein, Nicholls, and Thomalla (2003, p. 38) discuss five primary objectives of local adaptation planning, including (1) increasing the resilience of physical infrastructure to climate change impacts; (2) enabling adaptability of threatened natural systems; (3) design of flexible policies to adjust to changing conditions; (4) addressing existing conditions that increase vulnerability of the coupled natural-human systems; and (5) promoting public outreach to raise awareness and facilitate preparedness and mitigation.

In conceptualizing local climate adaptation, many recent studies emphasize the key differences between *incremental* and *transformative adaptation* (Füssel, 2007; Roggema, Vermeend & van den Dobbelsteen, 2012; Torabi, Dedekorkut-Howes & Howes, 2017; Alexandra, 2017). Roggema, Vermeend, and van den Dobbelsteen (2012) argue that current practices of spatial planning and design are ill-equipped to address the challenges that lie ahead. The authors suggest three strategic "pathways" to climate adaptation planning, including *incremental adaptation*, which emphasizes the need for small incremental steps; *transition*, which provides opportunities for implementing alternatives that would facilitate future more radical change; and *transformation*, which requires innovative, far-reaching policies and practical adaptation actions. Furthermore, Roggema, Vermeend, and van den Dobbelsteen (2012, p. 2527) argue that "current spatial planning practice is very well equipped to accommodate incremental change and in some examples transitions, but is badly prepared for transformation." For example, the three most recent regional plans for the Dutch province of Gröningen would allow only 2% of the region to change permitted uses over the next 13 years while challenges posed by climate change may require a more fundamental long-term *transformational* change to address emerging threats and adapt to changing environmental conditions (Roggema, Vermeend & van den Dobbelsteen, 2012). Similarly, Torabi, Dedekorkut-Howes, and Howes (2017) and Alexandra (2017) highlight several challenges to transformational change faced by urban systems including short-term horizons of local plans, conflicting ideas forcing political processes into stark juxtaposition, fragmented institutions, sunk investment in fixed urban infrastructure, and differences in resources, capacities, and knowledge.

Several studies have analyzed the potential drivers of climate adaptation (Brandes & LeBlanc, 2013; Rauken, Mydske & Winsvold, 2015; Larsson, Keskitalo & Akermark, 2016; Webler, Tuler, Dow, Whitehead & Kettle, 2016; Dilling, Pizzi, Berggren, Ravikumar & Andersson 2017; Meerow & Mitchell, 2017). Drawing upon the literature on policy change, Dilling, Pizzi, Berggren, Ravikumar, and Andersson (2017) designed four hypotheses to test the drivers of climate policy adoption across 60 local governments in the states of Colorado, Utah, and Wyoming. The study found that external funding had a strong positive association with higher rates of adoption of policy change while the presence of policy champions, experience with natural hazards, and perceived risk had only a weak relationship to adaptation action. However, the number of adopted initiatives increased with the number of past hazardous events and higher levels of perceived risk (Dilling, Pizzi, Berggren, Ravikumar & Andersson, 2017). Larsson, Keskitalo, and Akermark (2016) analyzed planning documents in 29 municipalities in northern Sweden to understand the institutional barriers for the implementation of local climate adaptation measures. The study found that climate adaptation, apart from consideration in risk and vulnerability analyses, is not a well-resourced policy area. However, integration with other considerably more developed policy areas such as flood management and emergency response has proven beneficial for addressing climate change impacts (Larsson, Keskitalo & Akermark, 2016). In an assessment of 98 climate adaptation plans in the United States, Koski and Siulagi (2016) found that cities that integrate climate adaptation into hazard management adopt more extensive climate adaptation initiatives than cities that do not. The study corroborates some of the findings by Dilling, Pizzi, Berggren, Ravikumar, and Andersson (2017), namely a weak relationship to political partisanship and environmental advocacy, but finds support for the role of municipal financing of adaptation actions (Koski & Siulagi, 2016). Rauken, Mydske, and Winsvold (2015) assessed various approaches to "mainstream" climate adaptation into existing policy sectors in five Norwegian municipalities. The study identified three aspects of mainstreaming: (1) comprehensiveness achieved through inclusion of climate adaptation in several interrelated policy areas; (2) aggregation attained through the cumulative effects of cross-sectoral adaptation actions; and (3) consistency suggesting a lack of conflicting policy objectives (Rauken, Mydske & Winsvold, 2015).

There is a growing interest in the scholarly literature on identifying the barriers and limits to adaptation (Moser & Ekstrom, 2010; Dow et al., 2013; Barnett et al., 2015; McNamara, Westoby & Smithers, 2017). In the context of climate change, limits to adaptation are defined as system states or thresholds beyond which irreversible changes in physical, ecological, or social systems occur (Adger et al., 2009; Moser & Ekstrom, 2010; Dow et al., 2013; Barnett et al., 2015). Barriers to adaptation are impediments that can be overcome "with concerted effort, creative management, change of thinking, prioritization, and related shifts in resources, land uses, institutions, etc." (Moser & Ekstrom, 2010, p. 2). The boundaries between limits and barriers are not viewed in absolute terms as many ostensible limits, especially those related to social, regulatory, or institutional issues, may de facto be overcome through civic engagement, adequate policies, and allocation of sufficient resources (Adger et al., 2009).

Studies have also addressed social vulnerability as a barrier to climate adaptation (Angueloski et al., 2016; Kashem, Wilson & van Zandt, 2016). Angueloski et al. (2016) examined the relationship between climate adaptation and equity in eight cities: New Orleans and Boston (U.S.), Medellin (Colombia), Manila (the Philippines), Santiago (Chile), Jakarta (Indonesia), Surat (India), and Dhaka (Bangladesh). The study found that protective infrastructure is often prioritized in areas with higher property values and economic assets at the expense of already underserved communities. Oftentimes land use regulations enforced to mitigate climate hazards disproportionately affect poor neighborhoods, leading to displacement (Angueloski et al., 2016). Inadequate risk governance mechanisms and excessive reliance on private funding may also exacerbate social inequality in climate adaptation (Angueloski et al., 2016). Bin Kashem, Wilson, and Van Zandt (2016) demonstrated that the spatial distribution of socially vulnerable groups varies over time and that the dispersal of such groups does not necessarily mean reduced vulnerability. The study suggests that measures of social vulnerability be tailored to the local context (Bin Kashem, Wilson & Van Zandt, 2016). Overall, more research is needed to understand how climate change can mitigate or exacerbate social vulnerability.

References

Abeygunawardena, P., Vyas, Y., Knill, P., Foy, T., Harrold, M., Steele, P., Tanner, T., Hirsch, D., Oosterman, M., Rooimans, J., Debois, M., Lamin, M., Liptow, H., Mausolf, E., Verheyen, R., Agrawala, S., Caspary, G., Paris, R., Kashyap, A., Sharma, A., Mathur, A., Sharma, M., & Sperling, F. 2009. *Poverty and Climate Change: Reducing the Vulnerability of the Poor through Adaptation (English)*. Washington, DC: World Bank. Retrieved from http://documents.worldbank.org/curated/en/534871468155709473/Poverty-and-climate-change-reducing-the-vulnerability-of-the-poor-through-adaptation.

Adaptation Fund. 2018. *Governance*. Retrieved from https://www.adaptation-fund.org/about/governance/.

Adger, W.N. 2002. Social and Ecological Resilience: Are They Related? *Progress in Human Geography* 4(3), 347–364.

Adger, W.N., Huq, S., Brown, K., Conway, D., & Hulme, M. 2003. Adaptation to Climate Change in the Developing World. *Progress in Development Studies* 3, 179–195.

Adger, W.N., Hughes, T.P., Folke, C., Carpenter, S.R., & Rockström, J. 2005. Social-ecological Resilience to Coastal Disasters. *Science* 309, 1036–1039.

Adger, W.N. 2006. Vulnerability. *Global Environmental Change* 16(3), 268–281.

Adger, W.N., Dessai, S., Goulden, M., Hulme, M., Lorenzoni, I., Nelson, D.R., Naess, L.O., Wolf, J., & Wreford, A. 2009. Are there Social Limits to Adaptation to Climate Change? *Climatic Change* 93, 335–354.

Adger, W.N., Brown, I., & Surminski, S. 2018. Advances in Risk Assessment for Climate Change Adaptation Policy. *Philosophical Transactions A, Royal Society A* 376, 20180106.

Alexandra, J. 2017. The City as Nature and the Nature of the City – Climate Adaptation Using Living Infrastructure: Governance and Integration Challenges. *Australasian Journal of Water Resources* 21(2), 63–76.

Anguelowski, I., Shi, L., Chu, E., Gallagher, D., Goh, K., Lamb, Z., Reeve, K., & Teicher, H. 2016. Equity Impacts of Urban Land Use Planning for Climate Adaptation: Critical Perspectives from the Global North and South. *Journal of Planning Education and Research* 36(3), 333–348.

Barnett, J., Evans, L.S., Gross, C., Kiem, A.S., Kingsford, R.T., Palutikof, J.P., Pickering, C.M., & Smithers, S.G. 2015. From Barriers to Limits to Climate Change Adaptation: Path Dependency and the Speed of Change. *Ecology and Society* 20(3), 5–15.

Becker, E. 2012. Social-ecological Systems as Epistemic Objects. In Glaser, M., Gesche, K., Ratter, B., & Welp, M. (Eds.) *Human-Nature Interactions in the Anthropocene: Potentials of Social-Ecological Systems Analysis*, 37–60. London, UK: Routledge.

Beichler, S.A., Hasibovic, S., Davidse, B.J., & Deppisch, S. 2014. The Role Played by Social-ecological Resilience as a Method of Integration in Interdisciplinary Research. *Ecology and Society* 19(3), 4.

Berke, P.R. 2014. Rising to the Challenge: Planning for Adaptation in the Age of Climate Change. In Glavovic, B.C., & Smith, G.P. (Eds.) *Adapting to Climate Change: Lessons from Natural Hazards Planning*. Dordrecht, Netherlands: Springer.

Berrang-Ford, L., Ford, J.D., & Paterson, J. 2011. Are We Adapting to Climate Change? *Global Environmental Change* 21(1), 25–33.

Biagini, B., Bierbaum, R., Stilts, M., Dobardzic, S., & McNeeley, S.M. 2014. A Typology of Adaptation Actions: A Global Look at Climate Adaptation Actions Financed through the Global Environment Facility. *Global Environmental Change* 25, 97–108.

Bin Kashem, S., Wilson, B., & Van Zandt, S. 2016. Planning for Climate Adaptation: Evaluating the Changing Patterns of Social Vulnerability and Adaptation Challenges in Three Coastal Cities. *Journal of Planning Education and Research* 26(3), 304–318.

Birkmann, J., & von Teichman, K. 2010. Integrating Disaster Risk Reduction and Climate Change Adaptation: Key Challenges – Scales, Knowledge, and Norms. *Sustainability Science* 5(2), 171–184.

Bloetscher, F., Polsky, C., Bolter, K., Mitsova, D., Pablicke Garces, K., King, R., Cosio Carballo, I., & Hamilton, K. 2016. Assessing Potential Impacts of Sea Level Rise on Public Health and Vulnerable Populations in Southeast Florida and Providing a Framework to Improve Outcomes. *Sustainability* 8, 315. doi:10.3390/su8040315

Bloetscher, F., & Romah, T. 2015. Tools for Assessing Sea Level Rise Vulnerability. *Journal of Water and Climate Change* 6(2), 181–190.

Bloetscher, F., Romah, T., Berry, L., Hammer, N.H., & Cahill, M.A. 2012. Identification of Physical Transportation Infrastructure Vulnerable to Sea Level Rise. *Journal of Sustainable Development* 5(12), 40–51.

Brandes, U., & LeBlanc, A. 2013. *Risk & Resilience in Coastal Regions*. Washington, DC: Urban Land Institute, ISBN: 978-0-87420-280-9.

Brooks, N., Adger, W.N., & Kelly, P.M. 2005. The Determinants of Vulnerability and Adaptive Capacity at the National Level and the Implications for Adaptation. *Global Environmental Change* 15, 151–163.

Bulkeley, H., Castan Broto, V., Hodson, M., & Marvin, S. (Eds.). 2010. *Cities and Low Carbon Transitions*. London: Routledge.

Bulkeley, H., Castan Broto, V., Hodson, M., & Marvin, S. 2011. Cities and the Low Carbon Transition. *The European Financial Review*, August 2011, 24–27.

Bulkeley, H., & Betsill, M.M. 2003. *Cities and Climate Change: Urban Sustainability and Global Environmental Governance*. London, UK: Routledge.

Davoudi, S., Crawford, F., & Mehmood, A. 2009. *Planning for Climate Change – Strategies for Mitigation and Adaptation for Spatial Planners*. London, UK: Earthscan.

Deyle, R.E., Bailey, K.C., & Matheny, A. 2007. *Adaptive Response Planning to Sea-level Rise in Florida and Implications for Comprehensive and Public-Facilities Planning*. Florida Planning and Development Lab, Department of Urban and Regional Planning, Florida State University (FSU).

Dilling, L., Pizzi, E., Berggren, J., Ravikumar, A., & Andersson, K. 2017. Drivers of Adaptation: Responses to Weather- and Climate-related Hazards in 60 Local Governments in the Intermountain Western U.S. *Environment and Planning A* 49(11), 2628–2648.

Dow, K., Berkhout, F., Preston, B.L., Klein, R.J.T., Midgley, G., & Shaw, M.R. 2013. Limits to Adaptation. *Nature Climate Change* 3(4), 305–307.

Edwards, M., & Haines, A. 2007. Evaluating Smart Growth Implications for Small Communities. *Journal of Planning Education and Research* 27(1), 49–64.

Flato, G.M., & Boer, G.J. 2001. Warming Asymmetry in Climate Change Experiments. *Geophys Research Letters* 28, 195–198.

Ford, J.D., Berrang-Ford, L., & Paterson, J. 2011. A Systematic Review of Observed Climate Change Adaptation in Developed Nations. *Climatic Change* 106(2), 327–336.

Füssel, H.M. 2007. Adaptation Planning for Climate Change: Concepts, Assessment Approaches, and Key Lessons. *Sustainability Science* 2, 265–275.

Girardet, H. 2008. *Cities People Planet: Urban Development and Climate Change*. Chichester, West Sussex, UK: John Wiley & Sons.

Gordon, C., Cooper, C., Senior, C.A., Banks, H.T., Gregory, J.M., Johns, T.C., Mitchel, J.F.B., & Wood, R.A. 2000. The Simulation of SST, Sea Ice Extents and Ocean Heat Transports in a Version of the Hadley Centre Coupled Model Without Flux Adjustments. *Climate Dynamics* 16, 147–168.

Hunt, A., & Watkiss, P. 2011. Climate Change Impacts and Adaptation in Cities: A Review of the Literature. *Climatic Change* 104, 13–49.

Huntjens, P., Lebel, L., Pahl-Wostl, C., Camkin, J., Schulze, R., & Kranz, N. 2012. Institutional Design Propositions for the Governance of Adaptation to Climate Change in the Water Sector. *Global Environmental Change* 22, 67–81.

Intergovernmental Panel on Climate Change (IPCC). 2007. Climate Change 2007: Impacts, Adoption, and Vulnerability. In Palutikof, J.P., van der Linden, P.J., & Hanson, C.E. (Eds.) *Contribution of Working Group II to the Fourth Assessment Report of the Intergovernmental Panel on Climate Change*. Cambridge, UK: Cambridge University Press.

International Council for Local Environmental Initiatives—World Wildlife Fund (ICLEI-WWF). 2015. *Measuring Up 2015: How US Cities Are Accelerating Progress Toward National Climate Goals*. Retrieved from http://icleiusa.org/wp-content/uploads/2015/08/Measuring_Up_2015.pdf.

Klein, R.J.T., Nicholls, R.J., & Thomalla, F. 2003. Resilience to Natural Hazards: How Useful is this Concept? *Global Environmental Change Part B: Environmental Hazards* 5(1), 35–45.

Knapp, A.K., Avolio, M.L., Beier, C., Carroll, C.J.W., Collins, S.L., Dukes, J.S., Fraser, L.H., Griffin-Nolan, R.J., Hoover, D.L., Jentsch, A., Loik, M.E., Phillips, R.P., Post, A.K., Sala, O.E., Slette, I.J., Yahdjian, L., & Smith, M.D. 2016. Pushing Precipitation to the Extremes in Distributed Experiments: Recommendations for Simulating Wet and Dry Years. *Global Change Biology* 23(5), 1774–1782.

Knutson, T.R., Delworth, T.L., Dixon, K.W., & Stouffer, R.L. 1999. Model Assessment of Regional Surface Temperature Trends (1949–1997). *Journal of Geophysical Research* 104(D24), 30 981–30 996.

Koski, C., & Siulagi, A. 2016. Environmental Harm or Natural Hazard? Problem Identification and Adaptation in U.S. Municipal Climate Action Plans. *Review of Policy Research* 33(3), 270–290.

Larsson, L., Keskitalo, E.C.H., & Akermark, J. 2016. Climate Change Adaptation and Vulnerability Planning within the Municipal and Regional System. *Journal of Northern Studies* 10(1), 67–90.

Lobell, D.B., Schlenker, W., & Costa-Roberts, J. 2011. Climate Trends and Global Crop Production Since 1980. *Science* 333(6042), 616–620. 10.1126/science.1204531.

McGuffie, K., & Henderson-Sellers, A. 2001. Forty Years of Numerical Climate Modelling. *International Journal of Climatology* 21, 1067–1109.

McNamara, K.E., Westoby, R., & Smithers, S.G. 2017. Identification of Limits and Barriers to Climate Change Adaptation: Case Study of Two Islands in Torres Strait, Australia. *Geographical Research* 55(4), 438–455.

McNamara, K.E., Bronen, R., Fernando, N., & Klepp, S. 2018. The Complex Decision-making of Climate-induced Relocation: Adaptation and Loss and Damage. *Climate Policy* 18(1), 111–117.

Meerow, S., & Mitchell, C.L. 2017. Weathering the Storm: The Politics of Urban Climate Change Adaptation Planning. *Environment and Planning A* 49(11), 2619–2627.

Millner, A., & Dietz, S. 2011. *Adaptation to Climate Change and Economic Growth in Developing Countries*. Grantham Research Institute on Climate Change and the Environment working paper no. 60, London School of Economics, London.

Mimura, N., Pulwarty, R.S., Duc, D.M., Elshinnawy, I., Redsteer, M.H., Huang, H.Q., Nkem, J.N., & Sanchez Rodriguez, R.A. 2014. Adaptation Planning and Implementation. In Field, C.B., Barros, V.R., Dokken, D.J., Mach, K.J., Mastrandrea, M.D., Bilir, T.E., Chatterjee, M., Ebi, K.L., Estrada, Y.O., Genova, R.C., Girma, B., Kissel, E.S., Levy, A.N., MacCracken, S., Mastrandrea, P.R., & White, L.L. (Eds.) *Climate Change 2014: Impacts, Adaptation, and Vulnerability. Part A: Global and Sectoral Aspects. Contribution of Working Group II to the Fifth Assessment Report of the Intergovernmental Panel on Climate Change*, 869–898. Cambridge, UK/New York, NY: Cambridge University Press.

Moser, S.C., & Ekstrom, J.A. 2010. A Framework to Diagnose Barriers to Climate Change Adaptation. *PNAS (Proceedings of the National Academy of Sciences of the United States)* 201007887. https://doi.org/10.1073/pnas.1007887107.

National Research Council (NRC). 2010. *Adapting to the Impacts of Climate Change*. Washington, DC: The National Academies Press.

Nemakonde, L.D. & van Niekerk, D. 2017. A Normative Model for Integrating Organizations for Disaster Risk Reduction and Climate Change Adaptation within SADC Member States. *Disaster Prevention and Management: An International Journal* 26(3), 361–376.

Neumann, J.E., Yohe, G., Nicholls, R., & Manion, M. 2000. *Sea Level Rise and Global Climate Change: A Review of Impacts to U.S. Coasts*. Pew Center on Global Climate Change. Retrieved from http://www.pewclimate.org/docUploads/env_sealevel.pdf.

Nicholls, R.J., Hoozemans, F.M.J., & Marchand, M. 1999. Increasing Flood Risk and Wetland Loss Due to Global Sea-level Rise: Regional and Global Analyses. *Global Environmental Change* 9, S69–S87.

Nozawa, T., Emori, S., Takemura, T., Nakajima, T., Numaguti, A., Abe-Ouchi, A., & Kimoto, M. 2000. *Coupled Ocean-Atmosphere Model Experiments of Future Climate Change Based on IPCC SRES Scenarios*. 11th Symposium on Global Change Studies, Long Beach, USA.

Obeysekera, J., & Park, J. 2013. Scenario-based Projection of Extreme Sea Levels. *Journal of Coastal Research* 29, 1–7.

Office of Management and Budget (OMB). 2006. *Proposed Risk Assessment Bulletin*. Released January 9, 2006. Washington, DC: Office of Management and

Budget, Executive Office of the President [online]. Retrieved from http://www.whitehouse.gov/omb/ inforeg/proposed_risk_assessment_bulletin_010906.pdf.

Organisation for Economic Co-operation and Development (OECD). 2006. *Adaptation to Climate Change: Key terms. OECD/IEA Project for the Annex I Expert Group on the UNFCCC.* Retrieved from https://www.oecd.org/env/cc/36278739.pdf.

Rauken, T., Mydske, P.K., & Winsvold, M. 2015. Mainstreaming Climate Change Adaptation at the Local Level. *Local Environment* 20(4), 408–423.

Roggema, R., Vermeend, T., & van den Dobbelsteen, A. 2012. Incremental Change, Transition or Transformation? Optimizing Change Pathways for Climate Adaptation in Spatial Planning. *Sustainability* 4(10), 2525–2549.

Seneviratne, S.I., Nicholls, N., Easterling, D., Goodess, C.M., Kanae, S., Kossin, J., Luo, Y., Marengo, J., McInnes, K., Rahimi, M., Reichstein, M., Sorteberg, A., Vera, C., & Zhang, X. 2012. Changes in Climate Extremes and Their Impacts on the Natural Physical Environment. In Field, C.B., Barros, V., Stocker, T.F., Qin, D., Dokken, D.J., K Ebi, L., Mastrandrea, M.D., Mach, K.J., Plattner, G-K, Allen, S.K., Tignor, M., & Midgley, P.M. (Eds.) *Managing the Risks of Extreme Events and Disasters to Advance Climate Change Adaptation.* A Special Report of Working Groups I and II of the Intergovernmental Panel on Climate Change (IPCC), 109–230. Cambridge, UK/New York, NY: Cambridge University Press.

Schelling, T.C. 1992. Some Economics of Global Warming. *The American Economic Review* 82(1), 1–14.

Strauss, B.H., & Kulp, S. 2014. *New Analysis Shows Global Exposure to Sea Level Rise, Research Report by Climate Central.* Retrieved from http://www.climatecentral.org/news/new-analysis-global-exposure-to-sea-level-rise-flooding-18066.

Streck, C. 2015. The Paris Agreement Summary. *Climate Focus.* Client Brief on the Paris Agreement III. Retrieved from http://www.climatefocus.com/sites/default/files/20151228%20COP%2021%20briefing%20FIN.pdf.

Thomalla, F., Downing, T., Spanger-Siegfried, E., Han, G., & Rockström, J. 2006. Reducing Hazard Vulnerability: Towards a Common Approach between Disaster Risk Reduction and Climate Adaptation. *Disasters* 30(1), 39–48.

Torabi, E., Dedekorkut-Howes, A., & Howes, M. 2017. Not Waving, Drowning: Can Local Government Policies on Climate Change Adaptation and Disaster Resilience make a Difference? *Urban Policy and Research* 35(3), 312–332.

United Nations (UN). 2015. *Sendai Framework for Disaster Risk Reduction 2015–2030.* Adopted at the Third United Nations World Conference on Disaster Risk

Reduction, Sendai, UN General Assembly, A/CONF.2224/L.2, 14-18 March, Geneva.

United Nations Development Programme (UNDP). 2008. *UNDP's Monitoring Framework for Climate Change Adaptation*. Retrieved from http://www.un.org/esa/sustdev/natlinfo/indicators/15Oct_2008/presentations_pdf/Bo%20 Lim.pdf.

United Nations Department of Economic and Social Affairs (UN DESA). 2017. *World Population Prospects: The 2017 Revision, Volume I: Comprehensive Tables., ST/ESA/SER.A/399*. UN-DESA Population Division.

United Nations Framework Convention on Climate Change (UNFCCC). 2015.

Vachette, A. 2017. Integrating Disaster Risk Reduction and Climate Change Adaptation in Vanuatu: The Art and Practice of Building Resilience to Hazards. In Leal Filho, W. (Ed.) *Climate Change Adaptation in Pacific Countries*. Springer, Cham: Climate Change Management.

Wamsler, C., & Brink, E. 2018. Mindsets for Sustainability: Exploring the Link between Mindfulness and Sustainable Climate Adaptation. *Ecological Economics* 151, 55–61.

Washington, W.M., Weatherly, J.W., Meehl, G.A., Semtner Jr., A.J., Bettge, T.W., Craig, A.P., Strand Jr., W.G., Arblaster, J.M., Wayland, V.B., James, R., & Zhang, Y. 2000. Parallel Climate Model (PCM): Control and Transient Simulations. *Climate Dynamics* 16(10–11), 755–774.

Webber, S. 2016. Climate Change Adaptation as a Growing Development Priority: Towards Critical Adaptation Scholarship. *Geography Compass* 10(10), 401–413.

Webler, T., Tuler, S., Dow, K., Whitehead, J., & Kettle, N. 2016. Design and Evaluation of a Local Analytic-deliberative Process for Climate Adaptation Planning. *Local Environment* 21(2), 166–188.

While, A., Jonas, A.E.J., & Gibbs, D. 2010. From Sustainable Development to Carbon Control: Eco-state Restructuring and the Politics of Urban and Regional Development. *Transactions of the Institute of British Geographers* 35(1), 76–93.

4 Addressing Climate Change

Initiatives and Coping Strategies from Across the Globe

Chapter Objectives

This chapter seeks to answer the following questions:

- How do cities address synergies and trade-offs between climate mitigation and adaptation planning?
- What are the challenges faced by the Small Island Developing States (SIDS) in addressing the impacts of climate change?
- What are the linkages between climate adaptation, food security, and development in the Global South?
- What coping strategies are being adopted by smallholder farming communities in the Global South?
- What is the role of climate services and geospatial technologies in facilitating climate adaptation planning?

Climate Adaptation Pathways: A Global Perspective

Climate adaptation and mitigation actions vary within and between nations and regions of the globe. Economic, social, political, and institutional arrangements, often predicated on historical processes and path dependencies, determine societal dynamics and possible pathways of climate adaptation action (Betzold, 2015; Fazey et al., 2016; McNamara, Westoby & Smithers, 2017; Beyerl, Mieg & Weber, 2018). Drawing on a geographically diverse set of local and regional approaches to climate adaptation, we examine how vulnerabilities, risk prioritization, institutional arrangements, and participatory processes are shaping climate adaptation initiatives and coping strategies across the globe. In doing so, this chapter focuses on five overarching themes:

- Synergies and trade-offs between climate mitigation and adaptation strategies
- Actions undertaken by coastal regions and megacities to cope with the threats of sea level rise
- Capacity building and participatory processes used by small island nations to prioritize and strengthen adaptation efforts
- Coping strategies to address food insecurity in the Global South in the face of increasing climate variability
- Contributions of climate information services and geospatial technologies to climate adaptation decision-making

For many cities and regions, climate mitigation and adaptation actions are intertwined with urban sustainability and resilience initiatives. In 2014, the International Council for Local Environmental Initiatives (ICLEI)–Local Governments for Sustainability in collaboration with the Massachusetts Institute of Technology surveyed 350 cities, members of ICLEI worldwide, to better understand the scope and implementation of climate policies (Aylett, 2014). Over 70% of the cities included in the survey reported efforts to achieve progress in both climate mitigation and climate adaptation. Nearly 80% reported on developing emission inventories and implementing specific greenhouse gas (GHG) emission reduction targets. Overall, 85% of the cities sought to reduce emissions from municipal waste, residential energy consumption, municipal buildings, and transportation (Aylett, 2014). In addition, vulnerable coastal cities and regions are increasingly adopting climate adaptation measures to strengthen coastal defenses and guide future development and infrastructure investments. The selection of feasible adaptation options is often constrained by the uncertainty associated with estimated future risks as well as anticipated costs.

Small Island Developing States (SIDS) are amongst the most threatened by climate change in their long-term survival. The most prominent among multiple environmental stressors faced by these SIDS are sea level rise, erosion, higher storm surge, increased cyclone activity, drought, saltwater intrusion, and significant uncertainty with respect to rainfall variability (Van Aalst, Ebi, Githeko, Yohe & Jones, 2004; Gaillard & Maceda, 2009; Barnett & O'Neill, 2012; Piccolella, 2013; DeGraff & Ramlal, 2015; Betzold, 2015; McNamara, Westoby & Smithers, 2017; Beyerl, Mieg & Weber, 2018). A study conducted by the World Bank in the late

1990s—*Cities, Seas, and Storms*—indicated that "the impacts of climate change are likely to be pervasive" (World Bank, 2000, p. 3). In order to "vastly decrease or downsize costs should climate change scenarios materialize," the Bank suggested the inclusion of adaptation objectives in future infrastructure expenditure planning and highlighted the importance of supporting community-based adaptation to prepare island nations for the future given the uncertainty in climate projections and the unknowns associated with timescale and magnitude of change (World Bank, 2000, p. x).

Climate change is expected to intensify drought conditions in many parts of the Global South, increasing the prevalence of moderate to severe food insecurity (FAO, 2017). Sub-Saharan Africa (SSA), known for its frequent and severe droughts and devastating famines, is highly dependent on rainfed crop and livestock production, which is particularly vulnerable to climate variability and change (Sidibe, Foudi, Pascual & Termansen, 2018, p. 588). In July 2016, the Food and Agriculture Organization (FAO) estimated that more than 40 million people in eastern and southern Africa, and nearly 20 million in other parts of the continent, endured severe food shortages because of the 2015/2016 El Niño–related droughts. It is expected that, by 2020, yields from rainfed agriculture in some parts of Africa could decrease by up to 50%, further reducing access to food and exacerbating malnutrition (FAO, 2017). Furthermore, climate models project an increase of 5% to 8% of arid and semi-arid lands in Africa by 2080 (FAO, 2017, p. 24).

In Southeast Asia, food security is highly dependent on monsoonal rainfall. By the end of the century, the average monsoonal rainfall is projected to increase by 5%–10% due to the warming of the Indian Ocean, which will carry higher amounts of moisture over Southeast Asia and India (Turner & Annamalai, 2012), amplifying flood risks for megacities such as Bangkok (Thailand), Jakarta (Indonesia), and Mumbai (India) (Swiss Re, 2013). Despite this general trend, a break in the monsoon rainfall during the growing season may cause water deficit and agricultural drought and affect the livelihoods of nearly one billion people (Turner & Annamalai, 2012). The slow movement of the monsoon rains in the summer in 2018, caused by a heat wave with temperatures of over 40°C (104°F) across northeastern India (India Meteorological Department, 2018), left many areas in northeast India, from Andhra Pradesh to West Bengal, with below-normal rainfall, jeopardizing expected yields from agriculture.

Reflecting on adaptation actions across geographic regions, this chapter examines how world cities, highly urbanized coastal regions, small island nations, and smallholder farming communities adapt to climate variability and change. The chapter begins with examples of how cities have addressed both greenhouse gas emission reductions and climate adaptation challenges. Various structural and non-structural approaches to address the impacts of sea level rise are also discussed. This is followed by a section which highlights the importance of participatory and consensus-building approaches to climate adaptation, including community-based disaster risk management to improve community resilience in the Small Island Developing States. The chapter also discusses how smallholder farming communities in the Global South adapt to the changing environmental conditions, highlighting the linkages between climate adaptation, food security, and development. The chapter concludes with a brief discussion of the role of climate services and geospatial technologies in facilitating climate adaptation planning and decision-making.

Mitigation and Adaptation: Synergies and Trade-offs

Managing the synergies and trade-offs between mitigation and adaptation is key to the successful implementation of climate change policies (Kane & Shogren, 2000; Thornton & Comberti, 2017; Schwirplies, 2018). Kane and Shogren (2000, p. 76) point out that "effective climate protection needs to use an integrated portfolio of mitigation and adaptation strategies," acknowledging the importance of adaptation in achieving emission abatement targets. Undoubtedly, the existing climate action plans initially developed to promote voluntary commitments to reduce greenhouse gas emission, provide a pathway for leveraging policy and planning experience, partnerships, institutional capacity, and financial resources in the design and implementation of climate adaptation actions. The Climate Action Planning Framework (CAPF) established by the C40 Cities initiative in 2018 highlights these synergies and their role in "developing climate action plans that are aligned with the objectives of the Paris Agreement" (C40 Cities, 2018, online). The framework incorporates four key components: (1) achieving greenhouse gas emissions neutrality or net-zero emissions by 2050 based on emission inventories, modeling, and GHG reduction commitments; (2) achieving climate resilience in the short-, medium-, and long-term by setting climate adaptation milestones based on climate scenarios and risk assessments; (3) ensuring that social, economic, and environmental benefits from climate action plans are equitably distributed and inclusive; and (4) establishing effective collaborations, partnerships, and climate action governance mechanisms (C40 Cities, 2018). A brief discussion of selected case studies highlighting various strategies in developing and implementing mitigation and adaptation actions is provided here.

The city of **London** maintains one of the most technologically advanced inventories with the London Energy and Greenhouse Gas Inventory (LEGGI). According to the *Mayor's Climate Change Mitigation and Energy Annual Report,* in 2013–2014, London reduced its CO_2 emissions by 20% compared to 2008 levels despite population growth of 8% (Greater London Authority [GLA], 2015). In addition to diversification of energy supply, London's mayoral office adopted a climate change adaptation strategy focusing on three pillars aimed at making the city more sustainable and resilient to climate change impacts: retrofitting building, greening the city, and addressing flood risk (GLA, 2011). As part of a comprehensive strategy, the city has initiated a home retrofitting program which helped improve the energy efficiency of 100,000 residential units across the capital. Combined with broader stakeholder participation and market incentives, the program has resulted in nearly 500,000 retrofitted homes in the Greater London area and accrued savings from electricity bills totaling £4.7 million (GLA, 2015). Furthermore, the city has undertaken measures to increase green space, improve air quality, reduce heat stress, and address flood risk posed by fluvial and pluvial floods, tidal inundation, and sea level rise (GLA, 2011).

In 2012, **Copenhagen**'s city council approved a plan to transform the Danish capital into the world's first "carbon-neutral" capital city by 2025 (City of Copenhagen, 2012; Gerdes, 2013). The city had already reduced its carbon emissions by 21%

between 2005 to 2011 and envisioned other sweeping changes in building design, innovative cooling and heating systems, public transportation, bicycle infrastructure, and open space (City of Copenhagen, 2012). Copenhagen created nearly 250 miles of bike lanes, and, as a result, 36% of all trips made by Danish capital residents to work or school involve the use of the bike lane system (Gerdes, 2013).

Over the past decade, the city of **Sao Paolo, Brazil**, has made concerted efforts to systematically integrate planning for urban sustainability with strategies to reduce aggregate municipal GHG emissions and address climate adaptation needs (Di Giulio, Bedran-Martins, Vasconcellos, Ribeiro & Lemos, 2018). An active member of ICLEI–Local Governments for Sustainability and the C40 Cities Climate Leadership Group, the city has spearheaded a number of initiatives to develop GHG emissions inventories, expand green areas, and improve urban mobility by offering sustainable transportation alternatives. In 2009, the city approved a Municipal Climate Law which became model legislation motivating other Brazilian cities and states to adopt similar measures (Setzer, Macedo & Rei, 2015, as cited in Di Giulio, Bedran-Martins, Vasconcellos, Ribeiro & Lemos, 2018). Sao Paolo's 2014 Strategic Master Plan further strengthens commitments to policy actions by addressing many of the megacity's current and future challenges. Reflecting on the goals and interventions proposed by the plan, Di Giulio, Bedran-Martins, Vasconcellos, Ribeiro, and Lemos (2018, p. 239) note that although the master plan does not explicitly recognize synergies and trade-offs between urban sustainability and climate mitigation/adaptation, it provides "an opportunity to address climate change actions in the context of other urban and environmental concerns, development pressures and goals, and a range of often conflicting values and priorities" (Di Giulio, Bedran-Martins, Vasconcellos, Ribeiro & Lemos, 2018, p. 239). A major strength of the plan is its focus on broadening stakeholder engagement and participatory processes. In 2014, Brazil's São Paulo State was recognized by the United Nations Office for Disaster Risk Reduction (UNISDR) for its commitment to urban resilience and broad participation of local municipalities in the Making Cities Resilient Campaign (UNISDR, 2014).

In the United States, over 130 municipalities have established GHG emissions reduction targets (ICLEI–World Wildlife Fund [WWF], 2015). The local governments of **Annapolis, Maryland**; **Seattle, Washington**; **Austin, Texas**; and **Northfield, Minnesota** have pledged to attain 100% reduction in CO_2 emissions by 2050 (ICLEI-WWF, 2015). In addition, 33 municipalities have established commitments to decrease GHG emissions by at least 80% by 2050 (ICLEI-WWF, 2015). Among them is **Portland, Oregon**, a city that has been universally recognized for its pioneering initiatives in support of urban sustainability and climate action (ICLEI-WWF, 2015). Portland has adopted an ambitious plan to cut down its emissions by 40% by 2030 and 80% by 2050. Moreover, the city has introduced an innovative consumption-based inventory to track progress toward cutting down GHG emissions. The city's climate action plan seeks to achieve these goals by expanding public transit and bikeways networks, increasing the number of certified green buildings and net-zero buildings, boosting the use of low-carbon fuel, promoting the use of electric vehicles, and expanding the number of solar energy systems (ICLEI-WWF, 2015). After joining the U.S. Conference of Mayors Climate Protection Plan in 2006, **Atlanta, Georgia**, established the first GHG emissions inventory for the

State of Georgia. Working closely with a broad range of stakeholders, the city has initiated programs for retrofitting buildings to improve energy efficiency, setting requirements for city-owned buildings to meet Leadership in Energy and Environmental Design (LEED) certification standards, and promoting the implementation of Property Assessed Clean Energy (PACE) financing (ICLEI-WWF, 2015).

In response to Executive Order S-13-08, the California Natural Resources Agency (CNRA) developed a comprehensive climate adaptation strategy for the **State of California** encompassing seven strategic sectors: management of ocean and coastal resources, infrastructure, water management, agriculture, forestry, biodiversity, and public health (CNRA, 2009). The 2009 California Climate Adaptation Strategy (CAS) prioritizes water resources management, highlighting the need to reduce the per capita water consumption, increase surface and groundwater storage, adopt measures for agricultural water use efficiency, and improve ecosystem health (CNRA, 2009, p. 7). The strategy recommends considering potential climate change impacts in future development and infrastructure projects. Furthermore, it emphasizes:

> California's ability to manage its climate risks through adaptation depends on a number of critical factors including its baseline and projected economic resources, technologies, infrastructure, institutional support and effective governance, public awareness, access to the best available scientific information, sustainably-managed natural resources, and equity in access to these resources.
>
> (CNRA, 2009, p. 4)

Three legislative measures adopted in 2015 further strengthen California's commitment to managing climate change: SB 246 (Chapter 606, Statutes of 2015), which creates a central hub for information and tools for local stakeholders; AB 1482 (Chapter 603, Statutes of 2015), which establishes a framework for coordination among state agencies; and SB 379 (Chapter 608, Statutes of 2015), which requires adoption of climate adaptation measures in local planning documents (California Legislature–Senate Committee on Environmental Quality, 2015).

These examples highlight a broad range of interventions across local and regional scales and how individual actions coalesce to form synergies in support of climate risk management and achievement of emission reduction goals. The implementation of these initiatives demonstrates that climate adaptation does not amount to a set of discrete actions undertaken in response to short-term conditions and objectives but rather a dynamic and continually evolving process in setting an active agenda of policy learning for transformational change.

Climate Change Adaptation in Urbanized Coastal Regions

Hallegatte (2009) emphasized that climate adaptation planning should benefit local communities in the long-term by considering strategic priorities, establishing "safety margins" for new investments, and adopting flexible design options. In this context, policies and practical actions often revolve around *no-regrets* or *low-regrets* measures that are expected to yield benefits even without considering the anticipated impacts of climate change (Hallegate, 2009; Berke, 2014). This section outlines three climate adaptation strategies to address the impacts of coastal flooding and sea level rise: (1) adaptive waterfront redesign through retrofitting, floodproofing, and nature-based infrastructure; (2) construction of storm surge barriers and tidal floodgates; and (3) efforts to engage coastal communities in adaptive climate governance. In many instances, these three approaches are intertwined in both the decision-making process and the outcome.

The U.S. Urban Land Institute (ULI) identified several areas of possible future action to reduce vulnerability of infrastructure to climate change such as retrofitting structures, redesigning public spaces, developing new engineering standards for climate-proof buildings, promoting "context-sensitive" local planning and urban design strategies, reducing the heat island effects, and aligning disaster preparedness and recovery with new realities and threats (Brandes & LeBlanc, 2013, p. 20). The **City of New York** developed a compendium of adaptive measures operating across multiple scales: at the *site*, along *shoreline* segments, *upland*, and *in-water* (NYC Department of City Planning, 2013, p. 4). The study outlined a range of shoreline stabilization measures for each specific combination of landform, armoring, presence/absence of habitat, and land use (NYC Department of City Planning, 2013). Newman et al. (2013) provided a detailed overview of the **City of Boston** hazard exposure to future climate threats and outlined a framework for floodproofing the city by implementing climate adaptive design. The study focused on several applications of "hazard-resilient landscape design" (Newman et al., 2013, p. 47) in which nature-based infrastructure provides a range of ecosystem services to help moderate urban heat islands and enhance flood protection and erosion control.

Developing structural and non-structural measures to address the impacts of tidal flooding and sea level rise in coastal regions faces multiple challenges, including explicit consideration of uncertainty in future infrastructure investments, adjustment of existing and creation of new design standards, and securing financial and institutional resources. Many coastal cities are in the process of conceptualizing comprehensive strategies for waterfront climate adaptive redesign. In the face of multiple threats posed by sea level rise, riverine floods, precipitation extremes, and rising groundwater levels, the Dutch City of **Rotterdam** has developed a comprehensive approach to a proactive "green" adaptation strategy. The strategy highlights the importance of small-scale landscape design solutions embedded in "blue-green" networks (Voskamp & van den Ven, 2015). "Blue-green" infrastructure integrates design and ecosystem functions synergistically to moderate microclimate, improve aquifer recharge, and enhance floodwater storage capacities (Voskamp & van den Ven, 2015). Several neighborhoods and districts in Rotterdam have benefitted from the "blue-green" redesign adaptation options in addition to over 3,600 "green" jobs (Rotterdam Climate Initiative, 2012). Some of the redesigned Rotterdam neighborhoods also offer a glimpse to the future of "floating communities" (Rotterdam Climate Initiative, 2012, p. 28).

Rising sea levels increase the vulnerability of coastal defenses to storm surge and storm-induced high tides. In many densely populated coastal areas, existing levees or other coastal infrastructure could be jeopardized by extreme events. In some instances, even raising the height of the coastal defense structures would not provide sufficient protection to coastal communities. Leveraging advances in science, technology, and engineering, some of the most vulnerable cities have opted to construct movable floodgate barriers to address such challenges. The Dutch Delta Programme, a government-led initiative for adaptive delta management, pioneered the construction of one of the earliest movable floodgates with a project on the **Hollandse River** in the late 1950s. Currently, the Dutch **Oosterschelde barrier**, with a total length of 9 kilometers, is the largest storm surge barrier in the world. The barrier is composed of 65 concrete pillars and 62 movable steel gates designed to both maintain navigation through the shipping channels and protect the Netherlands from storm surge from the North Sea (Kohl, 1986). The **Thames Barrier**, completed in 1984, currently protects 1.25 million residents living in the Thames tidal floodplain, including sections of **London**, **North Kent**, and **South Essex** (UK Environment Agency, 2012). Large clusters of economic and financial activity, key government buildings, thousands of commercial and industrial properties, over 3,000 hectares of sensitive heritage sites, nearly 1,000 electrical substations, and 400 schools are located in the Thames Estuary floodplain (UK Environment Agency, 2012, p. 14). Over the past 34 years, the barrier's 10 steel gates have been closed 182 times to protect London from both tidal and combined tidal/fluvial flooding (The Thames Barrier, 2014).

Following the destruction caused by Hurricane Katrina in 2005, the U.S. Army Corps of Engineers began the construction of a complex Hurricane and Storm Damage Risk Reduction System (HSDRSS) to improve flood protection and facilitate the recovery of New Orleans and the surrounding areas (Miller, Desoto-Duncan & Hertzler, 2013). The HSDRSS strategic goals focus on addressing four major flood risks to the **Greater New Orleans metropolitan area**: (1) prevent a storm surge coming ashore; (2) reduce the risk of flooding from the Mississippi River; (3) minimize runoff; and (4) mitigate coastal erosion (Miller, Desoto-Duncan & Hertzler, 2013, p. 5). In addition to raising and strengthening the existing levees and seawalls, a key new feature of the coastal protection system is **Louisiana's Inner Harbor Navigation Canal (IHNC) Lake Borgne Surge Barrier**. The IHNC surge barrier, a 3,000-meter-long structure placed at the convergence of the Mississippi River Gulf Outlet and the Gulf Intercoastal Waterway, is designed to block 8-meter-high storm waves and withstand a 500-year storm event that could threaten the Greater New Orleans metropolitan area (Miller, Desoto-Duncan & Hertzler, 2013, pp. 8–9).

Some of the most populous coastal megacities in Southeast Asia, including Tokyo (Japan), Manila (the Philippines), Jakarta (Indonesia), Taipei (Taiwan), and the Pearl River Delta (China), are developed on large coastal plains exposed to typhoons, storm surge, and heavy rains. With a total population of over 10 million, **Jakarta** is considered one of the most vulnerable coastal cities in the world due to the compounding effects of both natural and anthropogenic factors (Takagi, Fujii, Esteban & Yi, 2017). Located in a low-lying coastal plain draining 13 rivers (Sherwell, 2016), the long-term survival of Jakarta is threatened by rising waters caused primarily by high rates of land subsidence (Takagi, Fujii, Esteban & Yi, 2017). Over the past 40 years, some areas of the city have sunk by nearly 4 meters (Win, 2017). Caused by unregulated groundwater extraction, land subsidence is exposing millions of residents to tidal and seasonal flooding (Takagi, Fujii, Esteban & Yi, 2017).

In 2014, the National Capital Integrated Coastal Development (NCICD) master plan, developed jointly by the governments of Indonesia and the Netherlands, envisioned the construction of a 32-kilometer seawall anchored approximately 2.5 kilometers from the current coastline to help reduce the impacts of flooding from the sea (NCICD, 2014). The project drew criticism for not sufficiently addressing the causes of land subsidence and a number of anticipated environmental impacts, including the bay water quality (Sherwell, 2016). Besides, the construction of the giant sea wall could potentially displace tens of thousands of residents and threaten their livelihoods (Win, 2017). Many in Indonesia believe that the most effective measure to mitigate future coastal floods is to address the causes of land subsidence (Sherwell, 2016; Win, 2017). The current rates of groundwater extraction could potentially sink vulnerable areas an additional 5 to 6 meters by 2100 (Takagi, Fujii, Esteban & Yi, 2017), increasing the city's vulnerability to storm surge, high tides, and sea level rise despite the promises of costly infrastructure improvements.

The unprecedented flood event in Thailand in 2011 demonstrated the vulnerability of tropical megacities to climate extremes. The 2011 Thailand floods affected 13.6 million people nationwide and caused damages in the amount of 10% of the national gross domestic product (GDP) (Haraguchi & Lall, 2015), ranking as the country's most destructive flood event to date (Gale & Saunders, 2013). The floods were triggered by anomalously high (over 40% above average) monsoonal rainfall between May and October, influenced by La Nina[1] (Gale & Saunders, 2013; Haraguchi & Lall, 2015). The stormwater storage in dams and reservoirs had already reached a peak capacity in late July when typhoons Nock-ten and Muifa made landfall in the northern part of Thailand (Haraguchi & Lall, 2015). The excess stormwater from the north reached the Chao Phraya River delta and flooded nearly 30,000 square kilometers in and around **Bangkok**, causing unparalleled losses and disruption of economic activities (Saito, 2014; Haraguchi & Lall, 2015). In addition to fluvial flooding, Bangkok is vulnerable to rising waters from the sea. Tidal surges can reverse the flow of the Chao Phraya River and inundate large portions of the city, particularly when interacting with floodwaters reaching the mouth of the river from the north (Saito, 2014). To address the flood risk from storms and high tides, Bangkok has implemented several measures, including raising flood barriers along the Chao Phraya River, improving natural infrastructure and groundwater management, changing zoning regulations and building codes, implementing early warning systems, strengthening emergency management, and reducing social vulnerability (Saito, 2014). Although some of these measures remain "stand-alone" interventions that do not explicitly consider climate change impacts (Saito, 2014), they engage decision-makers at all levels in developing regulatory and planning tools to strengthen flood risk management in both the near and longer term. Improving climate forecasts (Haraguchi & Lall, 2015) and good governance (Saito, 2014) could facilitate better water management decisions and mitigate some of the worst impacts.

Building regional capacity to address the challenges posed by climate change is paramount to adaptation decision-making. Apart from technical and design solutions, adaptation efforts often require new forms of adaptive climate governance (Birkmann, Garschagen, Kraas & Quang, 2010). Vella, Butler, Sipe, Chapin, and Murley (2016) discussed the complex multi-level voluntary collaborative arrangements of the Southeast Florida Regional Climate Change Compact (SFRCCC).

South Florida faces multiple challenges in the years ahead as one of the most vulnerable regions in the world to sea level rise due to its low-lying topography and porous limestone geology (Obeysekera & Park, 2012). Bordered by the ocean and the freshwater ecosystem of the Everglades, South Florida's ecological and economic vitality depends profoundly on its coastal and eco-marine systems. Growing populations and current patterns of development have exposed some of the region's most valuable assets and economic activities to the impacts of erosion, storm surge, flooding, and sea level rise. Since its establishment in 2009, the four-county compact[2] led by a nonpartisan Steering Committee had guided policy development and facilitated implementation of climate adaptation actions. As Vella, Butler, Sipe, Chapin, and Murley (2016, p. 372) note, "the combination of voluntary approaches, data sharing, and scientific coordination has helped achieve some key aspects of adaptive governance, particularly related to community-based action and social learning at the local government level."

Climate Adaptation in Small Island Developing States: A Focus on Capacity Building

The importance of participatory and capacity building approaches to climate adaptation and community resilience has been strongly emphasized by the Hyogo Framework of Action (HFA) (2005–2015), which called for strengthening community-based disaster risk management (CBDRM) (United Nations Office for Disaster Risk Reduction, 2007). Initially promoted by nongovernmental organizations (NGOs) providing humanitarian assistance to developing countries threatened by natural hazards (Benson, Twigg & Myers, 2001), CBDRM has become a standard practice in a wide range of community resilience building efforts supported by international organizations, multilateral banks, national and international aid agencies, and national institutions (Maceda, Gaillard, Stasiak, Le Masson & Le Berre, 2009, Gaillard & Maceda, 2009; Maskrey, 2011).

The Kiribati Adaptation Program (KAP) is one of the first climate adaptation projects in the Small Island Developing States with a strong emphasis on national consultations and participatory processes (Van Aalst, Ebi, Githeko, Yohe & Jones, 2004; Elrick & Kay, 2009; Webber, 2015). The **Republic of Kiribati** is an atoll nation in the Central Pacific region consisting of 33 low-lying coral islands of which 21 are inhabited (Van Aalst, Ebi, Githeko, Yohe & Jones, 2004; World Bank, 2011). The highest elevation point on the island chain does not exceed 6 meters. Most of the land of the densely populated areas in and around the capital Tarawa is below 3-meter elevation (Elrick & Kay, 2009). Crop yields are poor due to the arid climate and low-productivity soils. The economy of the islands is strongly dependent on fisheries, as a source of revenue and subsistence, and on foreign aid (Van Aalst, Ebi, Githeko, Yohe & Jones, 2004; World Bank, 2011). The Kiribati Adaptation Program funded by the World Bank placed a strong emphasis on (1) protecting and enhancing scarce water resources through water and sanitation investments, infrastructure improvements, rainwater harvesting, and assessment of future changes in precipitation and drought conditions; (2) increasing coastal resilience to storms, sea level rise, and flooding; and (3) strengthening community resilience through raising public awareness and conducting national consultations (Kiribati Climate Change,

2016). The KAP experiment highlighted the limits to adaptation beyond which proposed solutions could be maladaptive or unproductive (Webber, 2015). Despite its limitations, the KAP project provided much-needed guidance to further investments in climate adaptation in the region.

In 2014, building on the KAP experience, the World Bank, in collaboration with the **Solomon Islands** government, initiated the Community Resilience to Climate and Disaster Risk Project (CRISP) (2014–2019) co-funded by the Global Environment Facility's Least Development Country Fund (LDCF) and the Global Facility for Disaster Reduction and Recovery (GFDRR) (World Bank, 2014; Webber, 2015). The project aimed at improving the capacity of select rural communities to address risks from natural hazards based on four resiliency pillars: (1) developing policies, institutional capacity, and governance mechanisms for integration of climate change adaptation (CCA) and disaster risk reduction (DRR); (2) forming a national risk information platform and strengthening the natural hazards early warning systems; (3) investing in both structural and non-structural adaptation measures and infrastructure improvements; and (4) monitoring and evaluation of project activities (World Bank, 2014).

Deeply seated human and social assets and strong resistance to relocation as an adaptation option were found by a recent study of climate risk perceptions among the indigenous people of two islands in **Torres Strait, Australia** (McNamara, Westoby & Smithers, 2017). A significant driver for climate adaptation in the two island communities—**Boigu** and **Erub**—was rooted in the traditional knowledge systems, including "strong social connections, strong ties to culture and tradition, and access to abundant natural resources" (McNamara, Westoby & Smithers, 2017, p. 445). The study revealed several limits and barriers to adaptation in the two island communities, including a need for further external assistance for coastal protection and infrastructure improvements and better water resources management (McNamara, Westoby & Smithers, 2017, p. 452). Similarly, Beyerl, Mieg, and Weber (2018, p. 39) found that climate adaptation in the **Pacific small island states of Tuvalu, Samoa, and Tonga** is shaped by strong community norms and the Pacific culture of "sharing and caring." The study revealed that the people of the small island states were aware of the impacts of existing unsustainable practices, including logging, sand mining, and fossil fuel consumption. The local residents viewed climate adaptation as a comprehensive strategy to address the threats emanating from climate change along with further efforts to achieve the goals of sustainable development, including ecosystem protection, increase in income-generating opportunities, and poverty reduction (Beyerl, Mieg & Weber, 2018). Relocation was not considered a viable climate adaptation option by the majority of the study participants (Beyerl, Mieg & Weber, 2018).

Climate Adaptation in the Global South: A Focus on Food Insecurity

Smallholder farming systems in Africa are particularly vulnerable to decreased annual rainfall and droughts. Sidibe, Foudi, Pascual, and Termansen (2018) highlighted the limitations of future investments in irrigation systems using "blue water" from rivers and lakes due to financial and market constraints and potential environmental impacts. The study highlighted the important role of soil biodiversity as a conservation mechanism of "green water" in rainfed agroecosystems. The proposed hydrological agronomic model demonstrated how the application of agroecological practices of soil moisture management and conservation enhanced the ecosystem services essential for agricultural productivity (Sidibe, Foudi, Pascual & Termansen, 2018). The study identified optimal soil conservation practices and socio-ecological conditions to support small farming systems' coping strategies against the effects of climate variability and change (Sidibe, Foudi, Pascual & Termansen, 2018).

Bawakyillenuo, Yaro, and Teye (2016) evaluated autonomous and planned adaptation strategies in six villages in the northern savannah region of **Ghana, West Africa**, based on focus groups, interviews, and a survey of residents. The study found that the rural communities of the **West Mamprusi**, **Savelugu Nanton**, and **Kassena Nankana East** districts adopted a host of autonomous adaptive strategies to address the challenges of changing rainfall patterns and soil moisture conditions. The villagers indicated that yields from rainfed agriculture are insufficient to support the families' nutritional needs (Bawakyillenuo, Yaro & Teye, 2016, p. 371). Discussion in women's focus groups revealed the importance of hand-dug wells for crop irrigation where shallow water table allows for such practices. In some villages, irrigation water was extracted from the **White Volta River** tributaries using diesel pumps. Other adaptation strategies, particularly access to the Nyoglo dam, were found less feasible due to the lack of financial resources and institutional capacity (Bawakyillenuo, Yaro & Teye, 2016).

Nigussie, van der Werf, Zhu, Simane, and van Ierland (2018) sought the perspectives of experts and smallholder farmers in the **upper Blue Nile River basin** to assess the applicability and feasibility of proposed climate adaptation measures. The study employed Multi-Criteria Analysis (MCA) to extract, evaluate, score, and rank stakeholder preferences and decision options against a range of criteria related to natural resource management, soil and land conservation, and water management (Nigussie, van der Werf, Zhu, Simane & van Ierland, 2018). Afforestation and improving water retention ranked high among the climate adaptation options for natural resource management while crop rotation and new fertilization practices such as composting scored higher among measures related to soil and land management. The water management adaptation options included erosion control, drip irrigation, and river diversions (Nigussie, van der Werf, Zhu, Simane & van Ierland, 2018, p. 143). When comparing experts' and farmers' preferences, the study found divergent opinions in several areas, highlighting the importance of participatory and consensus-building decision-making processes to bring together conflicting viewpoints and values (Nigussie, van der Werf, Zhu, Simane & van Ierland, 2018, p. 143).

Funder, Mweemba, and Nyambe (2018) examined the role of climate adaptation politics in rural development. The authors argued that the customary land rights system in many parts of Africa favored autonomous adaptation through smallholder farmers' coping strategies. However, climate adaptation through "planned" interventions caused a fundamental change in social organization, agroecosystems-based livelihoods, and resource management (Funder, Mweemba & Nyambe, 2018). Studying two semi-arid rural districts—**Kazungula and Sesheke**, located along the **Zambezi River** in **Zambia**—the authors noted that "government supported climate adaptation interventions also serve as a vehicle for central state agencies to assert and legitimize central state authority and natural resource control in a rural setting where state authority is otherwise fragmented" (Funder, Mweemba & Nyambe, 2018, p. 31). Waldman and Richardson (2018) examined agricultural decisions among male and female farmers in **Mali, West Africa**, focusing on three regions with varied (high to low) annual precipitation—**Sikasso**, **Koulikoro**, and **Segou**. A series of choice experiments provided further insight into farmers' preferences for cultivation of hybrid and perennial sorghum, known for its resiliency to climate variability such as higher temperatures and dry spells as well as other changing environmental conditions such as increased soil salinity (Waldman & Richardson, 2018).

Participatory mapping increasingly facilitates climate adaptation decision-making processes. Velempini, Smucker, and Clem (2018) examined the role of indigenous knowledge and local institutions in community-based adaptation (CBA) in the context of water availability and water resource management in the **North Pare Mountains** in **Tanzania**. The 5-year study was part of the Local Knowledge and Climate Change Adaptation Project (LKCCAP) in the **Kilimanjaro region** spearheaded by the government of Tanzania and U.S. research institutions. The study included a survey and a series of community workshops on drought history and water management to address the challenges in sustaining traditional livelihoods in a changing climate (Velempini, Smucker & Clem, 2018). The study participants generated sketch maps of existing water resources, identified small-scale water sources most vulnerable to drought conditions, and established local priorities for improving water access and availability as part of the community-based adaptation planning process (Velempini, Smucker & Clem, 2018, p. 41).

Adapting to climate variability and change, especially amongst smallholder farmers, has the potential to improve food security in Asia. Delaporte and Maurel (2018) examined a range of adaptation practices in response to climate extremes among the smallholder farmers in **Bangladesh**. The study found that the loss of agricultural income by one percentage point increases the probability of seeking alternative options by three percentage points. Income and level of development limited the options available to farmers. Farmers with access to electricity and resources were more likely to adopt measures that require fixed investment and to adapt proactively to changing conditions than those who lacked such resources. The results highlighted the critical role of development in climate adaptation. Drawing on a survey of 720 farming households in Nepal, Khanal, Wilson, Lee, and Hoang (2018) found that adaptation practices have the potential to improve food productivity.

Recent studies have also sought to identify key factors influencing farmers' decisions to adopt adaptation measures. Some of the most commonly cited factors include (1) local knowledge about soil and water management best practices, (2) access to climate information, (3) experience with past weather extremes, and (4) participation in farmers' associations and other support networks (Khanal et al., 2018; Funder, Mweemba & Nyambe, 2018; Nigussie, van der Werf, Zhu, Simane & van Ierland, 2018).

Implications for Data Collection and Geospatial Technologies

The increasing cost of climate-related disasters, and the extent to which they could hinder development and adversely affect local coping strategies, has focused the attention of multilateral organizations; international humanitarian aid agencies; national, regional, and local governments; and the private sector on the need to provide timely scientific information to users at all levels of society (Vaughan & Dessai, 2014). Climate services have emerged as a tool to deliver such information in a format that is tailored to the needs of a broad range of stakeholders. According to the World Meteorological Organization (WMO, 2011, p. 37), "a climate service is a process of developing and delivering climate information in such a way as to meet a user's need." In 2011, the World Meteorological Organization introduced the Global Framework for Climate Services (GFCS) to capitalize on existing knowledge, expertise, data, and databases (WMO, 2011). Climate services, for instance, provide farmers in **Africa**, **China**, and **Australia** with seasonal consensus outlooks and forecasts to facilitate planning of agricultural activities. In **Africa**, a consensus seasonal forecast, prepared by the Regional Climate Outlook Forum (RCOF) consisting of over 200 scientists and agricultural experts from around the world, is made available to farmers through each country's agricultural department to ensure that the farmers possess the information and the tools they need to select crop species and rotation plans, adjust irrigation needs, and make market and personal decisions (WMO, 2011).

Climate services combined with geospatial technologies are increasingly used to mobilize assistance from the international community and support global risk governance. Two early warning systems—the Famine Early Warning Systems Network (FEWS NET), sponsored by the United States Agency for International Development (USAID), and Global Monitoring for Food Security (GMFS), funded by the European Space Agency (ESA)—combine remote sensing data with regional analysis of food availability and sociopolitical factors to estimate the number of people in need of humanitarian assistance. In 2017, FEWS NET determined that nearly 80 million people in **Africa** are threatened by food insecurity and a high prevalence of malnutrition among children (FEWS NET, 2017). FEWS NET issues food security alerts, demonstrating the value of geospatial technologies as decision support for early warning systems and mobilization of resources for rapid response. The Working Group Climate (WGClimate) of the Committee on Earth Observation Satellites (CEOS) and the Climate Change Initiative (CCI) of the European Space Agency also provide climate data services to ensure timely delivery of climate information to stakeholders and end users (Giuliani, Nativi, Obregon, Beniston & Lehmann, 2017). Chapter 8 provides additional details on the application of remote sensing for climate adaptation planning.

There is a growing interest in developing geographic information system (GIS)-based climate adaptation decision-support tools (Neset, Opach, Lion, Lilja & Johansson, 2016). Building upon the decision support capabilities of GIS, Kunapo, Fletcher, Ladson, Cunningham, and Burns (2018) introduced a spatially explicit Integrated Climate Adaptation Model (ICAM) to facilitate the development of climate scenarios and management practices. The tool, tested in **Melbourne, Australia**, integrates various types of risk assessment including river flooding, excessive runoff, sea level rise, extreme temperatures, and droughts (Kunapo, Fletcher, Ladson, Cunningham & Burns, 2018). While decision-support tools can play an essential role in climate adaptation planning, they are often limited by input data quality, trade-offs between user-friendliness and computational tractability, and lack of input from key stakeholders (Kunapo et al., 2008).

Several studies have explored the integration of community-based disaster risk management and participatory three-dimensional modeling (P3DM) to facilitate participatory planning and community engagement in climate adaptation decisions (Gaillard & Maceda, 2009; Maceda, Gaillard, Stasiak, Le Masson & Le Berre, 2009; Rambaldi, 2010; Piccolella, 2013; Caribbean Natural Resources Institute [CANARI], 2013; Bobb-Prescott, 2014; DeGraff & Ramlal, 2015). Rambaldi (2010, p. 3) defines P3DM as "a mapping method based on extracting topographic information" to develop a stand-alone physical model based on inputs from community stakeholders. P3DM embodies the principles of participatory learning and action (PLA), GIS, and cartography (Rambaldi, 2010). The method has found several applications in rural communities and small island nations around the world where advanced technologies such as computer-based geographic information systems are not readily available due to the lack of resources and skilled workforce (Gaillard & Maceda, 2009; Maceda, Gaillard, Stasiak, Le Masson & Le Berre, 2009; Rambaldi, 2010; Piccolella, 2013). Upon completion, the content of the 3D physical models is scaled and integrated with other layers of geographic information using GIS and global positioning system (GPS) (Gaillard & Maceda, 2009; Rambaldi, 2010). See Chapter 5 for more details on citizen science and participatory mapping.

Notes

1 The anomaly was positively associated with the Southern Oscillation Index (Gale & Saunders, 2013).

2 The Southeast Florida Regional Climate Change Compact includes Miami-Dade, Monroe, Broward, and Palm Beach counties.

References

Aylett, A. 2014. *Progress and Challenges in the Urban Governance of Climate Change: Results of a Global Survey*. Cambridge, MA: MIT Press.

Barnett, J., & O'Neill, S.J. 2012. Islands, Resettlement and Adaptation. *Nature Climate Change* 2, 8–10.

Bawakyillenuo, S., Yaro, J.A., & Teye, J. 2016. Exploring the Autonomous Adaptation Strategies to Climate Change and Climate Variability in Selected Villages in the Rural Northern Savannah Zone of Ghana. *Local Environment* 21(3), 361–382.

Benson, C., Twigg, J., & Myers, M. 2001. NGO Initiatives in Risk Reduction: An Overview. *Disasters* 25(3), 199–215.

Berke, P.R. 2014. Rising to the Challenge: Planning for Adaptation in the Age of Climate Change. In Glavovic, B.C., & Smith, G.P. (Eds.) *Adapting to Climate Change: Lessons from Natural Hazards Planning*. Dordrecht, Netherlands: Springer.

Betzold, C. 2015. Adapting to Climate Change in Small Island Developing States. *Climatic Change* 133(3), 481–489.

Beyerl, K., Mieg, H.A., & Weber, E. 2018. Comparing Perceived Effects of Climate-related Environmental Change and Adaptation Strategies for the Pacific Small Island States of Tuvalu, Samoa, and Tonga. *Island Studies Journal* 13(1), 25–44.

Birkmann, J., Garschagen, M., Kraas, F., & Quang, N. 2010. Adaptive Urban Governance: New Challenges for the Second Generation of Urban Adaptation Strategies to Climate Change. *Sustainability Science* 5(2), 185–206.

Bobb-Prescott, N. 2014. *Case Study on the Use of Participatory Three-dimensional Modeling to Facilitate Effective Contribution of Civil Society in the Caribbean Islands in Planning for Action on Climate Change*. CANARI Technical Report 401, Laventille, Trinidad.

Brandes, U., & LeBlanc, A. 2013. *Risk & Resilience in Coastal Regions*. Washington, DC: Urban Land Institute, ISBN: 978-0-87420-280-9.

C40 Cities. 2018. *Climate Action Planning Framework*. Retrieved from https://resourcecentre.c40.org/climate-action-planning-framework-home.

California Climate Adaptation Strategy. 2008. A Report to the Governor of the State of California in Response to Executive Order S-13-2008. Retrieved from http://resources.ca.gov/docs/climate/Statewide_Adaptation_Strategy.pdf.

California Legislature–Senate Committee on Environmental Quality. 2015. *Climate Change Impacts and Adaptation Efforts in California: Summary of Oversight Committee*. Retrieved from https://senv.senate.ca.gov/sites/senv.senate.ca.gov/files/Climate%20Change%20Adaptation_California%20State%20Senate_2015.pdf.

California Natural Resources Agency (CNRA).

Caribbean Natural Resources Institute (CANARI). 2013. *Using traditional knowledge for decision-making on climate change in the Caribbean*. Policy Brief 15. Laventille, Trinidad: CANARI.

City of Copenhagen. 2012. *Copenhagen–Carbon Neutral by 2025*. Copenhagen Climate Adaptation Plan. Retrieved from http://kk.sites.itera.dk/apps/kk_pub2/index.asp?mode=detalje&id=983.

DeGraff, A., & Ramlal, B. 2015. *Participatory Mapping: Caribbean Small Island Developing States*. St. Augustine, Trinidad: Forum on the Future of the Caribbean.

Delaporte, I., & Maurel, M. 2018. Adaptation to Climate Change in Bangladesh. *Climate Policy* 18(1), 49–62.

Di Giulio, G.M., Bedran-Martins, A.M.B., Vasconcellos, M.P., Ribeiro, W.C., & Lemos, M.C. 2018. Mainstreaming Climate Adaptation in the Megacity of Sao Paolo, Brazil. *Cities* 72, 237–244.

Elrick, C., & Kay, R. 2009. *Mainstreaming of An Integrated Climate Change Adaptation Based Risk Diagnosis and Response Process into Government of Kiribati: Final Report*. Report prepared for the KAP Project, Phase II. Coastal Zone Management Pty Ltd, Perth.

Famine Early Warning Systems Network (FEWS NET). 2017. *Already Unprecedented Food Assistance Needs Grow Further: Risk of Famine Persists*. Retrieved from http://www.fews.net/global/alert/june-2017.

Fazey, L., Wise, R.M., Lyon, C., Câmpeanu, C., Moug, P., & Davis, T.E. 2016. Past and Future Adaptation Pathways. *Climate and Development* 8(1), 26–44.

Food and Agriculture Organization (FAO). 2017. *Regional Overview of Food Security and Nutrition in Africa 2016. The Challenges of Building Resilience to Shocks and Stresses*. Accra, Ghana: Food and Agriculture Organization of the United Nations. Retrieved from http://www.fao.org/3/a-i6813e.pdf.

Funder, M., Mweemba, C., & Nyambe, I. 2018. The Politics of Climate Change Adaptation in Development: Authority, Resource Control and State Intervention in Rural Zambia. *The Journal of Development Studies* 54(1), 30–46.

Gaillard, J.-C., & Maceda, E.A. 2009. Participatory 3-Dimensional Mapping for Disaster Risk Reduction. *Participatory Learning and Action* 60, 109–118.

Gale, E.L., & Saunders, M.A. 2013. The 2011 Thailand Flood: Climate Causes and Return Periods. *Weather* 68(9), 233–237.

Gerdes, J. 2013. Copenhagen's Ambitious Push To Be Carbon Neutral by 2025. *Yale Environment 360*. Retrieved from http://e360.yale.edu/features/copenhagens_ambitious_push_to_be_carbon_neutral_by_2025.

Giuliani, G., Nativi, S., Obregon, A., Beniston, M., & Lehmann, A. 2017. Spatially Enabling the Global Framework for Climate Services: Reviewing Geospatial Solutions to Efficiently Share and Integrate Climate Data and Information. *Climate Services* 8, 44–58.

Greater London Authority (GLA). 2011. *Managing Risks and Increasing Resilience: The Mayor's Climate Change Adaptation Strategy*. Retrieved from https://www.london.gov.uk/sites/default/files/gla_migrate_files_destination/Adaptation-oct11.pdf.

Greater London Authority (GLA). 2015. *The Mayor's Climate Change Mitigation and Energy Annual Report, 2013-14, Greater London Authority City Hall.* Retrieved from https://www.london.gov.uk/WHAT-WE-DO/environment/environment-publications/mayors-climate-change-mitigation-and-energy-annual.

Hallegatte, S. 2009. Strategies to Adapt to An Uncertain Climate Change. *Global Environmental Change* 19(2), 240–247.

Haraguchi, M., & Lall, U. 2015. Flood Risks and Impacts: A Case Study of Thailand's Floods in 2011 and Research Questions for Supply Chain Decision Making. *International Journal of Disaster Risk Reduction* 14, 256–272.

India Meteorological Department. 2018. *Advance of Southwest Monsson 2018.* Ministry of Earth Sciences, Government of India. Retrieved http://www.imd.gov.in/pages/monsoon_main.php.

International Council for Local Environmental Initiatives—World Wildlife Fund (ICLEI-WWF). 2015. *Measuring Up 2015: How US Cities Are Accelerating Progress Toward National Climate Goals.* Retrieved from http://icleiusa.org/wp-content/uploads/2015/08/Measuring_Up_2015.pdf.

Kane, S., & Shogren, J.F. 2000. Linking Adaptation and Mitigation in Climate Change Policy. In Kane, S.M., & Yohe, G.W. (Eds.) *Societal Adaptation to Climate Variability and Change.* Dordrecht, Netherlands: Springer Verlag.

Khanal, U., Wilson, C., Lee, B.L., & Hoang, V.-N. 2018. Climate Change Adaptation Strategies and Food Productivity in Nepal: A Counterfactual Analysis. *Climatic Change* 148, 575–590.

Kiribati Climate Change. 2016. *Kiribati Adaptation Program.* Office of the President, Republic of Kiribati. Retrieved from http://www.climate.gov.ki/category/action/adaptation/kiribati-adaptation-program/.

Kohl, L. 1986. The Oosterschelde Barrier – Man Against the Sea. *National Geographic* 170(4), 526–537.

Kunapo, J., Fletcher, T.D., Ladson, A.R., Cunningham, L., & Burns, M.J. 2018. A Spatially Explicit Framework for Climate Adaptation. *Urban Water Journal* 15(2), 159–166.

Maceda, E.A., Gaillard, J.-C., Stasiak, E., Le Masson, V., & Le Berre, I. 2009. Experimental Use of Participatory Three-dimensional Models in Island Community-based Disaster Risk Management. *Shima: The International Journal of Research into Island Cultures* 3(1), 72–84.

Maskrey, A. 2011. Revisiting Community-based Disaster Risk Management. *Environmental Hazards* 10(1), 42–52.

McNamara, K.E., Westoby, R., & Smithers, S.G. 2017. Identification of Limits and Barriers to Climate Change Adaptation: Case Study of Two Islands in Torres Strait, Australia. *Geographical Research* 55(4), 438–455.

Miller, D., Desoto-Duncan, A., & Hertzler, B. 2013. Hurricane Katrina and the Inner Harbor Navigation Canal-Lake Borgne Surge Barrier. *La Houille Blanche* 2, 5–11.

National Capital Integrated Coastal Development (NCICD). 2014. *Master Plan of National Capital Integrated Coastal Development 2014*. Retrieved from https://www.bureauanl.nl/files/MP-final-NCICD-LR.pdf.

Neset, T.-M., Opach, T., Lion, P., Lilja, A., & Johansson, J. 2016. Map-based Web Tools Supporting Climate Change Adaptation. *The Professional Geographer* 68(1), 103–114.

Newman, J., Springer, M., Sheehan, T., John Gravelin, J., Trouche, L., Slaughter, S., & Wilson, A. 2013. *Building Resilience in Boston*. Prepared by Linnean Solutions, the Built Environment Coalition, and the Resilient Design Institute for the Boston Green Ribbon Commission Climate Preparedness Working Group. Retrieved from https://www.architects.org/sites/default/files/Building_Resilience_in_Boston_SML_0.pdf.

Nigussie, Y., van der Werf, E., Zhu, X., Simane, B., & van Ierland, E.C. 2018. Evaluation of Climate Change Adaptation Alternatives for Smallholder Farmers in the Upper Blue-Nile Basin. *Ecological Economics* 151, 142–150.

NYC Department of City Planning. 2013. *Urban Waterfront Adaptive Strategies: A Guide to Identifying and Evaluating Potential Strategies for Increasing the Resilience of Waterfront Communities to Coastal Flooding and Sea Level Rise*. Retrieved from http://www1.nyc.gov/site/planning/plans/sustainable-communities/climate-resilience.page?tab=2.

Obeysekera, J., & Park, J. 2012. Scenario-based Projection of Extreme Sea Levels. *Journal of Coastal Research* 29, 1–7.

Piccolella, A. 2013. Participatory Mapping for Adaptation to Climate Change: The Case of Boe Boe, Solomon Islands. *Knowledge Management for Development Journal* 9(1), 24–36.

Rambaldi, G. 2010. *Participatory Three-dimensional Modelling: Guiding Principles and Applications* (2010 edition). Wageningen, the Netherlands: CTA.

Rotterdam Climate Initiative. 2012. *Rotterdam Climate Change Adaptation Strategy*. City of Rotterdam, the Netherlands. Retrieved from http://www.rotterd

amclimateinitiative.nl/documents/2015-en-ouder/Documenten/20121210_RAS_EN_lr_versie_4.pdf.

Saito, N. 2014. Challenges for Adapting Bangkok's Flood Management Systems to Climate Change. *Urban Climate* 9, 89–100.

Schwirplies, C. 2018. Citizens' Acceptance of Climate Change Adaptation and Mitigation: A Survey in China, Germany, and the U.S. *Ecological Economics* 145, 308–322.

Setzer, J., Macedo, L.V., & Rei, F. 2015. Combining Local and Transnational Action in Adaptation of Climate Policies in the City of Sao Paolo. In Johnson, C., Toly, N., & Schroeder, H. (Eds.) *The Urban Climate Challenge: Rethinking the Role of Cities in the Global Climate Regime*, 101–118. New York, NY: Routledge.

Sherwell, P. 2016. $40bn to Save Jakarta: The story of the Great Garuda. *The Guardian*, 22 November 2016. Retrieved from https://www.theguardian.com/cities/2016/nov/22/jakarta-great-garuda-seawall-sinking.

Sidibe, Y., Foudi, S., Pascual, U., & Termansen, M. 2018. Adaptation to Climate Change in Rainfed Agriculture in the Global South: Soil Biodiversity as Natural Insurance. *Ecological Economics* 146, 588–596.

Swiss Re. 2013. *Mind the Risk—A Global Ranking of Cities under Threat from Natural Disasters*. Zurich, Switzerland: Swiss Reinsurance Company.

Takagi, H., Fujii, D., Esteban, M., & Yi, X. 2017. Effectiveness and Limitation of Coastal Dykes in Jakarta: The Need for Prioritizing Actions Against Land Subsidence. *Sustainability* 9(4), 619. https://doi.org/10.3390/su9040619.

The Thames Barrier. 2014. How the Thames Barrier works, and when it is scheduled to close (last updated 8 October 2018). https://www.gov.uk/guidance/the-thames-barrier.

Thornton, T.F., & Comberti, C. 2017. Synergies and Trade-offs between Adaptation, Mitigation and Development. *Climatic Change* 140(1), 5–18.

Turner, A.G., & Annamalai, H. 2012. Climate Change and the South Asian Monsoon. *Nature Climate Change* 2, 587–595. doi:10.1038/nclimate1495.

United Kingdom Environment Agency (UK Environment Agency). 2012. *Thames Estuary 2100: Managing flood risk through London and Thames Estuary*. London, UK: Thames Estuary Programme. Retrieved from https://assets.publishing.service.gov.uk/government/uploads/system/uploads/attachment_data/file/322061/LIT7540_43858f.pdf.

United Nations Office for Disaster Risk Reduction (UNISDR). 2007. *Hyogo Framework for Action 2005-2015: Building the Resilience of Nations and Communities to Disasters*. Extract from the final report of the World Conference on Disaster Reduction (A/CONF.206/6). Retrieved from https://www.unisdr.org/we/inform/publications/1037.

United Nations Office for Disaster Risk Reduction (UNISDR). 2014. *Brazil's São Paulo State Recognized by UN for Pioneering Work in Building Urban Resilience to Disasters*. Retrieved from https://www.unisdr.org/archive/40966.

Van Aalst, M., Ebi, K.L., Githeko, A., Yohe, G., & Jones, R. 2004. Case Studies. In Lim, B., Spanger-Siegfried, E., Burton, I., Malone, E., & Huq, S. (Eds.) *Adaptation Policy Frameworks for Climate Change: Developing Strategies, Policies and Measures*. Cambridge, UK: Cambridge University Press.

Vaughan, C., & Dessai, S. 2014. Climate Services for Society: Origins, Institutional Arrangements, and Design Elements for an Evaluation Framework. *Wiley Interdisciplinary Reviews: Climate Change* 5(5), 587–603.

Velempini, K., Smucker, T.S., & Clem, K.R. 2018. Community-based Adaptation to Climate Variability and Change: Mapping and Assessment of Water Resource Management Challenges in the North Pare Highlands, Tanzania. *African Geographical Review* 37(1), 30–48.

Vella, K., Butler, W.H., Sipe, N., Chapin, T., & Murley, J. 2016. Voluntary Collaboration for Adaptive Governance: The Southeast Florida Regional Climate Change Compact. *Journal of Planning Education and Research* 36(3), 363–376.

Voskamp, L.M., & van den Ven, F.H.M. 2015. Planning Support System for Climate Adaptation: Composing Effective sets of Blue-green Measures to Reduce Urban Vulnerability to Extreme Weather Events. *Building and Environment* 83, 159–167.

Waldman, K.B., & Richardson, R.B. 2018. Confronting Tradeoffs between Agricultural Ecosystem Services and Adaptation to Climate Change in Mali. *Ecological Economics* 150, 184–193.

Webber, S. 2015. Mobile Adaptation and Sticky Experiments: Circulating Best Practices and Lessons Learned in Climate Change Adaptation. *Geographical Research* 53(1), 26–38.

Win, T.L. 2017. In Flood-prone Jakarta, will 'Giant Sea Wall' Plan Sink or Swim? *Reuters*, September 14, 2017. Retrieved from https://www.reuters.com/article/us-indonesia-infrastructure-floods/in-flood-prone-jakarta-will-giant-sea-wall-plan-sink-or-swim-idUSKCN1BP0JU.

World Meteorological Organization (WMO). 2011. *Climate Knowledge for Action: A Global Framework for Climate Services–Empowering the Most Vulnerable*, WMO 1965. Geneva, Switzerland: WMO. Retrieved from https://library.wmo.int/pmb_ged/wmo_1065_en.pdf.

World Bank. 2000. *Cities, Seas and Storms: Managing Change in Pacific Island Economies*. Washington, DC: World Bank.

World Bank. 2011. *Project Appraisal Document – Kiribati Adaptation Program Phase III*. Report No: 63874-KI. East Asia and Pacific Region Sustainable Development Department, Timor-Leste, Papua New Guinea/The Pacific Islands.

World Bank. 2014. *Community Resilience to Climate and Disaster Risk in Solomon Islands Project*. Retrieved from http://www.worldbank.org/en/news/loans-credits/2014/03/06/solomon-islands-community-resilience-to-climate-and-disaster-risk-project.

PART 2
Geospatial Technologies
Fundamentals and Terminology

5 Natural Hazards

Visualization and Basic Mapping

Chapter Objectives

This chapter seeks to answer the following questions:

- What are climate-related natural hazards?
- What is the range of data sources and geospatial technologies that are used for vulnerability hazard assessments?
- What role do citizens play in participatory mapping?

Common Natural Hazards

> *Floods, droughts, heat waves, and storms have always been part of human lives because they are a normal part of climate variability. However, the observed trends and projected changes in global climate have the potential to alter patterns of these climatic hazards and extreme weather events.*
>
> —Wilhelmi and Morss (2013), p. 49

Hazard exposure arises from people's occupancy of hazard-prone areas. At a global level, floods are the most widespread problem and are among the costliest disasters worldwide (Lott & Ross, 2006; Brody et al., 2007; Smith & Katz, 2013; Smith & Matthews, 2015; National Oceanic and Atmospheric Administration–National Centers for Environmental Information [NOAA-NCEI], 2017). Hurricanes, as we saw in 2017 with Hurricanes Harvey, Irma, and Maria, can have disastrous and deadly implications for entire communities and islands. Summertime hurricanes are an obvious example, but moisture in the winter months can spawn nor'easters as well as giant storms, such as Superstorm Sandy, which affected several countries, including the **United States** and **Cuba**. In addition to floods and hurricanes, this chapter provides a brief overview of several natural hazards (storm surge, tsunamis, tornadoes, landslides, wildfire, and drought) as context for discussing basic mapping and related datasets for understanding hazard exposure at multiple scales.

Even with knowledge of natural hazards, planners and policy-makers face the difficult challenge of balancing between: (i) promotion of economic development, (ii) management of growth, and (iii) reduction/minimization of vulnerability to coastal hazards. Hazard mitigation and adaptation policy actions must be informed by accurate vulnerability assessments at the local level and, in turn, accurate spatial representation of demographic, socioeconomic, and biophysical variables. Scholars and practitioners, such as coastal managers, have embraced geographical investigations (Fletcher, 2007) to inform their work. Application of geospatial technologies, including basic mapping, is now a fundamental component of a hazard mitigation and adaptation planning toolkit.

Floods

Floods are described in terms of their statistical frequency (e.g., 50-year flood, 100-year flood, 500-year flood). The 100-year flood (base flood) is an arbitrary standard of safety that reflects a compromise between the goals of providing long-term safety and developing economically valuable land. It does not mean that a flood of that size happens only once every 100 years. It is the level of floodwater expected to be equaled or exceeded every 100 years based on statistical average. It is also referred to as the 1% flood, meaning that there is a 1% chance that a flood this size will happen during any year (Holmes & Dinicola, 2010). Floods create secondary impacts such as pollution of drinking water supplies, disruption of gas and electrical service, and disruption of transportation systems, as highlighted most recently by the catastrophic floods caused by Hurricane Harvey.

Nuisance flooding—recurrent flooding that takes place at high tide—is becoming more common and is more widely reported and tracked in the United States. This type of flooding causes public inconveniences such as traffic delays and business closures, as well as hastening of roadway deterioration due to saltwater corrosion. According to Sweet and Marra (2015, p. 2), nuisance flooding in the United States has become a "sunny day" event (i.e., not necessarily linked to storms or heavy rain). The top 10 areas with nuisance flooding in the United States span the east, southeast, and west coasts, with cities on the east coast ranking among the top 10 (Metcalfe, 2014; Sweet & Marra, 2015). Additionally, extreme precipitation events and resulting flash floods present a constant threat to lives and property despite flood management infrastructure (Wilhelmi & Morss, 2013).

Flood maps show areas susceptible to flooding and delineate a range of flood risk areas, from areas outside flood zones to those subject to high-velocity wave action. Flood maps are created using historical data on river stages and discharge of previous flood, topographic data,[1] and aerial photos and satellite images[2] of prior floods. For example, the National Flood Hazard Layer created by the United States Federal Emergency Management Agency (FEMA) designates flood risk areas in several parts of the United States. The features indicating the extent of the 100- and 500-year floodplain can be overlaid with the census boundaries and used by local emergency management teams and hazard mitigation planners to estimate how many people currently live in such high-hazard areas. These estimates provide critical information as part of vulnerability assessments, but they must be mapped appropriately. One study modeled the floodplain in Iran utilizing Cartosat-1 images and GIS (geographic information system) for terrain mapping (Sarhadi, Soltani & Modarres, 2012). It is important to have updated flood maps, especially in rapidly developing areas.

Hurricanes

A hurricane is a tropical storm with maximum sustained winds greater than 74 miles per hour (mph). They form over tropical and subtropical waters but originate in the Atlantic basin (NOAA, 2016b). Hurricanes are particularly destructive because they can induce damage in several ways: high winds, heavy rainfall, floods, and the storm surge. The *Saffir-Simpson Hurricane Wind Scale* (SSHWS), formerly the Saffir-Simpson Hurricane Scale (SSHS), classifies hurricanes into five categories based on the intensities of their sustained winds. Figure 5.1 provides details about the type of damage that can result from these five categories.

Storm Surge and Shoreline Erosion

Hurricanes and other extreme storms generate storm surge and large waves that reshape and transform coastal landscapes (United States Geological Survey [USGS], 2015). Storm-induced erosion is rapid and dynamic, and predicting their impacts is essential for coastal planning and management (Passeri, Hagen, Medeiros, Bilskie, Alizad & Wang, 2015). Shoreline erosion is the removal of sediment from the shoreline, and it is measured by linear retreat (feet of erosion per year) or volumetric loss (e.g., cubic yards of eroded sediment or foot of shoreline per year). Shoreline erosion can be categorized as: (i) long-term erosion, caused by natural changes (e.g., tidal inlets) and human activities (e.g., dredging, damming rivers, alteration of vegetation); and (ii) short-term erosion, which can cause the equivalent shoreline change of several decades of long-term erosion.

LiDAR (Light Detection and Ranging) is a remote sensing method[3] used to measure the surface of the earth (NOAA, 2016c). LiDAR is based on infrared (1.1 μm) laser pulses that go back and forth below a flight path, scanning the ground and improving upon vertical accuracy (International Hurricane Research Center [IHRC], 2004). LiDAR data support activities such as inundation and storm surge modeling, hydrodynamic modeling, shoreline mapping, emergency response, hydrographic surveying, and coastal vulnerability analysis (NOAA, 2016c).

Category	Sustained Winds	Types of Damage Due to Hurricane Winds
1	74–95 mph 64–82 kt 119–153 km/h	**Very dangerous winds will produce some damage:** Well-constructed frame homes could have damage to roof, shingles, vinyl siding and gutters. Large branches of trees will snap and shallowly rooted trees may be toppled. Extensive damage to power lines and poles likely will result in power outages that could last a few to serveral days.
2	96–110 mph 83–95 kt 154–177 km/h	**Extremely dangerous winds will cause extensive damage:** Well-constructed frame homes could sustain major roof and siding damage. Many shallowly rooted trees will be snapped or uprooted and block numerous roads. Near-total power loss is expected with outages that could last from several days to weeks.
3 (major)	111–129 mph 96–112 kt 178–208 km/h	**Devastating damage will occur:** Well-built framed homes may incur major damage or removal of roof decking and gable ends. Many trees will be snapped or uprooted, blocking numerous roads. Electricity and water will be unavailable for several days to weks after the storm passes
4 (major)	130–156 mph 113–136 kt 209–251 km/h	**Catastrophic damage will occur:** Well-built framed homes can sustain severe damage with loss of most of the roof structure and/or some exterior walls. Most trees will be snapped or uprooted and power poles downed. Fallen trees and power poles will is olate residential areas. Power outages will last weeks to possibly months. Most of the area will be uninhabitable for weeks or months.
5 (major)	157 mph or higher 137 kt or higher 252 km/h or higher	**Catastrophic damage will occur:** A high percentage of framed homes will be destroyed, with total roof failure and wall collapse. Fallen trees and power poles will isolate residential areas. Power outages will last for weeks to possibly months. Most of the area will be uninhabitable for weeks or months.

Figure 5.1 Saffir-Simpson Hurricane Wind Scale

Source: NOAA National Weather Service (NWS), www.nhc.noaa.gov/aboutsshws.php

A wide variety of datasets are used to map the potential extent of sea level rise under various scenarios and to conduct vulnerability and adaptation planning assessments. Some of the most commonly used datasets include:

- Shoreline change over time—historical photos georeferenced against stable features in high-resolution satellite images as a proxy for shoreline change (Albert, Leon, Grinham, Church, Gibbes & Woodroffe, 2016)
- Wind and wave hindcast data for estimating of wave energy
- Digital elevation data generated from airborne LiDAR to develop sea level rise scenarios
- Data on sea level trends, maximum tidal range, maximum significant wave tide, and mean long-term erosion trend
- Output from the SLOSH (Sea, Lake, and Overland Surge from Hurricanes) model of the U.S. National Hurricane Center to map storm surge risk zones
- Historical land use and future land use datasets
- Land parcel and tax appraisal data
- Other datasets including cost and extent of beach nourishment projects, a number of permits for seawalls, and other hard structures for shoreline protection

Tsunamis

A tsunami is a series of waves caused by earthquakes or undersea volcanic eruptions (NOAA, 2017). These sea waves can be generated by earthquakes and by volcanic eruptions or landslides that usually occur undersea. For example, the Indonesian eruption of **Mount Krakatoa** in 1883 generated 130-foot tsunamis across the Indian Ocean (Kurczy, 2010). By far, the leading cause of tsunamis are earthquakes. In the European Mediterranean area, 83% of tsunamis are caused by submarine earthquakes (Maramai, Brizuela & Graziani, 2014). Tsunamis are largely invisible in the open ocean, but as they approach the shallow continental shelf, wave run-up can rise several meters and submerge low-lying coastal areas.

Tornadoes

Tornadoes are rotating columns of air, extending from the base of a storm cloud to the ground. The mechanism behind tornado formation is not fully understood, but they typically form in association with supercell thunderstorms. Supercells are characterized by large-scale internal rotation. In a supercell, vertical wind shear generates horizontal rotation that when tilted vertically by updrafts results in a vertical

Figure 5.2 Tornado formation

Source: Courtesy of AccuWeather Inc., 385 Science Park Road, State College, Pennsylvania, 16803, 814-237–0309, www.accuweather.com/en/weather-news/watch-erupting-volcano-creates/23751780

rotating column of air that accelerates as it converges and is sucked upward (Markowski & Richardson, 2014). The rotating air is invisible unless it is filled with water droplets, dirt, or debris. Their formation leads to a drop in pressure inside the vortex and an increase in wind speed. Debris trapped in wind fields carries immense destructive power—leading to the destruction of buildings, vehicles, and large trees. Tornadoes are measured on the Fujita scale, ranging from 0 to 5. A Fujita score of 5 exhibits gusts of over 300 miles per hour (NOAA, 2007).

According to Markowski and Richardson (2014) and Jones (2017), more tornadoes occur in the United States on an annual basis than anywhere else on the planet. The highest incidence of tornadoes is in the central region of the country in the spring from April to June (NOAA, 2016a). A three-fold increase in tornado impact activity is expected over the coming century, with a potential 50% increase in tornadic activity due to climate change (Jones, 2017).

Residents of mobile homes tend to be most impacted given that these structures are built substantially less sturdily than site-built homes.

Landslides

Masses of rock, earth, or debris move down a slope during landslides, which can be triggered by heavy rains, earthquakes, volcanic eruptions, and fire, as well as modification of land by humans (USGS, 2014). Landslides can also occur in coastal and onshore environments (USGS, 2014). In 2010, 38 people were killed on the **Portuguese island of Madeira** as a result of a landslide. That same year, 1,500 people were evacuated as a result of mudslides (fast-moving landslides) in **Pemberton, British Columbia, Canada** (Assilzadeh, Levy & Wang, 2010).

Landslides are diffuse and unpredictable (i.e., they do not occur in locations that are simple to define), thus making it difficult to produce conclusive maps. Maps are therefore reduced to inventories and susceptibility assessments. In the **United States**, landslide susceptibility maps are available from the United States Geological Survey. Assilzadeh, Levy, and Wang (2010) have also reported on a framework of landslide management and monitoring system. In a test case of the framework, the authors used GIS to forecast potential landslides by computing and mapping slope stability indices and landslide risk for **Penang Island** in the **Straits of Malacca**, situated in the northwest of Peninsular **Malaysia**. Seventeen layers of variables were constructed and overlaid on a map, including administrative boundaries, transportation networks, population distributions, river networks, slope, and previous hazards. The result was a landslide risk map that delineated four risk categories: very high, high, moderate, and low (Assilzadeh, Levy & Wang, 2010).

Wildfire

Wildfires are natural processes in forestland and remain one of the most ignored natural hazards. Dry conditions and droughts resulting from higher temperatures can create explosive wildfires. Although land use and fire suppression could also be contributing factors, drier conditions caused by warmer spring and summer seasons are thought to be the primary explanation for the increase in wildfires (Westerling, Hidalgo, Cayan & Swetnam, 2006)

Suburban and exurban development have brought wildfire hazard to our backyards. Even with recurrent wildfires in some regions, people often rebuild in the same area without vegetative cleanup or improved construction, thereby subjecting themselves to future disasters. Between 1980 and 2004, **Spain** was one of the most affected countries in Europe in the number of fires and land area burned (Martínez, Vega-Garcia & Chuvieco, 2009). In 2016, a wildfire near **Fort McMurray, Canada**, displaced over 80,000 residents and destroyed 2,400 homes—about 10% of the city.

Wildland fire threat maps, albeit not common, show the fire threat rating for specific areas and regions. For example, a fire threat map was developed in the **United States** by the California Department of Forestry and Fire Protection. That map delineates five threat classes: extreme, very high, high, moderate, and little or no threat. These threat areas are usually buffered to identify the people and property in extreme and very high fire threat areas. Monitoring stations, forecasting models, and remote sensing technology are also instrumental in mitigating the impact of wildfire and forest fire threats. The National Oceanic and Atmospheric Administration provides a Hazard Mapping System (HMS) fire and smoke product that combines multiple satellite imagery to produce smoke plume forecasts. The BlueSky Western Canada Wildfire Smoke Forecasting Framework developed by the U.S. Forest Service produces particulate matter forecasts as many as 48 hours in advance (Larkin et al., 2008; Yao et al., 2013).

Droughts

Drought is an extended period of time with deficient precipitation relative to normal and generally affects large areas (NWS, 2012). In drier regions, after soil moisture evaporates, the cooling provided by evapotranspiration slows and temperatures can increase, leading to exacerbated droughts. The "megadroughts" typical of California's earlier history could come again. **California** experienced serious drought between 2011 and 2014, as shown in Figure 5.3, but the state has a history of prolonged drought periods, some lasting more than a decade, during the past 1,000 years (Rogers, 2014). As noted by Ghose (2015), research suggests that conditions will only worsen in the coming decades. Drought has far-reaching direct and indirect effects because it can affect the water and food supply. The effects of drought are going to be exacerbated as groundwater basins across the globe become depleted (National Aeronautics and Space Administration [NASA], 2015).

Hazard and Vulnerability Assessments

At the basic level, GIS technology is used for mapping natural hazards and for assessing vulnerability and providing insight into the unprotected nature of the exposure (Randolph, 2004). GIS facilitates the entry, storage, manipulation, analysis, visualization, and presentation of spatially referenced information, connects information stored in a database or attribute table to features on a map, and combines *layers* of spatial and attribute information to build and analyze relationships of spatially referenced information. The ability to integrate both spatial and nonspatial information that is otherwise difficult to associate through other means is also important given myriad data sources and formats—physical maps (which can be digitized or scanned), spreadsheets (which can be geocoded), aerial photos, orthophotos, satellite imagery, global positioning system (GPS) data points, and local knowledge.

> Vulnerability is the propensity of exposed elements to suffer adverse effects when impacted by hazard events (Cardona et al., 2012, p. 69) and relates to pre-disposition, susceptibilities, fragilities, weaknesses, deficiencies, or lack of capacities due to adverse effects of exposed elements (Cardona et al., 2012, p. 70).

Vulnerability assessments range from basic identification of natural hazards to more complex formulation of algorithms, models, and indices of vulnerability, risk, and even population displacement potential to inform planning and policy-making (Bolin, 1985; Bolin,1994; Clark et al., 1998; Davidson & Lambert, 2001; Cutter, 2003; Boruff, Emrich & Cutter, 2005; Jain, Davidson & Rosowsky, 2005; Miles & Chang, 2006; FEMA, 2017; Kates, Colten, Laska & Leatherman, 2006; Simpson, 2006; French, Feser & Peacock, 2008; Zahran, Brody, Peacock, Vedlitz & Grover,

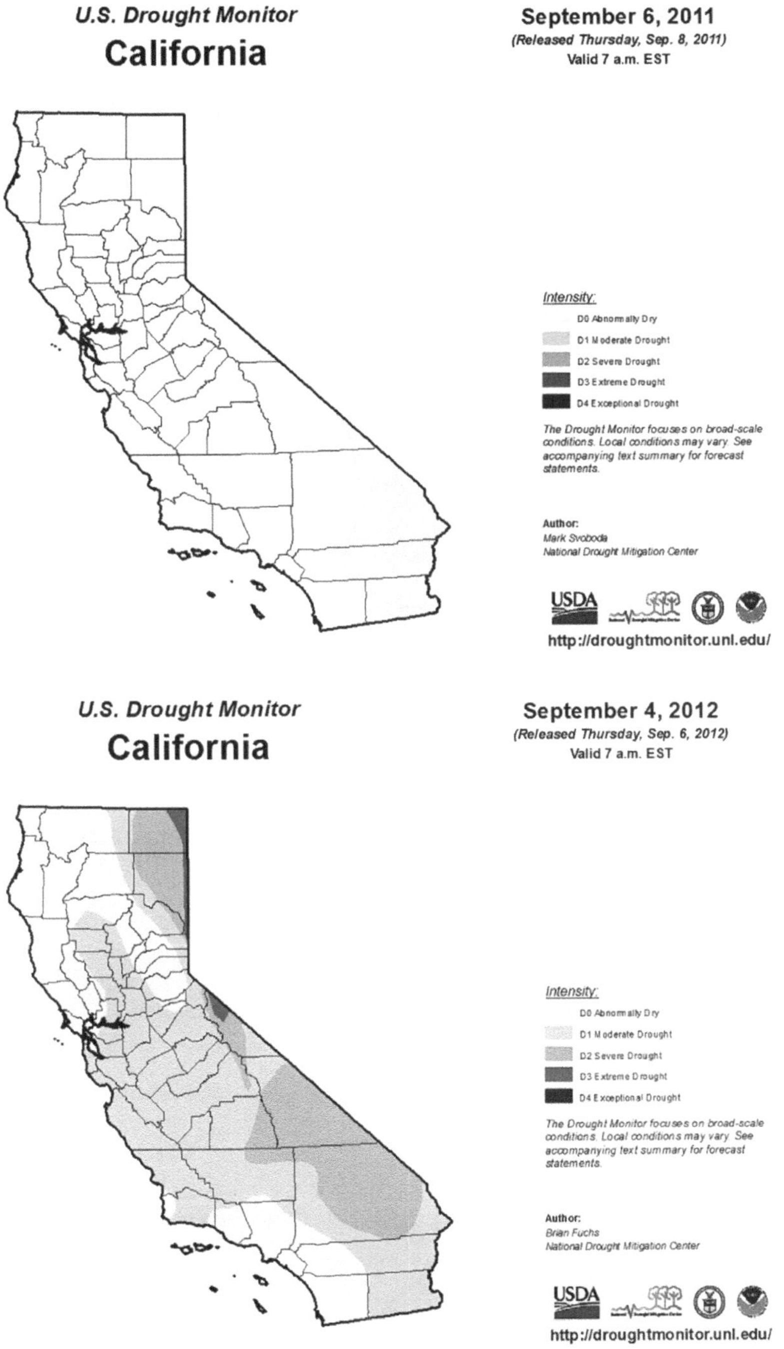

Figure 5.3 Drought extent in California for four consecutive years: September 6, 2011; September 4, 2012; September 3, 2013; and September 2, 2014

Source: The U.S. Drought Monitor is jointly produced by the National Drought Mitigation Center at the University of Nebraska–Lincoln (NDMC-UNL), the United States Department of Agriculture, and the National Oceanic and Atmospheric Administration. Map courtesy of NDMC-UNL, http://droughtmonitor.unl.edu/MapsAndData/WeeklyComparison.aspx

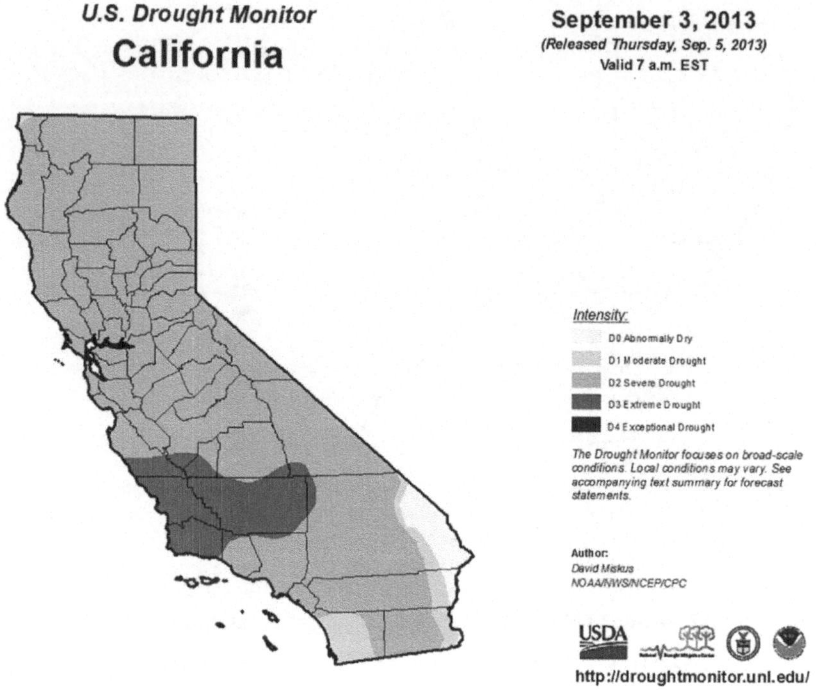

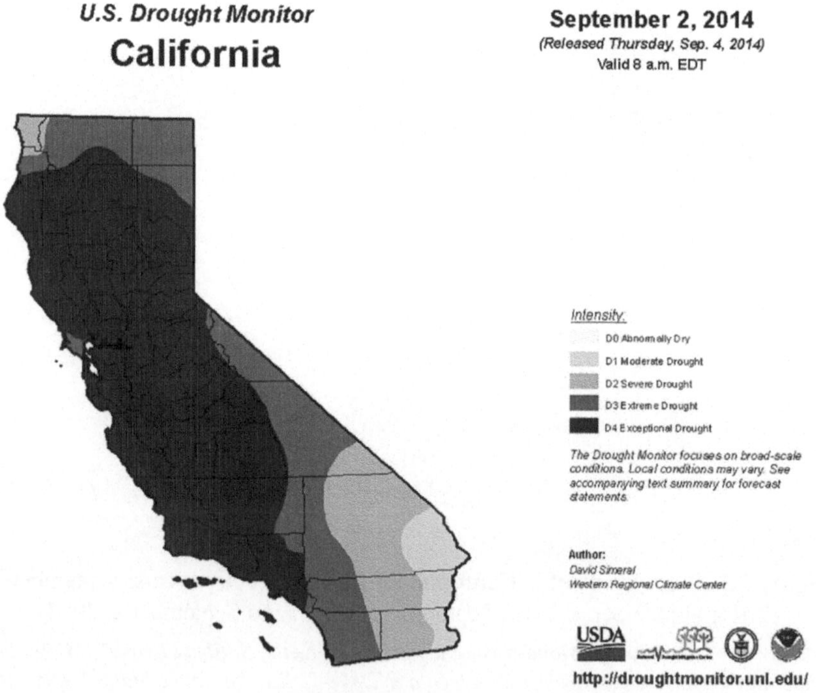

Figure 5.3 (Continued)

2008; Cutter, Burton & Emrich, 2010; Esnard, Sapat & Mitsova, 2011). These efforts are well informed by scholars from multiple disciplines (geography, demography, engineering, planning, sociology, anthropology, and science) who have advanced our knowledge of vulnerability as a spatial multi-scalar phenomenon with physical, social, policy, and governance dimensions.

Vulnerability assessments have a range of applications, which include:

- Provision of information on who and what is vulnerable, including estimation of damage and casualties that can result from various intensities of the hazard
- Delineation of the spatial extent of vulnerability and potential impact of disasters on people and infrastructure
- Evaluation and planning of evacuation routes and identification of gaps in services and shelters
- Definition of areas (e.g., landslide-prone areas) in which detailed studies are required prior to development
- Informing development regulations and property acquisition programs, as well as resource allocation
- Provision of information for formulation of mitigation and adaptation plans
- Development of targeted interventions and preparedness and response plans

In 1992, the European Union mandated GIS-based coastal and marine spatial planning (CMSP). This type of spatial planning is also utilized in other countries, including the United States, for management of coastal marine environments. Successful uses of CMSP programs involve stakeholders at early stages, in a bottom-up approach. GIS is used to disseminate data in a context suitable for each stakeholder audience (Becker, Burnell & Ron, 2012).

International institutions such as the World Bank have also launched important clearinghouses and portals. One example is the Climate Change Knowledge Portal (CCKP), which is a web-based platform that facilitates the integration of frameworks, information, data, and tools for climate change assessments at various scales (global, regional, and country), as well as capacity development and knowledge development (The World Bank Group, 2017). Part of the information framework includes spatially referenced environmental, disaster risk, and socioeconomic datasets that can be visualized on a Google Maps interface. One key product is the Climate Adaptation Country profiles, which can be used to explore, evaluate, synthesize, and learn about climate-related vulnerabilities and risks. See Chapters 7 and 15 for descriptions of other web-based portals and platforms.

Citizen Science and Participatory Mapping

Participatory mapping and public participation GIS (PPGIS) have been defined in multiple ways, but central tenets remain a focus on community mapping and the involvement and engagement of the broader public given their local knowledge of their communities. Tomaszewski (2015, p. 174) defines PPGIS as a "series of techniques for utilizing GIS as a means of involving the public in decisions related to planning and decision-making." Such techniques are one means for citizens to merge indigenous knowledge with available science and technology and to express local knowledge through visual tools such as maps. As a result, the public contributes to decision-making processes that affect them.

For example, an orientation and project planning workshop titled "Participatory Mapping and Community Empowerment for Climate Change Adaptation, Planning and Advocacy" held in Honiara, Solomon Islands, in 2012 was attended by participants from 17 countries. A participatory 3D modeling (P3DM) exercise allowed villagers from one community to produce a scaled and georeferenced replica of their community to facilitate monitoring, mitigation, and adaptation decisions in response to gradual impacts of climate change (see PPGIS/PGIS Blog at http://participatorygis.blogspot.com/2012/05/day-1-participatory-mapping-and.html).

P3DM has the advantage of combining indigenous knowledge systems with scientific information to facilitate participatory and consensus-building approaches for developing and implementing climate change adaptation policies and plans (Piccolella, 2013). The small island community of **Boe Boe** in the **Choiseul province of Solomon Islands** used P3DM to assess the risks of natural hazards and develop future sea level rise scenarios (Piccolella, 2013). Funded by AusAID and facilitated by The Nature Conservancy (TNC), the participatory mapping process presented the local stakeholders with the opportunity to discuss and implement ecosystem-based approaches to conservation, adaptation, and community resilience (Piccolella, 2013, p. 31). Gaillard and Maceda (2009) and Maceda et al. (2009) reported on three projects integrating community-based disaster risk management (CBDRM) and P3DM in the **Philippines**. The P3DM project in **Divinubo,** a small island located off the coast of the city of Borongan (the capital of the province of Eastern Samar), aimed at improving the existing disaster risk reduction program "based on the participants' own knowledge and experience" (Gaillard & Maceda, 2009, p. 111). Using the tools of P3DM, the island community of Divinubo identified and mapped several critically important assets that sustain local livelihoods, including tourism, fishing, subsistence farming, cash crop agriculture, and transportation (Maceda et al., 2009, p. 78). P3DM projects were also conducted in the villages of **Masantol** and **Dagupan**. Reflecting on the lessons learned from these initiatives, Gaillard and Maceda (2009) outlined a five-step methodology integrating CBDRM and P3DM which includes: (1) building a physical topographic model; (2) symbolizing thematic geographic layers and landmark features; (3) delineating hazard-prone areas based on local knowledge; (4) identifying and mapping disaster risk reduction measures; and (5) integrating P3DM with GIS.

In 2012, the Caribbean Natural Resources Institute (CANARI) in collaboration with the University of the West Indies engaged community groups on the island of **Tobago** in participatory three-dimensional modeling to facilitate the development of climate adaptation policies and plans (CANARI, 2013; Bobb-Prescott,

2014; DeGraff & Ramlal, 2015). In Tobago, the P3DM project involved over 100 residents and resulted in more than 80 map layers (CANARI, 2013). Indigenous spatial knowledge facilitated the visualization of natural habitats and landslide- and erosion-prone areas (Bobb-Prescott, 2014). The process of participatory mapping also provided a forum for discussion of the impacts of climate change on local livelihoods. The residents of Tobago shared adaptation practices and participated in discussions to identify appropriate coping strategies (Bobb-Prescott, 2014). In 2013, the Sustainable Grenadines, Inc. in partnership with the Grenada Fund for Conservation and The Nature Conservancy organized a P3DM workshop on **Union Island, Saint Vincent and the Grenadines**, to obtain indigenous ecological knowledge and develop policies and plans for ecosystem-based climate adaptation (DeGraff & Ramlal, 2015). Participatory GIS and marine spatial planning projects and mapping initiatives were also spearheaded in **Saint Lucia, Saint Kitts and Nevis, Belize, Barbuda, and Haiti** (DeGraff & Ramlal, 2015).

Other projects and initiatives are also worth noting with respect to climate change adaptation. For example, the Landscape Values Institute developed a website that used PPGIS to map climate change observations from people around the world. These data points were displayed and compared with data collected from the scientific community. The Landscape Values and PPGIS Institute is a non-profit consortium of international researchers and planners interested in advancing knowledge about participatory mapping to improve land and marine planning and management efforts at multiple scales ranging from local, to regional, to national. See www.landscapemap2.org for more details.

Other tools serve a complementary purpose toward climate adaptation planning for use by planners, developers, master planners, local authorities, and urban forestry initiatives. For example, the GraBS (Green and Blue Space Adaptation for Urban Areas and Eco-Towns) adaptation action planning toolkit project ran from 2008 to 2011 as a means of facilitating the development of climate adaptation action plans. In particular, the STAR (surface temperature and runoff) tools, created as part of the GraBS project, can enhance one's understanding of the influence of urban greening on the local climate. In one application, the STAR tools can be used at a neighborhood scale (in the northwest of **England** and beyond) to test the impact of different land cover scenarios of greening and development on surface temperatures and runoff, under different temperature and precipitation scenarios (The Mersey Forest & the University of Manchester, 2011).

Notes

1 *Topographic maps* show contours, in addition to basic features such as roads, buildings, water features, vegetation, and boundaries. Contour lines are imaginary lines that join points of equal elevation on the surface of the land above or below a reference surface, such as mean sea level. Contours make it possible to measure height, depth, and slope steepness.

2 *Satellite imagery data* is captured by remote sensing. Images of the Earth's surface are obtained by orbiting satellites. Information is derived from measurements

of the amount of electromagnetic radiation reflected, emitted, or scattered from objects. Examples of products and sources of satellite imagery include:

- MODIS (Moderate Resolution Imaging Spectroradiometer): moderate resolution (250- to 1,000-meter) multispectral data from the Terra Satellite (2000 to present) and Aqua Satellite (2002 to present)
- TM (Thematic Mapper): 30- to 120-meter multispectral data from Landsat 4 and 5 (1982 to present)
- MSS (Multispectral Scanner): 80-meter multispectral data from Landsats 1 to 5 (1972 to 1992)
- Multi-Resolution Land Characteristics (MRLC): national dataset with a 30-meter resolution created with Landsat Thematic Mapper satellite imagery as the main data source

3 Indirect methods that rely on sensing devices for capturing data far removed from the observer are also collectively referred to as remote sensing. However, remote sensing in this instance refers to the use of satellites that obtain data about the surface of large regions of the planet, most often in the form of computer-compatible data (DeMers, 1997, p. 41). Similar instruments for remote sensing can operate from aircrafts or satellites (Clarke, 2003).

References

Albert, S., Leon, J.X., Grinham, A.R., Church, J.A., Gibbes, B.R., & Woodroffe, C.D. 2016. Interactions Between Sea-Level Rise and Wave Exposure on Reef Island Dynamics in the Solomon Islands. *Environmental Research Letters* 11.

Assilzadeh, H., Levy, J.K., & Wang, X. 2010. Landslide Catastrophes and Disaster Risk Reduction: A GIS Framework for Landslide Prevention and Management. *Remote Sensing* 2(9), 2259–2273.

Becker, P., Burnell, G., & Ron, T.B. 2012. Using GIS to Improve Coastal Marine Spatial Planning. *Sea Technology* 53(8), 29–35.

Bobb-Prescott, N. 2014. Case study on the use of participatory three-dimensional modeling to facilitate effective contribution of civil society in the Caribbean islands in planning for action on climate change. CANARI Technical Report 401. Laventille, Trinidad.

Bolin, R. 1985. Disasters and Long-Term Recovery Policy: A Focus on Housing and Families. *Review of Policy Research* 4(4), 709–715.

Bolin, R. 1994. *Household and Community Recovery After Earthquakes*. Boulder, CO: Institute of Behavioral Science, University of Colorado.

Boruff, B.B., Emrich, C., & Cutter, S.L. 2005. Erosion Hazard Vulnerability of US Coastal Counties. *Journal of Coastal Research* 21(5), 932–942.

Brody, S.D., Zahran, S., Maghelal, P., Grover, H., & Highfield, W.E. 2007. The Rising Cost of Floods: Examining the Impact of Planning and Development Decisions on Property Damage in Florida. *Journal of the American Planning Association* 73(3), 330–345.

Cardona, O.D., van Aalst, M.K., Birkmann, J., Fordham, M., McGregor, G., Perez, R., Pulwarty, R.S., Schipper, E.L.F., & Sinh, B.T. 2012. Determinants of Risk: Exposure and Vulnerability. In: Field, C.B., V. Barros, T.F. Stocker, D. Qin, D.J. Dokken, K.L. Ebi, M.D. Mastrandrea, K.J. Mach, G.-K. Plattner, S.K. Allen, M. Tignor, and P.M. Midgley (Eds.) *Managing the Risks of Extreme Events and Disasters to Advance Climate Change Adaptation*, 65–108. A Special Report of Working Groups I and II of the Intergovernmental Panel on Climate Change (IPCC). Cambridge, UK and New York, NY: Cambridge University Press.

Caribbean Natural Resources Institute (CANARI). 2013. Using traditional knowledge for decision-making on climate change in the Caribbean. Policy Brief 15. Laventille, Trinidad: CANARI.

Clark, G.E., Moser, S.C., Ratick, S.J., Dow, K., Meyer, W.B., Emani, S., Jin, W., Kasperson, J.X., Kasperson, R.E., & Schwarz, H.E. 1998. Assessing the Vulnerability of Coastal Communities to Extreme Storms: The Case of Revere, MA., USA. *Mitigation and Adaptation Strategies for Global Climate Change* 3, 59–82.

Clarke, K.C. 2003. *Getting Started with Geographic Information Systems* (4th edition). Upper Saddle River, NJ: Pearson Education Inc.

Cutter, S.L. 2003. The Vulnerability of Science and the Science of Vulnerability. *Annals Association of American Geographers* 93(1), 1–12.

Cutter, S.L., Burton, C.B., & Emrich, C.T. 2010. Disaster Resilience Indicators for Benchmarking Baseline Conditions. *Journal of Homeland Security and Emergency Management* 7(1), 1–22.

Davidson, R.A., & Lambert, K.B. 2001. Comparing the Hurricane Disaster Risk of U.S. Coastal Counties. *Natural Hazards Review* 2(3), 132–142.

DeGraff, A. & Ramlal, B. 2015. *Participatory Mapping: Caribbean Small Island Developing States*. St. Augustine, Trinidad: Forum on the Future of the Caribbean.

DeMers, M.N. 1997. *Fundamentals of Geographic Information Systems*. Hoboken, NJ: John Wiley and Sons.

Esnard, A-M., Sapat, A., & Mitsova, D. 2011. An Index of Relative Displacement Risk to Hurricanes. *Natural Hazards* 59(2), 833–859.

Federal Emergency Management Agency (FEMA). 2017. *HAZUS Overview*. Retrieved from www.fema.gov/hazus-mh-overview (accessed July 2017).

Fletcher, S. 2007. Introduction to the Theme Issue: The Role of Geography in Contemporary Coastal Management. *Coastal Management* 35(4), 413–417.

French, S.P., Feser, E., & Peacock, W.G. 2008. *Quantitative Models of the Social and Economic Consequences of Earthquakes and Other Natural Hazards*. Final Report, Mid-America Earthquake Center Project SE-2, University of Illinois at Urbana-Champaign, Urbana, IL, 1–50.

Gaillard, J-C & Maceda, E.A. 2009. Participatory 3-Dimensional Mapping for Disaster Risk Reduction. *Participatory Learning and Action*, 60, 109-118.

Ghose, T. 2015. What Record-Breaking Drought Means for California's Future. *Live Science*. Retrieved from www.livescience.com/50417-california-drought-future.html (accessed 09/2017).

Holmes, R.R., & Dinicola, K. 2010. *100-Year Flood: It's All About Chance Haven't We Already Had One This Century?* U.S. Geological Survey General Information Product 106. Retrieved from https://pubs.usgs.gov/gip/106/pdf/100-year-flood-handout-042610.pdf (accessed August 2017).

International Hurricane Research Center (IHRC). 2004. *Windstorm Simulation and Modeling Project: Airborne Airborne LIDAR Data and Digital Elevation Models in Broward County, Florida*. Broward County Emergency Management Division, Florida International University.

Jain, V.K., Davidson, R., & Rosowsky, D. 2005. Modeling Changes in Hurricane Risk Over Time. *Natural Hazards Review* (2), 88–96.

Jones, B. 2017. Cities Build Their Vulnerability. *Nature Climate Change* 7, 237–238.

Kates, R., Colten, C., Laska, S., & Leatherman, S. 2006. Reconstruction of New Orleans After Hurricane Katrina: A Research Perspective. *Proceedings of the National Academy of Sciences*. Retrieved from www.pnas.org/content/103/40/14653.full.pdf (accessed July 2017).

Kurczy, S. 2010, October. Japan Tsunami Is Small Compared to Five of the World's Biggest Tsunamis. *Christian Science Monitor*. Retrieved from www.csmonitor.com/World/Global-Issues/2010/1028/Japan-tsunami-is-small-compared-to-five-of-world-s-biggest-tsunamis/1883-Krakatoa-tsunami.

Larkin, N.K., Sullivan, D., O'Neill, S., Raffuse, S., Solomon, R., Krull, C., Rorig, K., & Strand, T. 2008. *Smoke Forecasting Using the BlueSky Smoke Modeling Framework*. 15th Joint Conference on the Applications of Air Pollution Meteorology with the Air and Waste Management Association, New Orleans, LA.

Lott, N., & Ross, T. 2006. *Tracking and Evaluating U.S. Billion-Dollar Weather Disasters, 1980–2005*. AMS Forum: Environmental Risk and Impacts on Society: Successes and Challenges, Atlanta, GA, Amer. Meteor. Soc. 1.2.

Maceda, E.A., Gaillard, J-C, Stasiak, E., Le Masson, V., Le Berre, I. 2009. Experimental use of participatory three-dimensional models in island community-based disaster risk management, Shima: The International Journal of Research into Island Cultures, 3(1), 72-84.

Maramai, A., Brizuela, B., & Graziani, L. 2014. The Euro-Mediterranean Tsunami Catalogue. *Annals of Geophysics* 57(4), S0435.

Markowski, P., & Richardson, Y. 2014. What We Know and Don't Know About Tornado Formation. *Physics Today* 67(9), 26–31.

Martínez, J., Vega-Garcia, C., & Chuvieco, E. 2009. Human-Caused Wildfire Risk Rating for Prevention Planning in Spain. *Journal of Environmental Management* 90(2), 1241–1252.

The Mersey Forest & the University of Manchester. 2011. *An Introduction to the STAR Tools: Surface Temperature and Runoff Tools for Assessing the Potential of Green Infrastructure in Adapting Urban Areas to Climate Change*. Retrieved from www.ppgis.manchester.ac.uk/grabs/pdf/STAR_tools.pdf.

Metcalfe, J. 2014, July. The U.S. Cities with the Worst Climate Change-Related Flooding. *The Atlantic CityLab*. Retrieved from www.citylab.com/weather/2014/07/the-us-cities-with-the-worst-climate-change-related-flooding/375212/

Miles, S.B., & Chang, S.E. 2006. Modeling Community Recovery from Earthquakes. *Earthquake Spectra* 22(2), 439–458.

National Aeronautics and Space Administration (NASA). 2015. *Study: Third of Big Groundwater Basins in Distress*. Retrieved from www.nasa.gov/jpl/grace/study-third-of-big-groundwater-basins-in-distress.

National Oceanic and Atmospheric Administration (NOAA). 2007. *Enhanced F Scale for Tornado Damage*. Retrieved from www.spc.noaa.gov/faq/tornado/ef-scale.html.

National Oceanic and Atmospheric Administration (NOAA). 2016a. *Severe Weather 101: Tornado Basics*. Retrieved from www.nssl.noaa.gov/education/svrwx101/tornadoes/.

National Oceanic and Atmospheric Administration (NOAA). 2016b. *What Is a Hurricane?* Retrieved from http://oceanservice.noaa.gov/facts/hurricane.html.

National Oceanic and Atmospheric Administration (NOAA). 2016c. *What Is LiDAR?* Retrieved from http://oceanservice.noaa.gov/facts/lidar.html.

National Oceanic and Atmospheric Administration (NOAA). 2017. *What Is a Tsunami?* Retrieved from https://oceanservice.noaa.gov/facts/tsunami.html.

National Oceanic and Atmospheric Administration National Centers for Environmental Information (NOAA-NCEI). 2017. *Billion Dollar Weather and Climate Disasters; Overview*. Retrieved from www.ncdc.noaa.gov/billions/.

National Weather Service (NWS). 2012. *Drought: Public Fact Sheet*. National Oceanic and Atmospheric Administration. Retrieved from www.nws.noaa.gov/om/csd/graphics/content/outreach/brochures/FactSheet_Drought.pdf.

Passeri, D.L., Hagen, S.C., Medeiros, S.C., Bilskie, M.V., Alizad, K., & Wang, D. 2015. The Dynamic Effects of Sea Level Rise on Low-Gradient Coastal Landscapes: A Review. *Earth's Future* 3, 159–181. DOI:10.1002/2015EF000298.

Piccolella, A. 2013. Participatory mapping for adaptation to climate change: The case of Boe Boe, Solomon Islands. *Knowledge Management for Development Journal*, 9(1), 24-36.

Rambaldi, G. 2010. *Participatory Three-dimensional Modelling: Guiding Principles and Applications*. 2010 edition. Wageningen, the Netherlands: CTA.

Randolph, J. 2004. *Environmental Land Use Planning and Management*. Washington, DC: Island Press.

Rogers, P. 2014, January. California Drought: Past Dry Periods Have Lasted More Than 200 Years, Scientists Say. *San Jose Mercury News*. Retrieved from www.mercurynews.com/science/ci_24993601/california-drought-past-dry-periods-have-lasted-more.

Sarhadi, A., Soltani, S., & Modarres, R. 2012. Probabilistic Flood Inundation Mapping of Ungauged Rivers: Linking GIS Techniques and Frequency Analysis. *Journal of Hydrology* 458, 68–86.

Simpson, D.M. 2006. *Indicator Issues and Proposed Framework for a Disaster Preparedness Index (dPi)*. Fritz Institute Disaster Assessment Project. The Fritz Institute, San Francisco, CA.

Smith, A., & Katz, R. 2013: U.S. Billion-Dollar Weather and Climate Disasters: Data Sources, Trends, Accuracy and Biases. *Natural Hazards* 67(2), 387–410.

Smith, A., & Matthews, J. 2015. Quantifying Uncertainty and Variable Sensitivity Within the U.S. Billion-Dollar Weather and Climate Disaster Cost Estimates. *Natural Hazards* 77(3), 1829–1851.

Sweet, W., & Marra, J. 2015. Understanding Climate: Billy Sweet and John Marra Explain Nuisance Flooding. *Cimage.gov*. National Oceanic and Atmospheric

Agency. Retrieved from www.climate.gov/news-features/understanding-climate/understanding-climate-billy-sweet-and-john-marra-explain.

The World Bank Group. 2017. *Climate Change Knowledge Portal*. Retrieved from http://sdwebx.worldbank.org/climateportal/.

Tomaszewski, B. 2015. *Geographic Information Systems (GIS) for Disaster Management*. Boca Raton, FL: CRC Press.

United States Geological Survey (USGS). 2014. *Landslide Types and Processes. Fact Sheet 2004–3072*. Retrieved from https://pubs.usgs.gov/fs/2004/3072/pdf/fs2004-3072.pdf (accessed August 2017).

United States Geological Survey (USGS). 2015. *Coastal Change Hazards: Hurricanes and Extreme Storms*. Retrieved from http://coastal.er.usgs.gov/hurricanes/coastal-change/.

Westerling, A.L., Hidalgo, H.G., Cayan, D.R., & Swetnam, T.W. 2006. Warming and Earlier Spring Increase Western US Forest Wildfire Activity. *Science* 313(5789), 940–943.

Wilhelmi, O.V., & Morss, R.E. 2013. Integrated Analysis of Societal Vulnerability in an Extreme Precipitation Event: A Fort Collins Case Study. *Environmental Science and Policy* 26, 49–62.

Yao, J., Brauer, M., & Henderson, S.B. 2013. Evaluation of a Wildfire Smoke Forecasting System as a Tool for Public Health Protection. *Environmental Health Perspectives* 121(10), 1142–1147.

Zahran, S., Brody, S.D., Peacock, W.G., Vedlitz, A., & Grover, H. 2008. Social Vulnerability and the Natural and Built Environment: A Model of Food Casualties in Texas. *Disasters* 32(4), 537–560.

6 Spatial Analysis and GIS Modeling

Chapter Objectives

This chapter provides a brief overview of a range of spatial analytical techniques and seeks to answer the following questions:

- How can geospatial analytical techniques inform vulnerability and hazard risk assessments to strengthen coastal management and climate and disaster planning?
- What are some of the most commonly used data-driven approaches applicable to exploring phenomena related to risk and vulnerability assessments?
- What are some of the model-driven approaches that can inform climate adaptation planning?

Introduction

There have been concerted efforts in the United States and elsewhere toward coastal management and climate and disaster planning informed by vulnerability assessments and hazard risk analyses. Geographic information systems (GIS), remote sensing, global positioning systems (GPS), and more recently, unmanned aircraft vehicles (UAVs) are fundamental components of hazards research and offer capabilities that can enhance our scientific approach to climate adaptation planning. Vulnerability assessments range from basic identification of natural hazards to more complex formulation of algorithms, models, and indices to inform planning and policy-making (Esnard, Sapat & Mitsova, 2011; Birkman et al., 2016; Cutter, 2016a, 2016b, 2016c; Ismail-Zadeh, Cutter, Takeuchi & Paton, 2017). These efforts are well informed by scholars from a multiplicity of disciplines (geography, demography, engineering, planning, sociology, anthropology, and science) who have advanced our knowledge of vulnerability as a spatial multi-scalar phenomenon with physical, social, policy, and governance dimensions.

There is little doubt that climate adaptation planning requires an interdisciplinary approach, incorporating concepts and methods of climate science, hazard analysis, spatial analysis, and statistics. This chapter provides a brief overview of the spatial analytical methods that apply to climate adaptation planning. It is beyond the scope of this book to include examples of all potential uses of these methods in climate studies. Nevertheless, our premise is to provide a working overview of the methodologies and their potential applications that can facilitate and guide the development of relevant adaptation strategies.

The methods presented in this chapter encompass a large subset of topics covered by the *Geographic Information Science & Technology Body of Knowledge* initiative (DiBiase et al., 2006). Geospatial data analysis is a type of data analysis that "seeks to understand both first-order (environmental) effects and second-order (interaction) effects" (DiBiase et al., 2006, p. 43). Further, a distinction is made between "data-driven" and "model-driven" approaches in geospatial analysis (DiBiase et al., 2006, p. 43). Data-driven approaches are exploratory (DiBiase et al., 2006). They include spatial and non-spatial queries of geospatial data and various types of overlay, proximity, and density analyses. They help develop research hypotheses and are often an essential requirement for developing model-driven approaches where knowledge of spatial phenomena is further advanced by simulating the spatial objects' properties and their interactions over various spatial and temporal scales (DiBiase et al., 2006). The applications of the spatial analytical methods briefly discussed in this chapter are further examined and supported by a review of the literature and case studies in the remaining chapters of this book.

Queries and Geoprocessing

Spatial data in GIS have three fundamental characteristics: an attribute (a thematic dimension), a time of occurrence (a temporal dimension), and a geographic location (a spatial dimension) (Heywood, Cornelius & Carver, 1998). Location can be relative (the cell placement on the *n*th column and *m*th row within the array of cells) or absolute (expressed as *x,y* coordinates or latitude/longitude) (Heywood, Cornelius & Carver, 1998, p.20). GIS functionality enables both spatial and non-spatial queries (Heywood, Cornelius & Carver, 1998, p. 13). Non-spatial queries such as **Select by Attribute** allow the user to explore the attributes of individual features or phenomena, select, and highlight on the map (and in the attribute table) the features that possess those attributes. The **Summarize** and **Statistics** functions produce summary statistics for the selected features and summary tables. The **Select by Location** tool provides numerous options to create spatial queries by selecting features within or at a particular distance from other features and explore their spatial and non-spatial characteristics.

Geoprocessing tools such as **Overlay**, **Proximity**, and **Generalization** offer a set of functions commonly used to analyze spatial relationships or create new datasets from existing data. In ArcGIS, overlay functions apply to both vector and raster data. The **overlay operators** for vector data include *Identity* (which splits the input feature using the overlay feature), *Intersect* (resulting in a new feature created from the overlapping geometry of all input features), *Union* (combining all input features), and *Update* (blending two datasets) (Environmental Systems Research Institute [ESRI], 2016b). The **generalization tools** (**Merge**, **Dissolve**) are used to combine and simplify input data layers. **Buffers** are commonly used in **proximity analysis** to identify features located within a certain distance from another feature. The **Near** tool measures the distance between points or points and polylines.

The overlay operations for raster data are powerful tools used to apply algebraic expressions to spatial data. They are often used in suitability analysis and derivation of exposure and vulnerability indices (Maantay & Ziegler, 2006). Both Clark Lab's TerrSet Monitoring and Modeling Software (formerly Idrisi) ***GIS Analysis Tools*** and ESRI's ***Spatial Analyst*** provide an expanded suite of mathematical, relational, and context operators to compose and execute overlay models. In ArcGIS, **Weighted Overlay** is a widely used tool in suitability and exposure analyses. The tool provides the opportunity to organize the raster overlay process by assigning scores (preferences) and level of influence (weights) for each map layer (ESRI, 2016a). These assumptions are incorporated in the computation of the final composite score. Another useful tool is **Zonal Statistics**, which yields a summary of raster values based on zones or polygons contained in another layer (ESRI, 2016a).

Almost any type of hazard identification analysis and mitigation planning begins with the application of these tools. For example, these operators are often used as first steps in flood analysis, where map layers displaying infrastructure assets, parcels, and population data are overlaid with maps of inundated areas. Raster map or geo-algebra applications are also widely used in urban growth modeling (Clark Labs, 2015). Torio and Chmura (2013) proposed a *Coastal Squeeze Index* to rank threats to coastal wetlands in Kouchibouguac National Park in New Brunswick, Canada, and Portland, Maine, in the United States. The proposed index is based on overlay of various physical factors such as slope and imperviousness weighted by fuzzy logic functions (Torio & Chmura, 2013)

Surface Analysis

Data structures used in surface analysis include *contour lines* (vector), *triangulated irregular networks* (TINs) (vector), *digital terrain models* (DTMs) (vector), and *digital elevation models* (DEMs) (raster). Contour lines connect points with similar height values *quantized*[1] at different interval values (Vieux, 2016, p. 31). Contour lines have long been used by civil engineers and hydrologists to determine water flow direction over land surfaces. DTMs are vector-based data structures consisting of regularly spaced points. DTMs are created using a variety of techniques including conversion of contour line data into points and stereophotogrammetry (Vieux, 2016). DEMs are created by interpolating the elevation points in a digital terrain model or converting a TIN model to a raster dataset. LiDAR (Light Detection and Ranging) point cloud data are used to derive two types of elevation surfaces: high-resolution DEMs, which represent a bare-earth model of land surfaces, and DSMs (*digital surface models*), which typically represent a 3D return surface with buildings, aboveground infrastructure, and vegetation.

DEMs are commonly used as inputs in an array of land surface analyses. Using the **Raster Surface** toolset in ArcGIS, various landform features and properties can be identified and extracted to prepare maps and visualizations or conduct further analysis. *Contours*, *contours with barriers*, *slope*, and *aspect* (the direction of the steepest downslope expressed in positive degrees from 0 to 360 clockwise) are commonly derived from DEMs to use as inputs in models and other applications (ESRI, 2016b). Other useful derivatives from digital elevation models include *hillshade* (a type of a shaded relief that factors in the direction of illumination), *curvature* (a new raster dataset indicating the degree of deflection of the land surface), and *cut fill* (a raster dataset that allows the user to estimate the volume of displaced material) (ESRI, 2016b). Additional tools for surface analysis are available in the **3D Features** toolset, including an option to add Z-values, create three-dimensional buffers around point and line features (Buffer 3D), and support the creation of 3D features from attribute information (ESRI, 2016b).

Watershed analysis is an essential component of hydrologic models. The analysis usually begins with a catchment area delineation. ESRI's **Hydrology** toolset for the ArcGIS **Spatial Analyst** toolset provides the functionality necessary to conduct watershed analysis including the derivation of stream networks from elevation data and delineation of catchment areas. Hydrology analysis using GIS requires preprocessing of the digital elevation data to remove spikes and sinks that may interfere with the output of the flow accumulation algorithm. The operation takes the average values of the cells surrounding a "sink" to derive a new value for the cells where the outliers are detected. The process of stream network delineation begins with creating the flow direction and flow accumulation grids. The flow accumulation grid is used to derive a raster stream network using **Map Algebra/Raster Calculator** (**Spatial Analyst Tools** in ArcToolbox). The user has to input a conditional statement using the operator CON to determine the number of cells over which cell value aggregation will occur. For example, if the expression *Con("DEM" > 5000,1)* is applied, the values of 5,000 or more cells will be aggregated to one value in a particular cell where the flow accumulates. The conditional statement controls the density of the stream network and the number and the size of the catchment areas in the watershed. Using the **Watershed** tool, the raster stream network can be converted to sub-watersheds to represent the hydrologic response units of the study area.

Spatial Statistics

Spatial statistics examine spatial dependence among geographic objects (Wong & Lee, 2005). The tools of spatial statistics describe spatial patterns in two fundamental ways: (i) using *global* measures to identify the attributes of the overall spatial distribution; and/or (ii) applying *local* measures to quantify the strength of the association between a particular observation and nearby sample data values (Wong & Lee, 2005; Cromley & McLafferty, 2012). Global measures seek to describe spatial arrangements of geographic objects in absolute terms, adopting the null hypothesis that the observed spatial pattern "is not statistically different from a random pattern" (Wong & Lee, 2005, p. 245). Thus, variations from the random pattern are characterized as either *clustered* or *dispersed* (Wong & Lee, 2005). Local measures use spatial weights to quantify the spatial lag between two observed values. Detecting a grouping of high values within a specified radius or across spatial units reveals the presence of a geographic *hotspot*, while clustering of low values shows a geographic *cold spot* (Cromley & McLafferty, 2012). The **nearest neighbor statistic** is a statistical measure of evaluating a hypothetical global pattern in a point distribution (Wong & Lee, 2005). **Ripley's K-function** modifies the ordered nearest neighbor statistic to detect spatial arrangements over various spatial scales (Wong & Lee, 2005; Kiskowski, Hancock & Kenworthy, 2009).

The concept of *spatial autocorrelation* reflects the basic principle that objects located in proximity to each other tend to have similar characteristics (Tobler, 1970; Griffith, Wong & Whitfield, 2003). In contrast to *R* and *K statistics*, the spatial autocorrelation coefficient (SAC) weighs the attributes of the geographic objects in relation to their locational characteristics. The two most common global measures of spatial autocorrelation are **Moran's I** and **Geary's C** (Wong & Lee, 2005). Both are inferential statistics that test the null hypothesis that the observed pattern is not significantly different from random. Moran's I is derived from the cross-product of the deviation of the observed attribute value from the mean while Geary's C is calculated using the difference between the attribute values of points x_i and x_j (Wong & Lee, 2005). Chen (2013) recently proposed new formulations of Moran's I and Geary's C using matrix algebra. Global measures of spatial autocorrelation provide statistical evidence of the observed pattern of spatial dependence (e.g., clustering, dispersion, or randomness), but they do not provide tools to map these patterns. In ESRI's ArcMap, the global measures of spatial autocorrelation can be found in the **Analyzing Patterns** toolset under Spatial Statistics Tools in ArcToolbox.

Mapping of spatial clusters is performed using two local measures of spatial autocorrelation: the G_i^* **statistic** (Getis & Ord, 1992) and **local indicators of spatial autocorrelation (LISA)** statistic (Anselin, 1995). Positive values of Getis and Ord's G_i^* statistics indicate a spatial clustering, or a *hotspot*, while negative values show a *cold spot*. Positive values of LISA specify a clustering of similar values. In epidemiology, a positive LISA value may indicate a clustering of high disease rates (a hotspot) or low disease rates (a cold spot) (Cromley & McLafferty, 2012). A negative LISA statistic reveals a clustering of dissimilar values where high values of the variable of interest are bordered by low values of the same variable (Cromley, McLafferty, 2012). The **Mapping Clusters** toolset under Spatial Statistics Tools in ArcToolbox provides several tools to map spatial clusters, and identify and represent outliers. The **Hot Spot Analysis** tool employs the Getis and Ord's G_i^* statistic to map spatial clustering of variables. Spatial autocorrelation has a wide range of applications in environmental exposure studies, spatial epidemiology, greenhouse gas (GHG) emission inventories and mapping, air quality, and analyzing spatial patterns of various socioeconomic variables and land-based activities that can increase vulnerability to climate change.

Spatial Interpolation Methods

Interpolation refers to a set of methods of fitting a function to a set of discrete point observations to predict unknown values (Johnston, Ver Hoef, Krivoruchko & Lucas, 2001). Interpolation methods can be used to estimate missing observations from a set of discrete values or to create continuous surfaces (e.g., raster datasets that consist of cells with numeric values) (Yu, 2010; Teegavarapu, 2012). Teegavarapu (2012) provides a detailed overview of spatial and temporal interpolation of missing rainfall data from gauge stations. Interpolation methods for generating continuous surfaces are commonly used to represent various geographic phenomena and processes. Variations in elevation, slope, rainfall, wind speed and direction, flow direction, atmospheric concentrations of pollutants, surface temperature, groundwater flow, and soil properties are commonly represented using interpolated surfaces (Johnston, Ver Hoef, Krivoruchko & Lucas, 2001; Yu, 2010; Vieux, 2016).

The **Geostatistical Analyst** toolset in ESRI's ArcGIS™ is a compendium of various interpolation methods (Childs, 2004). These methods are broadly defined as deterministic (based on distance decay or other mathematical functions) and stochastic (based on probability functions). Among the most commonly used deterministic interpolation methods are inverse distance weighting, Thiessen polygons, natural and nearest neighbor, local and global polynomial interpolation, and radial basis functions (Johnston, Ver Hoef, Krivoruchko & Lucas, 2001; De Smith, Goodchild & Longley, 2015). ***Inverse Distance Weighting*** (IDW) is a univariate method based on a simple distance decay function. It assigns higher weights to cells/pixels that are closer to the point observation of origin (Johnston, Ver Hoef, Krivoruchko & Lucas,2001). The IDW method is often critiqued for distortions caused by the so-called "*tentpole effect*" as the intensity of the interpolated values is higher at locations closer to the observed value and diminishes gradually as the distance increases (Vieux, 2016, p. 47).

Another deterministic method is ***Trend Surface Analysis*** (TSA), in which first-, second-, or third-order (cubic) polynomial functions are fitted to the observed point data to derive a surface (Johnston, Ver Hoef, Krivoruchko & Lucas, 2001). **Kernel density** tools calculate intensity (magnitude per unit area) from the point or polyline features, assigning a larger weight closer to the sample point or line. The intensity tapers as the distance increases, resulting in smoother surfaces (Scott, 2015). ***Thiessen polygons*** have long been used by hydrologists to interpolate precipitation data. Thiessen polygons are constructed using Delaunay's triangulation principle. Any point inside the Thiessen zone is closer to its point of origin than any other point in the network system (Webster & Oliver, 2007). Delaunay tessellation is often the first step in conducting Natural Neighbor interpolation (De Smith, Goodchild & Longley, 2015).

Radial basis functions (RBFs) are a set of deterministic interpolation methods based on neural networks (De Smith, Goodchild & Longley, 2015; ESRI, 2017). Standard radial basis functions include Gaussian, multi-quadratic, inverse multiquadric function, polyharmonic, and spline (Schaback, 2007; ESRI, 2017). RBF interpolators such as the ***thin-plate spline*** force the interpolation through each observed data point while simultaneously smoothing the overall surface (Vieux, 2016; ESRI, 2017). **Splines** are the preferred method for generating digital elevation models (Vieux, 2016).

Stochastic interpolation relies on statistical techniques. In **ordinary kriging**, the expected value μ and a random component $\varepsilon_{(h)}$ determine the spatially correlated component of the variation around an unknown mean using a model that takes the form of $Z(\lambda) = \mu + \varepsilon_{(\lambda)}$ (Johnston, Ver Hoef, Krivoruchko & Lucas, 2001). Value at the location of interest is predicted by using a semivariogram that quantifies the change in spatial variation as a function of distance (Young & Young, 1998, p. 250). Fitting an appropriate semivariogram model is "a key step between spatial description and spatial prediction" (Johnston, Ver Hoef, Krivoruchko & Lucas, 2001, p. 67). When there is a positive correlation between two points, the covariance between two points is positive, and the resulting semivariogram values will have a smaller squared difference. At a certain distance, when the points are no longer correlated (i.e., they are independent), the covariance becomes zero, and the fitted line levels. In such cases, the semivariogram has larger squared differences (Johnston, Ver Hoef, Krivoruchko & Lucas, 2001). The output of the kriging procedure in ESRI's **Geostatistical Analyst** can be evaluated using scatter plots of observed versus predicted values, error and standardized error plots, and Q-Q plots. In the validation procedure, the full dataset is split into two subsets: a *test* and a *training* set (Johnston, Ver Hoef, Krivoruchko & Lucas, 2001). An autocorrelation model is fitted to the *test* data points subset and then used to generate a surface for the training dataset. A scatter plot of measured versus predicted values helps examine how well the fitted model corresponds with the empirical semivariogram (Young & Young, 1998; Johnston, Ver Hoef, Krivoruchko & Lucas, 2001).

Sun, Le, Zidek, and Burnett (1998) developed a Bayesian multivariate spatial interpolator pooling together air quality data from different monitoring networks to examine environmental health impacts. Validation and cross-validation indicated that the multivariate interpolation resulted in a smaller mean square error compared to the univariate interpolation (Sun, Le, Zidek & Burnett, 1998). More recently, *Empirical Bayesian Kriging* (EBK) and *Moving Window Kriging* (MWK) methods were added to the **Geostatistical Analyst** toolset (Knotters, Heuvelink, Hoogland & Walvoort, 2010). Interpolation methods have multiple applications in climate adaptation planning. They can be used to create spatially explicit rainfall fields essential to improve the accuracy of distributed rainfall–runoff generation modeling (Vieux, 2016). Multivariate interpolation can be used to model the spatial patterns of GHG emissions using various types of monitoring data. Health impacts studies often rely on interpolation of point observations of exposure variables or health outcomes.

Remote Sensing and GPS-UAV Technologies

Remote sensing products have been available for over 40 years, beginning with the launch of the Landsat Multispectral Scanner (MSS) in 1972 (United States Geological Survey [USGS], 2015). Technological development and advancement in sensors, transmission, acquisition, data processing, and storage over the last decade have led to a wide variety of new products with great potential to facilitate scientific breakthroughs and applications (Roy et al., 2016). Advances in both instrumentation and retrieval algorithms as well as the assemblage of extensive satellite records have greatly improved our knowledge of the climate system and its variability (Yang, Gong, Fu, Zhang, et al., 2013). Data collected through satellite remote sensing (SRS) provides critical information about changes in global average air and ocean temperatures; mass loss in polar ice sheets, glaciers, and ice caps; and rising global average sea level. Land surface temperatures (LST) derived from thermal sensors play a key role in describing the characteristics of the urban heat islands (White-Newsome et al., 2014) and in soil-vegetation-atmosphere transfer modeling of terrestrial ecosystems (Vlassova, Perez-Cabello, Nieto, Martin, Riano & de la Riva, 2014).

In remote sensing, sensors are broadly classified as passive and active. Passive sensors detect radiation emitted or reflected by the observed objects (NASA, 2018). Passive sensors collect data in the visible, infrared, thermal infrared, and microwave portions of the electromagnetic spectrum (NASA, 2018). Multi- and hyperspectral radiometers sensing in a range of wavelengths have been used to detect, classify, and quantify changes in vegetative cover, measure temperature and humidity, evaluate crop stress, monitor point source greenhouse gas emissions, and assess water quality. High-spectral resolution sensors are often designed for remote sensing of specific geophysical parameters (NASA, 2018). An active sensor sends a signal toward an observed object and quantifies the backscattered radiation emitted by the object (NASA, 2018). Active sensors include laser altimeters, LiDAR, radar (Radio Detection and Ranging), InSAR (interferometric synthetic aperture radar), and variuos types of sounders (NASA, 2018).

Remote sensing data have multiple applications in mapping vegetative cover, water bodies, urban areas, and land and sea surface temperature and in detecting changes on the earth's surface, including areas affected by droughts, earthquakes, storm surges, and sea level rise (Stone & Norman, 2006; Xiao, Ouyang, Zheng, Li, Schienke & Wang, 2007; Kumar, Bhaskar & Padmakumari, 2012; Roy et al., 2016). Figure 6.1 displays some of the visible impacts of Hurricane Irma on the Lower Keys, Florida. Image (a) shows the Lower Keys three days before Irma's landfall on September 10, 2017, as a Category 4 hurricane. Image (b) reveals the extent of the nearshore sediment disturbance caused by the strong winds of Irma.

High-resolution optical remote sensing imagery is available from EROS, IKONOS 2, SPOT 5, and QuickBird 2 (Gillespie, Chu, Frankenberg, & Thomas, 2007; Joyce, Belliss, Samsonov, McNeil, & Glassey, 2009; Brakenridge, Syvitsky, Overeem, Higgins, et al., 2012). Thermal imaging sensors which provide essential information on land and sea surface temperatures and "hotspots" of wildfire activity are mounted on several satellite platforms including Landsat, Terra, and Aqua. Most thermal scanners including the Advanced Spaceborne Thermal Emission and Reflection Radiometer (ASTER) and Moderate Resolution Imaging Spectroradiometer (MODIS) are multispectral (i.e., they collect data across several wavelengths of the electromagnetic spectrum). Earth surface changes on a global scale can be evaluated using moderate resolution (200-meter to 1-kilometer) ASTER and MODIS scenes available through NASA's Land Processes Distributed Active Archive Center (LP DAAC) including *Data Pool* (https://lpdaac.usgs.gov/lpdaac/get_data/data_pool). Higher resolution (15- and 30-meter) surface reflectance products are available from Landsat 4–5 Thematic Mapper (TM) (1982–2012), Landsat 7 Enhanced Thematic Mapper (ETM+) (1999–present), and Landsat 8 OLI/TIRS (Operational Land Imager/Thermal Infrared

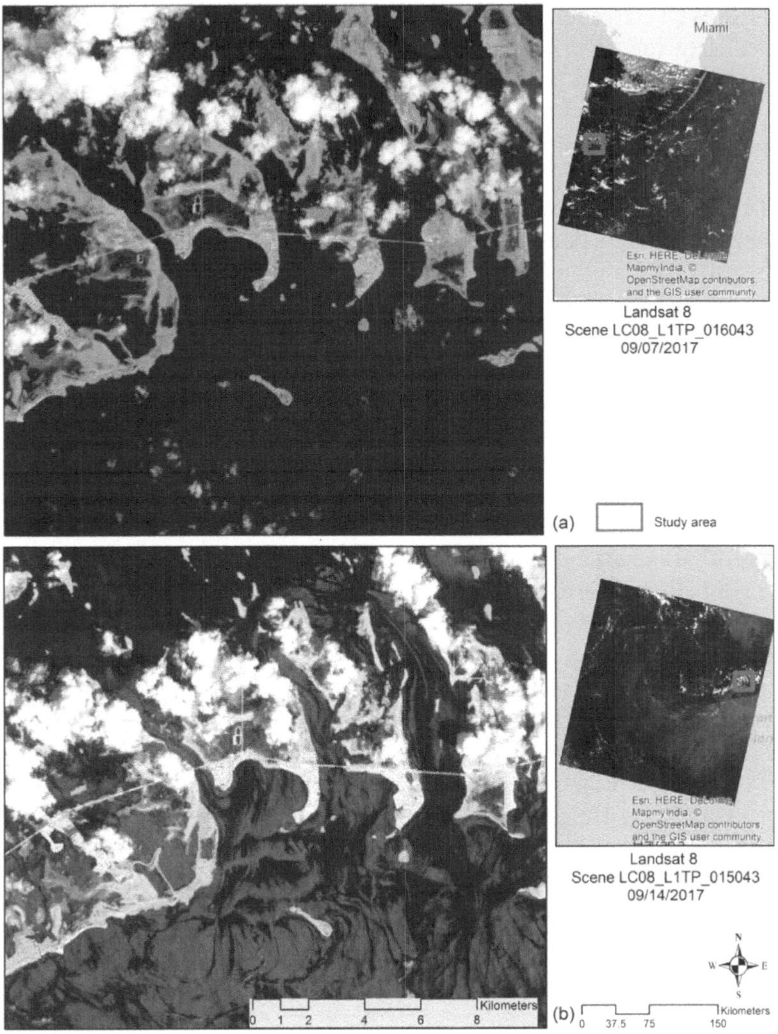

Figure 6.1 Before and after Hurricane Irma: Landsat 8 natural color image of Cudjoe Key, Florida, where Irma made landfall as a Category 4 storm on September 10, 2017

Source: USGS GloVis viewer, https://glovis.usgs.gov; map created by the authors

Sensor) (2013–present) (USGS, 2015). High-quality, high-resolution multispectral and hyperspectral data are available through Earth Observing-1 (O-1) (2000 to 2017) and the European Space Agency's *Sentinel-2* (https://lta.cr.usgs.gov/sentinel_2) (2015–present) (USGS, 2015). These products can be downloaded from several USGS-supported servers, including *GloVis* (http://glovis.usgs.gov), *EarthExplorer* (http://earthexplorer.usgs.gov), and *LandsatLook Viewer* (http://landsatlook.usgs.gov). The main limitation of using optical and thermal sensors is that imagery is often obstructed by cloud cover, smoke, or haze (Gillespie et al., 2007, Joyce et al. 2009).

LiDAR is an active remote sensing technique using an airborne system (usually mounted on an aircraft). Topographic LiDAR data are collected using near-infrared laser pulses that travel back and forth under the flight path while bathymetric LiDAR data are derived from green reflectance laser pulses that have the ability to penetrate water surfaces (National Oceanic and Atmospheric Administration National Ocean Service [NOAA-NOS], 2017). High-accuracy DEMs derived from LiDAR are widely used to map changes in shoreline configuration and delineate areas affected by floods, hurricanes, and sea level rise (NOAA-NOS, 2017).

In recent years, a growing body of research has documented the use of interferometric synthetic aperture radar (InSAR) for monitoring, detecting, and evaluating the consequences of natural hazards (Zhou, Zhou, Deng, Ai, et al., 2014). In contrast to passive optical and thermal sensors, data collected through radar instruments are not influenced by cloud cover, precipitation, or other environmental conditions (Pritchard, 2006). The most advanced InSAR data are currently available from Envisat, Sentinel-2, ALOS PALSAR, and TerraSAR. InSAR remote sensing techniques are increasingly used to examine ice sheet deformation and flows, develop high-resolution digital elevation models, and monitor floods and wildfires (Sanyal & Lu, 2004; Tralli, Blom, Zlotnicki, Donnellan, et al., 2005; Joyce et al., 2009). *Interferometric synthetic aperture radar* (InSAR) data are available through the Alaska Satellite Facility (ASF) (www.asf.alaska.edu) and UNAVCO (www.unavco.org/data/imaging/sar/sar.html).

Maps, indices, and measures derived from remote sensing data have multiple applications in climate adaptation planning. The *Normalized Difference Vegetation Index* (NDVI) is a widely used measure of classifying and quantifying vegetative cover. Plants absorb the sunlight red wavelengths and reflect near-infrared wavelengths (Qiu, Xu & He, 2014). These properties of the reflectance bands are used to derive NDVI from remote sensing data. The index is calculated as the difference between the *near-infrared* and *red* wavelength bands divided by the sum of the two wavelengths (*Equation 1*) (Qiu, Xu & He, 2014):

$$NDVI = \frac{R_{NIR} - R_{RED}}{R_{NIR} + R_{RED}} \tag{1}$$

where R_{NIR} denotes reflectance in the near-infrared band (0.85–0.88μm) and R_{RED} is reflectance in the red band (0.64–0.67μm), respectively. An NDVI index above 0.5 is considered an indication of healthy vegetation.

A useful measure of the extent of the impervious surfaces is the *Normalized Difference Built-up Index* (NDBI) (Zha, Gao & Ni, 2003; Bhatti & Tripathi, 2014). Impervious surfaces are associated with higher reflectance in the *short-wave infrared* wavelengths. Hence, NDBI is calculated as the difference between the *short-wave infrared* wavelength (1.57–1.65 μm) band and the *near-infrared* band (0.85–0.88μm) divided by the sum of the two wavelengths (Zha, Gao & Ni, 2003; Bhatti & Tripathi, 2014). The index is given by *Equation 2*:

$$NDBI = \frac{R_{SWIR} - R_{NIR}}{R_{SWIR} + R_{NIR}} \tag{2}$$

Bhatti and Tripathi (2014) developed the *Built-up Area Extraction Method* (BAEM) combining NDVI, the Modified Normalized Difference Water Index (MNDWI), and land surface temperature data to improve the accuracy of the built-up area extraction. The *Normalized Difference Water Index* (NDWI) is used to detect the moisture content in vegetation canopy. The index is calculated using the *near-infrared* and *short-wave infrared* reflectance bands as shown in *Equation 3* (Gao, 1996; Fang, Ju, Zhan, Cheng, Qiu & Wang, 2016):

$$NDWI(canopy) = \frac{R_{NIR} - R_{SWIR}}{R_{NIR} + R_{SWIR}} \quad (3)$$

Water bodies can be detected using the *near-infrared* and *green* reflectance bands (*Equation 4*):

$$NDWI(waterbody) = \frac{R_{GREEN} - R_{NIR}}{R_{GREEN} + R_{NIR}} \quad (4)$$

Gu, Hunt, Wardlow, Basara, Brown, and Verdin (2008) examined the relationship between NDVI and the NDWI canopy index to explore the relationship between soil moisture deficits and drought conditions in Oklahoma. Pinter, Santos, Hui, and Schaefer (2016) used NDWI (waterbody) to measure the inundation extent of the flooding in Louisiana triggered by a 1,000-year rainfall event (see also Chapter 8).

Emerging technologies such as GPS and UAV provide new opportunities to obtain image time series and other types of field data that can support a variety of applications in climate research and adaptation planning. Drones, as they are commonly known, have been used in disasters to improve situational awareness and emergency response (Sandvik & Lohne, 2014), sea level rise, coastal erosion, and distribution of mangrove and coral reef species (Casella et al., 2014) and in epidemiological studies to improve surveillance of infectious diseases and abate the spread of Zika and other viruses (Fornace, Drakeley, William, Espino & Cox, 2014). Chapter 8 outlines some of the recent applications of these techniques for monitoring and evaluation of global environmental change.

Geocomputation and Spatiotemporal Modeling with GIS

Simulation models are associated with an evolutionary framework where processes are time dependent. Geocomputation borrows ideas from statistical physics, fractal geometry, and complexity theory combined with advances in GIS (Atkinson & Martin, 2000). Geocomputation, as it emerged in the latter half of the 1990s, provides the theoretical and technological basis for a new generation of modeling systems capable of capturing the dynamism of processes and systems, including scenario planning, stakeholder participation, and policy adaptation. Advances in simulation modeling were facilitated by new computer technologies and development of spatial analysis tools based on GIS and remote sensing (RS). The new technologies encouraged improvements in three major ways: support for decision-making, incorporation of space-time dynamics, and a higher level of sophistication of the modeling techniques (Batty, 1994). These contributions radically improved the ability of urban simulation models to represent a wide variety of urban phenomena (Klosterman, 2001; Torrens, 2003). As a result, recent advances in dynamic urban modeling expanded the use of fractal-based models, cellular automata (CA), agent-based models (ABMs), and various statistical models using parametric and nonparametric techniques.

Can these powerful and computationally complex models assist communities in their climate change adaptation efforts, and how? The literature suggests that models of urban growth and land use/land cover change can inform communities about future patterns of development that can place people and property at risk from floods and sea level rise. As Lathrop, Auermuller, Trimble, and Bognar (2014, p. 409) indicate, "future adaptation to sea level rise should not only be an engineering issue but rather primarily a land use issue." Cities seeking to develop climate adaptation strategies need these tools to inform their citizens about future risks, develop alternative scenarios, and propose solutions to address the most significant threats (Lathrop, Auermuller, Trimble & Bognar, 2014).

An underlying assumption of both cellular automata and agent-based models is that urban systems are dynamic and their components are in a process of change driven by macro-level forces (e.g., molecular interactions) as well as by macro-scale dynamics such as population growth and economic activities (Allen, 1997). Despite their common conceptual basis, CA models and ABMs differ in their architecture. Cellular automata are pattern-based while agent-based models are process-based. CA models operate on a *lattice* similar to the raster data structure in GIS. Spatial dynamics in CA models are driven by *rules* that determine changes in a *cell state* depending on transition patterns occurring in *neighboring cells* or as a result of stochastic mutations. ABMs are object-oriented process models in which agents are "self-directed objects" who follow internal rules to achieve self-actualization through actions aimed at accomplishing specific goals within a given time frame (Brown, Riolo, Robinson, North & Rand, 2005, p. 30).

These models offer advanced capabilities for climate adaptation planning as they allow development of scenarios related to land use/land cover change and evaluation of the impacts associated with the projected changes in urban heat islands, hydrology, and open space conservation. Clarke's UGM (Urban Growth Model) and SLEUTH (Slope, Land use, Exclusion, Urban extent, Transportation, Hillshade) (Clarke & Gaydos, 1998) forecast land cover change using five coefficients: "slope resistance, road gravity, breed, diffusion and spread" (Clarke & Gaydos, 1998). The coefficients determine suitability (e.g., the *slope coefficient*) as well as the type of development (e.g., development along corridors, contiguous sprawling development, or a random scattering of new urban patches). Arthur-Hartranft, Carlson, and Clark (2003), for example, use Landsat data to derive *scaled radiant surface temperature (T*)* and *fractional vegetation cover (Fr)* indices. These indices provided the analytical tools to evaluate the outputs of SLEUTH and the *soil-vegetation-atmosphere transfer model* and examine the relationship between microclimate and future land use.

TerrSet offers several modules and models for dynamic simulation of land cover/land use change and climate adaptation planning (Clark Labs, 2015). The modules include CELLATOM, which is a purely cellular automata model, and CA-MARKOV, which combines cellular automata with Markov chain analysis. Land Change Modeler (LCM) and GEOMOD are extensions of TerrSet GIS software with enhanced urban simulation capabilities (Clark Labs, 2015). Newer additions to the software package include *GeOSIRIS* (a tool to estimate CO_2 emissions from deforestation), *Earth Trends Modeler* (a compendium of tools and data to track earth observation time series), and the *Climate Change Adaptation Modeler* (CCAM) (a module that allows users to access climatic datasets and derive several bioclimatic variables) (Clark Labs, 2015). The integration of ABMs with GIS has proven to be more challenging (Brown, Riolo, Robinson, North & Rand, 2005). Applications of ABMs and land use/land cover change have been explored by Torrens and O'Sullivan (2001); Parker, Manson, Janssen, Hoffmann, and Deadman (2003); Verburg, Nijs, van Eck, Visser, and Jong (2004); and Batty (2005). Chapter 14 provides a detailed overview of the use of urban simulation models in environmental impact assessments, analysis of the complex interactions between cities and climate change, and evaluation of proposed policy measures.

Decision Support Systems and Web-Based GIS

Efforts to make the decision-making processes more inclusive through communicative, collaborative, and participatory approaches led to the development of *decision support systems* (DSSs), which first appeared in the 1980s to lend technology-based, support-based improved capabilities for data management and analysis (Brail & Klosterman, 2001). *Planning support systems* (PSSs) came to the fore in the 1990s as a response to the challenge of developing a more effective decision-support system by incorporating public involvement in real-time assessment of alternative solutions (Klosterman, 2001). Decision/planning support systems consist of two basic components: *information technologies* integrated with specific *aims and tasks* (Klosterman, 2001; Geertman & Stillwell, 2004). They combine knowledge of urban dynamics with spatial analysis tools with the objective to improve capabilities for data compilation, visualization, and analysis using GIS. *Planning support systems* differ from *spatial decision support systems* (SDSSs) in that they tend to encompass a broader vision of long-term planning strategies. More recently, PSS have been used to integrate technology and decision-making to improve our ability to design and promote solutions that address the impact of climate stressors and support communities' long-term sustainability (Deal & Pan, 2017; Choi & Lee, 2017). Triantakonstantis and Mountrakis (2012) review the applicability of urban growth models in the planning practice and discuss their potential for improving decision-making on a variety of issues that require long-term planning.

Deal and Pan (2017) used the *Land-Use Evolution and Impact Assessment Model* (LEAM) (Sun, Deal & Pallathucherill, 2005) as a participatory tool to engage planning professionals and community stakeholders in designing alternative paths of development in Peoria, Illinois; St. Louis, Missouri; and McHenry County, Illinois. The focus of the planning process was to reveal hidden environmental costs and establish consensus on "preferred" rather than projected paths of development that take into consideration the trade-offs between economic growth and potential consequences for the environment and arable lands. The simulations incorporated intended policy measures and revealed some unanticipated future impacts. The participatory planning process provided a forum for discussion and corrective solutions that otherwise might have been omitted (Deal & Pan, 2017).

Among the most widely used GIS-based tools in hazard analysis and mitigation planning is **Hazus-MH**, which was first released by the Federal Emergency Management Agency (FEMA) in 1997 upon request from the National Institute of Building Sciences (FEMA, 2017). Hazus-MH provides a unified methodology for damage assessment and loss estimation in the wake of natural disasters such as hurricanes, storm surge, earthquakes, tsunamis, and floods (FEMA, 2017). Hazus provides decision support to emergency managers, local governments, and regional authorities in preparedness and mitigation planning and response and recovery operations (FEMA, 2017). Figure 6.2 shows a snapshot of a Hazus-MH project for Lehigh Valley, Pennsylvania.

Another commonly used decision support tool is ***CommunityViz*** (Janes & Kwartler, 2008). The tool has been available since 1994 and has served many

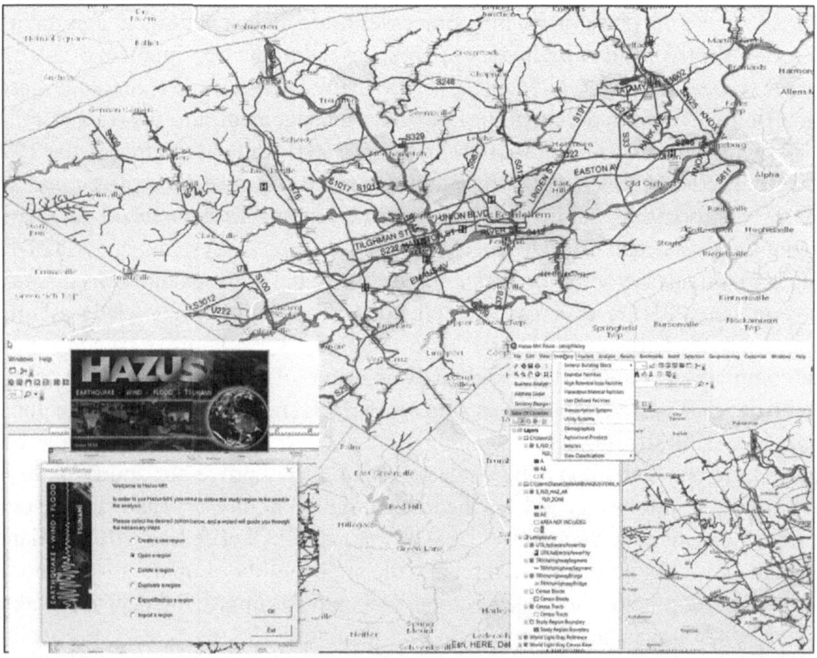

Figure 6.2 Visualization of Hazus-MH interface and a set of layers including infrastructure, buildings, and population data from the U.S. Census for Lehigh Valley, Pennsylvania

Source: Map created by the authors

planning and policy efforts, providing options for scenario development using the *Scenario Constructor* module, a 3D visual-simulation environment for real-time realistic visualizations using the *SiteBuilder* module, and simulation of future patterns of development through an agent-based model incorporated in the *Policy Simulator* (Janes & Kwartler, 2008). ***INDEX*** (Allen, 2008) and ***UrbanSim*** (Waddell, Liu & Wang, 2008) also offer several tools to design and compare scenarios, evaluate alternatives using a set of flexible indicators, and create a wide variety of visualizations.

Many of these tools are now available on a cloud platform combining the power of GIS to extract, process, and deliver spatial data with the flexibility provided by hosting web services that can facilitate access, enable queries by non-GIS users, and assist diffusion of spatial information (Peng & Tsou, 2003). Many of these decision support systems, including ***UrbanSim*** (www.urbansim.com/platform) and the ***EcoCities Spatial Portal*** (www.ppgis.manchester.ac.uk/ecocities), provide web-based services offering easy access to data and computational resources. Another form of delivering spatial data to the World Wide Web is through ArcGIS for Server and Portal for ArcGIS, which provide organizations with capabilities to share and manage geospatial information by transforming it into web-based services (ESRI, 2013). Desktop GIS provides a wide range of spatial analysis tools that are still not available through web-based GIS services. Recent efforts to deploy a GIS infrastructure that can address the needs of communities and organizations led to hybrid modes of integration where "ArcGIS for Server provides the behind-the-firewall data storage, sharing, and processing, while ArcGIS Online provides cloud-based sharing, dissemination, and collaboration" (ESRI, 2013). Chapter 15 provides a brief overview of a variety of climate adaptation tools and web resources.

Note

1 Convert continuous elevation data to discrete values by specifying an interval range.

References

Allen, E. 2008. Clicking Toward Better Outcomes: Experience with INDEX, 1994–2006. In: Brail, R.K. (Ed.) *Planning Support Systems for Cities and Regions*, 139–166. Cambridge, MA: Lincoln Institute of Land Policy.

Allen, P.M. 1997. *Cities and Regions as Self-Organizing Systems: Models of Complexity*. London & New York: Routledge.

Anselin, L. 1995. Local Indicators of Spatial Association—LISA. *Geographical Analysis* 27(2), 93–115.

Arthur-Hartranft, T., Carlson, T.N., & Clark, K.C. 2003. Satellite and Ground-Based Microclimate and Hydrologic Analyses Coupled with a Regional Urban Growth Model. *Remote Sensing of Environment* 86, 385–400.

Atkinson, P.M., & Martin, D. 2000. Introduction. In: Atkinson, P.M., & Martin, D. (Eds.) *Innovations in GIS VII: GIS and GeoComputation*. London: Taylor & Francis, 1–7.

Batty, M. 1994. A Chronicle of Scientific Planning: The Anglo-American Modeling Experience. *Journal of the American Planning Association* 60(1), 7–16.

Batty, M. 2005. Approaches to Modeling in GIS: Spatial Representation and Temporal Dynamics. In: Mcguire, David, Batty, Michael and Goodchild, Michael F. (Eds.) *GIS, Spatial Analysis, and Modeling*, 41–62. Redlands, CA: ESRI Press.

Bhatti, S., & Tripathi, N. 2014. Built-Up Area Extraction Using Landsat 8 OLI Imagery. *GIScience & Remote Sensing* 51(4), 445–467.

Birkman, J., Wenzel, F., Greiving, S., Garschagen, M., Vallée, D., Nowak, W., Welle, T., Fina, S., Goris, A., Rilling, B., Fiedrich, F., Fekete, A., Cutter, S.L., Düzgün, S., Ley, A., Friedrich, M., Kuhlmann, U., Novák, B., Wieprecht, S., Riegel, C., Thieken, A., Rhyner, J., Ulbrich, U., & Mitchell, J.K. 2016. Extreme Events, Critical Infrastructures, Human Vulnerability and Strategic Planning: Emerging Research Issues. *Journal of Extreme Events* 3(2), 1650017 (25 pages).

Brail, R.K. & Klosterman, R.E. 2001. *Planning Support System: Integrating Geographical Information Systems, Models and Visualization Tools*. Redlands, CA: ESRI Press.

Brakenridge, G.R., Syvitsky, J.P.M., Overeem, L., Higgins, S.A., Kettner, A.J., Stewart-Moore, J.A., Westerhoff, R. 2012. Global mapping of storm surges and the assessment of coastal vulnerability. *Natural Hazards*, 66, doi: 10.1007/s11069-012-0317-z.

Brown, D.G., Riolo, R., Robinson, D.T., North, M., & Rand, W. 2005. Spatial Process and Data Models: Toward Integration of Agent-Based Models and GIS. *Journal of Geographical Systems* 7, 25–47.

Casella, E., Rovere, A., Pedroncini, A., Mucerino, L., Casella, M., Cusati, L.A., Vacchi, M., Ferrari, M., & Firpo, M. 2014. Study of Wave Run-Up Using Numerical Models and Low-Altitude Aerial Photogrammetry: A Tool for Coastal Management. *Estuarine, Coastal & Shelf Science* 149, 160–167.

Chen, Y. 2013. New Approaches for Calculating Moran's Index of Spatial Autocorrelation. *PloS* 8(7), e68336. DOI: 10.1371/journal.pone.0068336.

Childs, C. 2004. Interpolating surfaces in ArcGIS. *Spatial Analyst. ArcUser*, July–September, 32–35.

Choi, H-S., & Lee, G-S. 2017. Planning Support Systems (PSS)-Based Spatial Plan Alternatives and Environmental Assessment. *Sustainability* 8, 286. DOI: 10.3390/su8030286.

Clark Labs. 2015. *TerrSet 18.2 Geospatial Monitoring and Modeling System*. Worcester, MA: Clark Labs, Clark University. Retrieved from https://clarklabs.org/wp-content/uploads/2016/03/TerrSet18-2_Brochure_WEB.pdf (accessed 07/26/17).

Clarke, K.C., & Gaydos, L. 1998. Loose-Coupling a Cellular Automaton Model and GIS: Long-Term Urban Growth Prediction for San Francisco and Washington-Baltimore. *International Journal of Geographic Information Science* 12, 699–714.

Cromley, E.K. & McLafferty, S.L. 2012. *GIS and Public Health* (2nd Edition). New York, NY: The Guilford Press.

Cutter, S.L. 2016a. The Landscape of Disaster Resilience Indicators in the USA. *Natural Hazards* 80(2), 741–758.

Cutter, S.L. 2016b. Resilience to What? Resilience for Whom? *The Geographical Journal* 182(2), 110–113.

Cutter, S.L. 2016c. Our Hazardous Environment: Four Decades of Progress or Retrenchment? *Environment: Science and Policy for Sustainable Development* 58(1), 2–4.

De Smith, M.J., Goodchild, M.F., & Longley, P.A. 2015. Geospatial Analysis: A Comprehensive Guide to Principles, Techniques and Software Tools, Online version. Retrieved from www.spatialanalysisonline.com/HTML/ (accessed 07/28/2017).

Deal, B., & Pan, H. 2017. Discerning and Addressing Environmental Failures in Policy Scenarios Using Planning Support System (PSS) Technologies. *Sustainability* 9, 13. DOI: 10.3390/su9010013.

DiBiase, D., DeMers, M., Johnson, A., Kemp, K., Luck, A.T., Plewe, B., & Wentz, E. 2006. Geographic Information Science &Technology Body of Knowledge, 1st edition. Washington, DC: Association of American Geographers.

Environmental Systems Research Institute (ESRI). 2013. Implementing Web GIS: Completely integrating ArcGIS for Server, Portal for ArcGIS, and ArcGIS Online. *ArcNews*. Retrieved from www.esri.com/esri-news/arcnews/fall13 articles/implementing-web-gis (accessed 07/25/2017).

Environmental Systems Research Institute (ESRI). 2016a. *Overlay Analysis*. Retrieved from http://desktop.arcgis.com/en/arcmap/10.3/analyze/commonly-used-tools/overlay-analysis.htm#GUID-BF7FF5E5-302F-4DC4-9108-7C 91795B5427 (accessed 07/28/2017).

Environmental Systems Research Institute (ESRI). 2016b. *An Overview of the Raster Surface Toolset*. Retrieved from http://desktop.arcgis.com/en/arcmap/10.4/tools/3d-analyst-toolbox/an-overview-of-the-surface-tools.htm (accessed 07/23/2017).

Environmental Systems Research Institute (ESRI). 2017. How Radial Basis Functions Work? Retrieved from http://pro.arcgis.com/en/pro-app/help/analysis/geostatistical-analyst/how-radial-basis-functions-work.htm (accessed 08/03/2017).

Esnard, A.M., Sapat, A., & Mitsova, D. 2011. An Index of Relative Displacement Vulnerability to Hurricanes. *Natural Hazards* 59(2), 833–859.

Fang, M., Ju, W., Zhan, W., Cheng, T., Qiu, F., & Wang, J. 2016. A New Spectral Water Index for the Estimation of Leaf Water Content from Hyperspectral Data of Leaves. *Remote Sensing of Environment* 196, 13–27.

Federal Emergency Management Agency (FEMA). 2017. *HAZUS-MH Overview.* Retrieved from www.fema.gov/hazus-mh-overview (accessed 09/20/2017).

Fornace, K.M., Drakeley, C.J., William, T., Espino, F., & Cox, J. 2014. Mapping Infectious Disease Landscapes: Unmanned Aerial Vehicles and Epidemiology. *Trends in Parasitology* 30(11), 514–519.

Gao, B-C. 1996. NDWI—a Normalized Difference Water Index for Remote Sensing of Vegetation Liquid Water from Space. *Remote Sensing of Environment* 58, 257–266.

Geertman, S., & Stillwell, J. 2004. Planning Support Systems: An Inventory of Current Practice. *Computers, Environment and Urban Systems* 28, 291–310.

Getis, A., & Ord, J.K. 1992. The Analysis of Spatial Association by Use of Distance Statistics. *Geographical Analysis* 24(3), 189–206.

Gillespie, T.W., Chu, J., Frankenberg, E., & Thomas, D. 2007. Assessment and Prediction of Natural Hazards from Satellite Imagery, *Progress in Physical Geography*, 31(5), 459-470.

Griffith, D.A., Wong, D.W.S., & Whitfield, T. 2003. Exploring Relationships Between the Global and Regional Measures of Spatial Autocorrelation. *Journal of Regional Science* 43(4), 683–710.

Gu, Y., Hunt, E., Wardlow, B., Basara, J.B., Brown, J.F., & Verdin, J.P. 2008. Evaluation of MODIS NDVI and NDWI for Vegetation Drought Monitoring Using Oklahoma Mesonet Soil Moisture Data. *Geophysical Research Letters* 35.

Heywood, I., Cornelius, S., & Carver, S. 1998. *An Introduction to Geographical Information Systems*, Upper Saddle River, NJ: Prentice Hall.

Ismail-Zadeh, A.T., Cutter, S.L., Takeuchi, K., & Paton, D. 2017. Forging a Paradigm Shift in Disaster Science. *Natural Hazards* 86(2), 969–988.

Ismail-Zadeh, A.T., Cutter, S.L., Takeuchi, K., & Paton, D. 2017. Forging a Paradigm Shift in Disaster Science. *Natural Hazards* 86(2), 969–988.

Janes, G.M., & Kwartler, M. 2008. Communities in Control: Developing Local Models Using CommunityViz®. In: Brail, R.K. (Ed.) *Planning Support Systems for Cities and Regions*, 167–184. Cambridge, MA: Lincoln Institute of Land Policy.

Johnston, K., Ver Hoef, J.M., Krivoruchko, K., & Lucas, N. 2001. *Using ArcGIS Geostatistical Analyst*. Redlands, CA: ESRI Press.

Joyce, K.E., Belliss, S.E., Samsonov, S.V., McNeil, S.J., & Glassey, P.J. 2009. A review of the status of satellite remote sensing and image processing techniques for mapping natural hazards and disasters, *Progress in Physical Geography*, 33(2), 193-207.

Kiskowski, M.A., Hancock, J.F., & Kenworthy, A.K. 2009. On the Use of Ripley's K-Function and Its Derivatives to Analyze Domain Size. *Biophysical Journal* 97(4), 1095–1103.

Klosterman, R. 2001. Planning Support Systems: A New Perspective on Computer-Aided Planning. In: Brail, Richard K., & Klosterman, Richard E. (Eds.) *Planning Support Systems*, 1–25. Redlands, CA: ESRI Press.

Knotters, M., Heuvelink, G.B.M., Hoogland, T., & Walvoort, D.J.J. 2010. *A Disposition of Interpolation Techniques*. Wageningen, The Netherlands: Wageningen University and Research Centre, Statutory Research Tasks Unit for Nature and the Environment.

Kumar, K., Bhaskar, P., & Padmakumari, K. 2012. Estimation of Land Surface Temperature to Study Urban Heat Island Effect Using Landsat ETM+ Image. *International Journal of Engineering Science and Technology* 4(2), 771–778.

Lathrop, R., Auermuller, L., Trimble, J., & Bognar, J. 2014. The Application of WebGIS Tools for Visualizing Coastal Flooding Vulnerability and Planning for Resiliency: The New Jersey Experience. *ISPRS International Journal of Geo-Information* 3, 408–429.

Longley, P.A., Goodchild, M.F., Maguire, D.J., & Rhind, D.W. (Eds). 2005. *Geographical Information Systems: Principles, Techniques, Management and Applications* (2nd edition). Abridged. ISBN: 978-0-471-73545-8, pp. 404.

Longley, P.A., Goodchild, M.F., Maguire, D.J., & Rhind, D.W. 2015. *Geographic Information Science and Systems*, 4th edition. Hoboken, NJ, Wiley, pp. 477(+xvi).

Maantay, J., & Ziegler, J. 2006. *GIS for the Urban Environment*. Redlands, CA: ESRI Press.

National Aeronautics and Space Administration (NASA). 2018. Remote Sensors: Overview. EOSDIS Earthdata. Retrieved from https://earthdata.nasa.gov/user-resources/remote-sensors.

National Oceanic and Atmospheric Administration National Ocean Service (NOAA-NOS). 2017. *What Is LiDAR?* Retrieved from https://oceanservice.noaa.gov/facts/lidar.html (accessed 07/30/2017).

Parker, D.C., Manson, S.M., Janssen, M.A., Hoffmann, M.J., & Deadman, P. 2003. Multi-Agent Systems for the Simulation of Land-Use and Land-Cover Change: A Review. *Annals of the Association of American Geographers* 93(2), 314–337.

Peng, Z-R., & Tsou, M-J. 2003. *Internet GIS: Distributed Geographic Information Services for the Internet and Wireless Networks*. Hoboken, NJ: John Wiley and Sons Inc.

Pinter, N., Santos, N., Hui, R., Schaefer, K. 2016. New Baton Rouge flood map show limits of current risk and planning methods. UC Davis Center for Watershed Sciences, available at https://californiawaterblog.com/2016/08/28/new-baton-rouge-flood-map-show-limits-of-current-risk-and-planning-methods/.

Pritchard, M.E. 2006. InSAR, a tool for measuring Earth's surface deformation. *Physics Today*, July 2006, 68-69.

Qiu, W., Xu, H., & He, Z. 2014. Study on the Difference of Urban Heat Island Defined by Brightness Temperature and Land Surface Temperature Retrieved by RS Technology. *Computer Modelling & New Technologies* 18(4), 222–225.

Roy, D.P., Kovalskyy, V., Zhang, H.K., Vermote, E.F., Yan, L., Kumar, S.S., & Egorov, A. 2016. Characterization of Landsat-7 to Landsat-8 Reflective Wavelength and Normalized Difference Vegetation Index Continuity. *Remote Sensing of Environment* 185, 57–70.

Sandvik, K.B., & Lohne, K. 2014. The Rise of the humanitarian Drone: Giving Content to an Emerging Concept. *Millennium—Journal of International Studies* 1–20. DOI: 10.1177/0305829814529470.

Sanyal, J. & Lu, X.X. 2004. Application of remote sensing in flood management with special reference to monsoon Asia: A review. *Natural Hazards*, 33, 283-301.

Schaback, R. 2007. A Practical Guide to Radial Basis Functions. In: Chen, C.S., Hon, Y.C., & Schaback, R.A. (Eds.) *Scientific Computing with Radial Basis Functions*. Retrieved from http://num.math.uni-goettingen.de/schaback/SCwRBF.pdf (accessed 08/02/2017).

Scott, D.W. 2015. *Multivariate Density Estimation: Theory, Practice, and Visualization* (2nd edition). New York, NY: John Wiley & Sons.

Stone, B., & Norman, J.M. 2006. Land Use Planning and Surface Heat Island Formation: A Parcel-Based Radiation Flux Approach. *Atmospheric Environment* 40, 3561–3573. http://dx.doi.org/10.1016/j.atmosenv.2006.01.015.

Sun, Z., Deal, B., & Pallathucherill, V.G. 2005. The Land-Use Evolution and Impact Assessment Model: A Comprehensive Urban Planning Support System. *URISA Journal*. Available at www.urisa.org/publications/journal/articles/the_land_use_evolution (accessed October 23, 2006).

Sun, W., Le, N.D., Zidek, J.V., & Burnett, R. 1998. Assessment of a Bayesian Multivariate Interpolation Approach for Health Impact Studies. *Environmetrics* 9(5), 565–586.

Teegavarapu, R.S.V. 2012. *Floods in Changing Climate: Extreme Precipitation*. Cambridge, UK: Cambridge University Press.

Tobler, W.R. 1970. A Computer Movie Simulating Urban Growth in the Detroit Region. *Economic Geography* 46 (supplement), 234–240.

Torio, D.D., & Chmura, G.L. 2013. Assessing Coastal Squeeze of Tidal Wetlands. *Journal of Coastal Research* 29(5), 1049–1061.

Torrens, P. 2003. Automata-Based Models of Urban Systems. In: Longley, Paul A., & Michael Batty (Eds.) *Advanced Spatial Analysis: The CASA Book of GIS*, 61–81. Redlands, CA: ESRI Press.

Torrens, P., & O'Sullivan, D. 2001. Cellular Automata and Urban Simulation: Where Do We Go from Here? *Environment and Planning B: Planning and Design* 28, 163–168.

Tralli, D.M., Blom, R.G., Zlotnicki, V., Donnellan, A., & Evans, D.L. 2005. Satellite remote sensing of earthquake, volcano, flood, landslide and coastal inundation hazards. *ISPRS Journal of Photogrammetry & Remote Sensing*, 59, 195-198.

Triantakonstantis, D., & Mountrakis, G. 2012. Urban Growth Prediction: A Review of Computational Models and Human Perceptions. *Journal of Geographic Information System* 4, 555–587.

United States Geological Survey (USGS). 2015. *Products*. Retrieved from https://lta.cr.usgs.gov/products_overview (accessed 07/30/2017).

Verburg, P., Nijs, T., van Eck, J.R., Visser, H., & Jong, K. 2004. A Method to Analyze Neighborhood Characteristics of Land Use Patterns. *Computers, Environment and Urban Systems* 28, 667–690.

Vieux, B.E. 2016. *Distributed Hydrologic Modeling Using GIS* (3rd Edition). Dordrecht, Netherlands: Springer.

Vlassova, L., Perez-Cabello, F., Nieto, H., Martin, P., Riano, D., & de la Riva, J. 2014. Assessment of Methods for Land Surface Temperature Retrieval from Landsat-5 TM Images Applicable to Multiscale Tree-Grass Ecosystem Modeling. *Remote Sensing* 6, 4345–4368.

Waddell, P., Liu, X., & Wang, L. 2008. UrbanSim: An Evolving Planning Support System for Evolving Communities. In: Brail, R.K. (Ed.) *Planning Support Systems for Cities and Regions*, 103–138. Cambridge, MA: Lincoln Institute of Land Policy.

Webster, R.A., & Oliver, M.A. 2007. *Geostatistics for Environmental Scientists* (2nd Edition). Hoboken, NJ: John Wiley & Sons.

White-Newsome, J.L., McCormick, S., Sampson, N., Miatta, A., Buxton, M.A., O'Neill, M.S., Gronlund, C.J., Catalano, L., Conlon, K.C., & Parker, E.A. 2014. Strategies to Reduce the Harmful Effects of Extreme Heat Events: A Four-City Study. *International Journal of Environmental Research and Public Health* 11, 1960–1988.

Wong, D.W.S., & Lee, J. 2005. *Statistical Analysis of Geographic Information With ArcView GIS and ArcGIS*. Hoboken, NJ: John Wiley & Sons.

Xiao, R-B., Ouyang, Z-Y., Zheng, H., Li, W-F., Schienke, E.W., & Wang, X-K. 2007. Spatial Patterns of Impervious Surfaces and Their Impact on Land Surface Temperature in Beijing, China. *Journal of. Environmental Sciences*, 19(2), 250–256.

Yang, J., Gong, P., Fu, R., Zhang, M., Chen, J., Liang, S., Xu, B., Shi, J., & Dickinson, R. 2013. The role of satellite remote sensing in climate change studies. *Nature Climate Change*, 3, 875-883.

Young, L.J., & Young, J.H. 1998. *Statistical Ecology: A Population Perspective*. Boston: Kluwer Academic.

Yu, H. 2010. GIS Design. In: Warf, Barney (Ed.) *Encyclopedia of Geography*. Thousand Oaks, CA: SAGE Publications, 1290–1293.

Zha, Y., Gao, J., & Ni, S. 2003. Use of Normalized Difference Built-Up Index in Automatically Mapping Urban Areas from TM Imagery. *International Journal of Remote Sensing* 24(3), 583–594.

Zhou, C., Zhou, Y., Deng, F., Ai, S., Wang, Z., Dongchen, E. 2014. Seasonal and interannual ice velocity changes of Polar Record Glacier, East Antarctica. *Annals of Glaciology*, 55(66), 45–51.

7 Climate Hazard Assessment and Adaptation Indicators

Chapter Objectives

The chapter focuses on the following questions:

- What are indicators and indices?
- Why are they useful for climate adaptation monitoring and planning?
- How are they used for climate adaptation monitoring and planning?
- How are they used for assessing climate vulnerability, risk, and resilience?
- What are the methodological challenges that hinder the development, selection, and applications of indicators for climate adaptation monitoring and planning?

What Are Indicators, and Why Are They Useful for Climate Hazard Assessment and Adaptation Monitoring?

Although used in the development context, the United States Agency for International Development (USAID)'s definition of an indicator as "a quantitative or qualitative variable that provides reliable means to measure a particular phenomenon or attribute" (USAID, 2009) is applicable to the domains of climate hazard assessment and climate adaptation monitoring. Indicators reveal the relative position of the phenomena being measured and, when evaluated over time, can inform mitigation, adaptation, and resilience strategies. Indicators have also been described as learning tools and signals given that "they deliver evaluative evidence for what has worked successfully in adaptation and what learned lessons should help guide future interventions or policy" (Climate-Eval Community of Practice, 2015, p. 33). In the context of climate change adaptation, indicators are largely seen as numerical expressions of information for operationalization purposes (Climate-Eval Community of Practice, 2015, p. 34) and as policy-relevant variables measurable over time and space (Dekker & Singer, 2011). Put another way, indicators are one approach to making theoretical concepts operational (Gallopin, 1997; Hinkel, 2011). Given the difficulty of quantifying phenomena that are not easily observable (e.g., social vulnerability), an argument has been made for a focus on operationalization rather than measurement (Holand, 2015).

To better capture multiple dimensions of concepts and phenomena related to climate adaptation (e.g., adaptive capacity), individual variables/indicators or thematic sets of variables/ indicators are often combined or aggregated as composite indicators or indices to produce a single measure (OECD, 2008; Baptista, 2014). In the climate evaluation context, some scholars have noted the small difference between indicators and indices mostly related to reference/baseline starting points (for indicators) compared to base times or values (for indices) (Climate-Eval Community of Practice, 2015).

Definitions: Indicator and Index

Indicator

- A direct measure, an indirect measure (proxy indicator), or a calculation used to represent an attribute of a system of interest (e.g., population, geographic region, socioeconomic sector, or coupled human-environment system) (Baptista, 2014, p. 1).

- A quantitative or a qualitative measure derived from a series of observed facts that can reveal relative positions (e.g., of a country) in a given area (Organisation for Economic Co-operation and Development [OECD], p. 8).

Composite Indicators/Indices

- Measures representing the integration of multiple individual indicators. Composite indices aim to capture complex realities and multidimensional concepts that cannot be adequately represented by an individual indicator or by an unstructured, disaggregated set of individual indicators (Baptista, 2014, p. 2).

- Formed when individual indicators are compiled into a single index on the basis of an underlying model. The composite indicator should ideally measure multidimensional concepts which cannot be captured by a single indicator alone (e.g., sustainability) (OECD, 2008, p. 13).

Indicators and indices are increasingly used for assessing and estimating social, environmental, ecological, physical, and economic vulnerabilities and resilience of communities, regions, and countries. They are also used to:

- Track conditions and trends
- Monitor the implementation and effectiveness of planned mitigation and adaptation policies, measures, and actions
- Assess risks and vulnerabilities to inform resiliency and planning for climate impacts
- Evaluate climate change strategies and adaptation frameworks laid out in plans that seek to make communities more resilient to the impacts of climate change
- Measure progress and performance in building adaptive capacity (process-based indicators) and delivering adaptation actions and outcomes (outcome-based indicators)
- Compare adaptation achievements across sectors, regions, and countries (United States Global Change Research Program [USGCRP]; Sniffer, 2012)

Overall, composite indicators and indices are attractive because they encapsulate a complex reality and summarize complex and often disparate technical information and data in a way that is easy to understand and apply by non-experts for climate hazard mitigation and climate adaptation decisions. Indicator projects and initiatives are carried out by academics and consultants in the private sector and are meant for audiences that span researchers, practitioners (in the public, private, and non-profit sectors), and policy-makers, as well as citizen scientists and stakeholders. The results, typically in the form of scores or rankings in tables or percentile scores on maps, make data and information more user-friendly for decision-makers. Maps also reveal spatial patterns and relationships. Whatever the application domain—sustainability, resilience, or climate adaptation—indicators and indices vary in purpose, approach, and complexity.

Applications: Vulnerability, Resilience, and Climate Adaptation Domains

> *Vulnerability, resilience, and adaptation are three fundamentally inter-related concepts among such research communities as global environmental/climatic change, social–ecological and disaster risk science.*
>
> —Lei, Wang, Yue, Zhou, and Yin (2014)

In various chapters of this book, we recognize the linkages between vulnerability, disaster risk reduction, sustainability, climate adaptation, and resilience. These domains are increasingly explored as part of climate adaptation assessment and risk reduction monitoring initiatives. This section presents a brief overview, rather than an exhaustive summary, of vulnerability and resilience indicators and indices since this is not the focus of this chapter.

Examples from the United States

Disaster scholars and vulnerability scientists have provided important insights on frameworks to *operationalize* community vulnerability and resilience (Cutter, Burton & Emrich, 2010; Peacock et al., 2010; Joerin, Shaw, Takeuchi & Krishnamurthy, 2014; DasGupta & Shaw, 2015; Siebeneck, Arlikatti & Andrew, 2015). The *Social Vulnerability Index* (SoVI) is a comparative metric that synthesizes 29 socioeconomic variables (including age, race, ethnicity, education, family structure, social dependence, occupation) that are standardized and reduced using a principal components analysis to create a smaller set of statistically independent factors (Cutter, Boruff & Shirley, 2003; Hazards Vulnerability Research Institute [HVRI]). Cutter, Burton, and Emrich (2010) also reported on the development of *baseline resilience indicators for communities* (BRIC) for southeastern U.S. counties prone to thunderstorms, flooding, tornados, and hurricanes. The variables include: (i) social resilience; (ii) economic resilience; (iii) institutional resilience; (iv) infrastructure resilience; and (v) community capital. To study community disaster resilience, Peacock et al. (2010) used as a case analysis 144 National Oceanic and Atmospheric Agency (NOAA) designated coastal counties also in the southeastern United States. Their *Community Disaster Resilience Index* incorporated 75 indicators categorized into four groups: (i) social capital; (ii) economic capital; (iii) physical capital; and (iv) human capital. In all these applications, GIS (geographic information system) was used to map the results to reveal spatial variations in patterns of the phenomena being studied at multiple units of analysis. In another example, Thatcher, Brock, and Pendleton (2013) developed a *Coastal Economic Vulnerability Index* (CEVI) to depict the vulnerability of each 1 kilometer of U.S. Gulf coastline segment to sea level rise. GIS layers (for physical coastline attributes, built environment, critical infrastructure, and population exposure) were created for prioritizing revenue allocation for restoration and protection of the coastline. The development of a *Climate Resilience Screening Index* is an ongoing initiative at the U.S. Environmental Protection Agency (EPA). With a focus on prioritizing activities to increase resilience, this index is described as a "much-needed measure for evaluating comparative adaptability and tracking progress toward climate resilience across the nation" (Summers, Smith, Harwell & Buck, 2017, p. 2).

The U.S. Department of Homeland Security (DHS) is developing indices—including the *Protective Measures Index*, the *Resilience Index*, and the *Criticality Index*—to better assist with risk management of critical infrastructure (Fisher & Norman, 2010). While their primary focus is on mitigating the effects of terrorist attacks on the nation's critical infrastructure and key facilities, the construction of the Resilience Index is based on three primary components—robustness, resourcefulness, and recovery—also used in other domains. For example, the World Economic Forum's Global Risks 2013 report provides useful insight into qualitative and quantitative indicators that can be used to assess national resilience. Five components (robustness, redundancy, resourcefulness, response, and recovery) are used to assess progress and identify gaps in five national-level subsystems: economic, environmental, governance, infrastructure, and social (World Economic Forum [WEF], 2013). In the context of infrastructure resilience, robustness and redundancy are relevant as cross-cutting issues during pre- and post-disaster planning and policy phases and because they lend themselves to actionable intervention strategies by government agencies. Robustness incorporates the concept of reliability and refers to the ability to absorb and withstand the effects of one or multiple hazards(s) without losing its function. Redundancy involves having excess capacity and backup systems or services that are able to replace or take over the function of another component (Bach, Gupta, Nair & Birkmann, 2013).

Beyond the United States

Siebeneck, Arlikatti, and Andrew (2015) reported on the application of the *Disaster Resilience of Place* (DROP) model developed by Cutter et al. (2008) using Thailand as the case study to compare the resilience of urban and rural areas. Over the 2000 to 2010 time frame, over 25 variables were measured to represent six dimensions of resilience: ecological, social, economic, institutional, infrastructure, and community competence. Principal component analysis was performed and four groupings were identified: (i) economic assets; (ii) community/response assets; (iii) household assets; and (iv) institutional structure. The results of overall resilience were mapped in five ranges from the average to one and two standard deviations below and above the average.

In a case study focused on in **Chennai, India**, a *Climate Disaster Resilience Index* (CDRI) was developed and consists of five dimensions—economic, institutional, natural, physical, and social—and provides a baseline assessment of resilience to climate-related hazards (Joerin, Shaw, Takeuchi & Krishnamurthy, 2014). The results from CDRI highlight the lower resilience values in the older northern sections of **Chennai**, compared to higher resilience values in thriving areas along the urban fringes (Joerin, Shaw, Takeuchi & Krishnamurthy, 2014). Asare-Kyei, Forkuor, and Venus (2015) reported on a *flood hazard index* for three areas in the **Sudanian Savanna ecological zone.** In that index, five flood hazard intensity levels were derived, and the maps were validated with local knowledge. Balica, Wright, and van der Meulen (2012) reported on the development of a *Coastal City Flood*

Vulnerability Index (CCFVI) for nine cities around the world. Based on hydrogeological, socioeconomic, and politico-administrative factors, the cities most vulnerable to coastal flooding were identified for more in-depth study, as well as for informing climate change adaptation measures.

Traditional economic resilience measures that work in urban coastal areas do not necessarily work in rural areas because they do not take into account the different ecosystems. DasGupta and Shaw (2015) developed a *resilience index for climate-related disasters in coastal communities in Indian Sundarbans*—islands in the low-lying deltas of northeastern **India**. The index was formulated based on five weighted factors: economic resilience, physical resilience, institutional resilience, resilience related to coastal zone management, and environmental and natural resilience. Based on these factors, a composite resilience profile was established, using a weighted mean, combining all the factors. The data were generated from a questionnaire disseminated to local leaders and experts. It covered five dimensions, for a total of 125 Likert scale variables. Five levels of susceptibility were mapped using the ArcGIS equal interval classification method. The results provide guidance for such eco-fragile regions where integrated economic development and planning for disaster and climate resilience is needed (DasGupta & Shaw, 2015).

Climate adaptation indicators capture a wider range of qualitative and quantitative approaches and applications. Most government agencies use a mix of qualitative and quantitative indicators to help monitor and evaluate progress. In some applications and tools, GIS is used to graphically display results and output, as is the case for vulnerability and resilience index outputs. Other tools have additional GIS functionality. For example, Liu, Timbal, Mo, and Fairweather (2011) developed a GIS-based *climate change adaptation strategy tool* (CCAST) with built-in GIS capability of map projection, boundary allocation, and interpolation, in addition to graphical display for visualizing climate change impacts on agricultural production. CCAST, developed for agricultural industries, was used in **Australia** to demonstrate how the selection of different genotypes of wheat (as an adaptation strategy) can mitigate the impact of climate change on wheat cropping.

Unlike the aforementioned examples, GIS is not always central to the development of adaptation indicators. Given that the outcomes of adaptation policy can, in most cases, be observed only in the far future, Hinkel (2011) has suggested that process indicators can be used to monitor the process of adaptation itself. For example, the development and implementation of an emergency management plan can be monitored as one aspect of the institutional phase of the adaptation process.

Global Indicator and Index Initiatives: Focus on Climate Change

Several decades of work by scholars and practitioners in the sustainability field provide lessons and insights on the pros and cons of indicators, composite indicators, and indices. Successful climate adaptation planning spans multiple scales as climate change impacts are manifested at all levels of human activity. So what tools do we have to track and measure changes? Time series data and geospatial techniques are used by scientists to develop indicators and techniques to track changes in the "vital signs" of the earth. They include indices to measure and map ocean heat, global land-ocean temperature, sea level rise, and ice loss. In this chapter, we are more interested in global-focused indicators and indices that seek to operationalize vulnerability, risk, resilience, and adaptation necessitated by these climate-related hazards. Several indices have been developed at national and global levels to monitor vulnerability, risk, and resilience, including the University of Notre Dame Global Adaptation Index (ND-GAIN), the Climate Change Vulnerability Index, and the World Risk Index.

Examples of Websites and Tools

- Climate Change Vulnerability Index: www.natureserve.org/conservation-tools/climate-change-vulnerability-index
- ND-GAIN Country Index: http://index.gain.org
- World Risk Index: www.uni-stuttgart.de/ireus/Internationales/WorldRiskIndex

Climate Change Vulnerability Index

The NatureServe Climate Change Vulnerability Index identifies plant and animal species that are vulnerable to the effects of climate change. The index can be used by resource managers, planners, managers, and conservationists to develop and prioritize actions that increase the resilience of species to climate change (Young, Byers, Hammerson, Frances, Oliver & Treher, 2015). The factors contributing to climate change impacts are categorized as (i) exposure to climate change; (ii) sensitivity to climate change; and (iii) capacity to adapt to change. The Andes version can be used in **Colombia**, **Ecuador**, **Peru**, and **Bolivia**. There is also a Canadian version tailored for use in Canadian assessment areas.

ND-GAIN Country Index

As noted on the ND-GAIN website, countries are ranked based on their vulnerability to climate change and other global challenges, as well as their readiness to improve resilience. The goal is to enhance understanding of the urgency for climate adaptation through targeted interventions for vulnerable countries. As of 2017, 45 indicators were being monitored for 192 countries. The indicators are categorized into: (i) *vulnerability*—exposure, sensitivity, and capacity for six supporting sectors, including food, water, health, ecosystem service, human habitat, and infrastructure; and (ii) *readiness*—comprised of economic readiness, governance readiness, and social readiness.

World Risk Index

The World Risk Index calculates climate hazard risk profiles of countries using multiple indicators of vulnerability—exposure, susceptibility, coping capacity, and adaptive capacity (Birkmann & Welle, 2015). Using the World Bank's income classification groups, the index (in map form as well) reveals global risk distribution. The output can inform risk reduction prioritization particularly for low-income and lower-middle-income countries that face a higher risk of being negatively affected by climate-related hazards (Birkmann & Welle, 2015).

Limitations and Methodological Issues

> *All existing indices of (social and outcome) vulnerability to climate change show substantial conceptual, methodological and empirical weaknesses including lack of focus, lack of a sound conceptual framework, methodological flaws, large sensitivity to alternative methods for data aggregation, limited data availability, and hiding of legitimate normative controversies.*
>
> —Füssel (2009), pp. 8–9

Referring to climate change vulnerability of households, regions, and countries as a complex state of affairs, Hinkel (2011) noted that indicators are a useful media given that single numbers can be used and interpreted for policy and related action tasks such as identification of mitigation targets and monitoring of adaptation policy. Despite the growing utility and applicability of indicators, there are a variety of challenges that must be acknowledged and addressed. Several of these caveats and limitations are discussed in the following.

Complexity, Value Judgment, and Attribution

In the earlier discussion, a distinction was made between measurement and operationalization given the difficulties of directly measuring, for example, a country's vulnerability to climate change (Füssel, 2009; Holand, 2015). Füssel also warned about the limitations of data-driven (inductive) approaches in developing indices of vulnerability to climate change. The limitation is that data-driven indices are developed and tested in the context of coping with short-term climate variability and extremes rather than adaptation to long-term climate change (Füssel, 2009).

Hinkel (2011) also refers to the *forward-looking aspect* of vulnerability. The thinking is that vulnerability (and adaptation) indicators "indicate a possibility, i.e. some state that might or might not come about in the future" (Hinkel, 2011, p. 201). Another complication is *value judgment* (Sniffer, 2012, p. 19). As such, the selected indicators may not meet the needs of all stakeholders. Government agencies such as the Department of Environment in Northern **Ireland** adopted a cross-departmental approach to adaptation and indicator development to address that problem (Sniffer, 2012).

The challenges presented by the *long timescales and dynamic baseline associated with climate change* have also been raised by Sniffer (2012)—specifically, indicators used to monitor progress and to design appropriate intervention strategies. Related is the complex issue of *attribution*. How does one develop indicators that can separate progress in adaptation from progress achieved due to other incremental activities, programs, initiatives, policies, or plans? Consequently, process-based indicators may be more appropriate for monitoring and evaluating adaptation (Sniffer, 2012, p. 14). Esnard, Sapat, and Mitsova (2011) similarly acknowledged that composite indicators and indices such as their Displacement Risk Index do not account for complex and cumulative interventions (e.g., preventative mitigation actions) over short and long time horizons. Sniffer (2012, p. 14) advises that the seriousness of the attribution problem depends on the end use. If used to monitor and report the status of the system and to observe trends, then attribution is less important compared to determining whether a policy, project, or investment has been worthwhile.

Aggregation Methods

Another common concern relates to the scientific soundness of indicators. For composite indicators and indices that stem from a predominantly quantitative approach, concerns linger about unit of analysis, scale, variable selection and data collection, *aggregation methods*, weighting, and validation (Füssel, 2009; Esnard, Sapat & Mitsova, 2011; Hinkel, 2011; Tate, 2013; Holand, 2015; Wisner, 2016).

The choice of *aggregation method* used is a *subjective process*, one that introduces additional uncertainty in the final results. Common concerns stem from the fact that: (i) individual information is lost in the aggregation process; and (ii) the unevenness of the indicators leads to a larger influence on the calculated index when there is large variation in one variable (Balica, Wright & van der Meulen, 2012). Index-based analyses and comparisons of climate change vulnerability and resilience can be more challenging when local geographic, ecological, and socioeconomic contexts vary widely even within a given area of interest (Baptista, 2014).

Beyond aggregation and generation of final index values, scores, or ranks, the composite indices should be internally and externally validated. Simpson (2006) identified the lack of simple unified scientific validation methods as one of the shortcomings of the use of composite indices. Füssel (2009) also warns that aggregated indices hide legitimate political or ethical controversies not possible to address even with better science. Overall, composite indicators and indices might best serve as a "big picture" starting point to initiate further discussion (OECD, 2008).

Resource for more technical details:

From the assessment and monitoring perspective, sustainability indicators assist policy-makers in identifying areas where the links between the economy, the environment, and society are weak (Sustainable Cities International, 2012). One well-known index is the *Environmental Sustainability Index* (ESI), which was developed as a comparative measure of a country's level of environmental sustainability for purposes of cross-national comparisons of environmental progress (Esty, 2001). Reports and articles on the ESI provide detailed computational insights about data selection, standardization, scaling, weighting, and aggregation methodology (Esty, 2001; Esty, 2002; Niemeijer, 2002; Esty, Levy, Srebotnjak & de Sherbinin, 2005). See also OECD (2008) for a detailed summary on imputation of missing data, normalization, weighting, and aggregation methodologies.

Uncertainty Analysis and Sensitivity Analysis

The robustness of composite indicators and indices can be evaluated using a combination of sensitivity analysis and uncertainty analysis. Sensitivity analysis is critical to the process of developing composite indicators and indices given subjective judgments (e.g., treatment of missing values, weights of the indicators) made during the process of index construction (OECD, 2008). Uncertainty analysis focuses on how uncertainty in the input factors propagates through the structure of the composite indicator and affects the composite indicator values (OECD, 2008, p. 34).

As Tate (2013, p. 526) warned, each step in the index development process is imbued with *uncertainty* because of the choices (between viable competing options) made by the index developer. The selection of variables for vulnerability indices involves what Wisner (2016) describes as circular reasoning, iterative thinking, and a *self-corrective cycle of deduction and induction* as literature and case studies become more available. The same pertains to climate adaptation indices. The case for field research and case studies is bolstered for such composite indicator and index initiatives that reveal aspects of vulnerability resilience and adaptive capacity that are otherwise hard to achieve (Wisner, 2016). When adaptation indices are solely focused on slow-onset hazards such as climate change, the introduction of new information and activities that shape the course of adaptation over long timescales needs to be built in as part of this self-corrective cycle referred to prior.

References

Asare-Kyei, D., Forkuor, G., & Venus, V. 2015. Modeling Flood Hazard Zones at the Sub-District Level with the Rational Model Integrated with GIS and Remote Sensing Approaches. *Water* 7(7), 3531–3564.

Bach, C., Gupta, A.K., Nair, S.S., & Birkmann, J. (2013). Critical Infrastructures and Disaster Risk Reduction. *National Institute of Disaster Management and Deutsche Gesellschaft für nternational Zusammenarbeit GmbH (GIZ)*, New Delhi, 72p.

Balica, S., Wright, N., & van der Meulen, F. 2012. A Flood Vulnerability Index for Coastal Cities and Its Use in Assessing Climate Change Impacts. *Natural Hazards* 64(1), 73–105.

Baptista, S.R. 2014. *Design and Use of Composite Indices in Assessments of Climate Change Vulnerability and Resilience: African and Latin American Resilience to Climate Change (ARCC)*. United States Agency for International Development. Retrieved from www.ciesin.org/documents/Design_Use_of_Composite_Indices.pdf.

Birkmann, J., & Welle, T. 2015. Assessing the Risk of Loss and Damage: Exposure, Vulnerability and Risk to Climate-Related Hazards for Different Country Classifications. *International Journal of Global Warming* 8(2), 191–212.

Climate-Eval Community of Practice. 2015. *Good Practice Study on Principles for Indicator Development, Selection, and Use in Climate Change Adaptation Monitoring and Evaluation*. Retrieved from www.climate-eval.org/study/good-practice-study-principles-indicator-development-selection-and-use-climate-change.

Cutter, S.L., Barnes, L., Berry, M., Burton, C.G., Evans, E., Tate, E.C., & Webb, J. 2008. A Place-Based Model for Understanding Community Resilience to Natural Disasters. *Global Environmental Change* 18(4), 598–606.

Cutter, S.L., Boruff, B., & Shirley, W. 2003. Social Vulnerability to Environmental Hazards. *Social Science Quarterly* 84(2), 242–261.

Cutter, S.L., Burton, C.G., & Emrich, C.T. 2010. Disaster Resilience Indicators for Benchmarking Baseline Conditions. *Journal of Homeland Security and Emergency Management* 7(1).

DasGupta, R., & Shaw, R. 2015. An Indicator Based Approach to Assess Coastal Communities' Resilience Against Climate Related Disasters in Indian Sundarbans. *Journal of Coastal Conservation* 19(1), 85–101.

Dekker, S., & Singer, J. 2011. *Integrated Community Sustainability Plans: Monitoring & Evaluating Success. Sustainable Cities, Canadian International Development Agency*. Retrieved from http://sustainablecities.net/wp-content/uploads/2015/04/icsps-monitoring-and-evaluating-success-final.pdf

Esnard, A-M., Sapat, A., & Mitsova, D. 2011. An Index of Relative Displacement Risk to Hurricanes. *Natural Hazards* 59(2),833–859.

Esty, D.C. 2001. Toward Data-driven Environmentalism: The Environmental Sustainability Index. *Environ Law Rep* 31(5),10603–10613.

Esty, D.C. 2002. Why Measurement Matters. In Esty, D.C., & Cornelius, P. (Eds.) *Environmental Performance Measurement: The Global 2001–2002 Report*. New York, NY: Oxford University Press.

Esty, D.C., Levy, M., Srebotnjak, T., & de Sherbinin, A. 2005. *2005 Environmental Sustainability Index: Benchmarking National Environmental Stewardship*. New Haven, CT: Yale Center for Environmental Law & Policy.

Fisher, R.E., & Norman, M. 2010. Developing Measurement Indices to Enhance Protection and Resilience of Critical Infrastructure and Key Resources. *Journal of Business Continuity & Emergency Planning* 4(3), 191–206.

Füssel, H-M. 2009. *Review and Quantitative Analysis of Indices of Climate Change Exposure, Adaptive Capacity, Sensitivity, and Impacts*. Background note to the World Development Report. Potsdam Institute for Climate Impact Research (PIK), Germany.

Gallopin, G.C. 1997. Indicators and Their Use: Information for Decision-Making. *SCOPE—Scientific Committee on Problems of the Environment International Council of Scientific Unions* 58, 13–27.

Hazard Vulnerability Research Institute (HVRI). *SOVI: Social Vulnerability Index for the United States—2010–2014*. Retrieved from http://artsandsciences.sc.edu/geog/hvri/sovi®-0.

Hinkel, J. 2011. Indicators of Vulnerability and Adaptive Capacity: Towards a Clarification of the Science-policy Interface. *Global Environmental Change* 21(1), 198–208.

Holand, I.S. 2015. Lifeline Issue in Social Vulnerability Indexing: A Review of Indicators and Discussion of Indicator Application. *Natural Hazard Review* 16(3).

Joerin, J., Shaw, R., Takeuchi, Y., & Krishnamurthy, R. 2014. The Adoption of a Climate Disaster Resilience Index in Chennai, India. *Disasters 38*(3), 540–561.

Lei, Y., Wang, J., Yue, Y., Zhou, H., & Yin, W. 2014. Rethinking the Relationships of Vulnerability, Resilience and Adaptation for a Disaster Risk Perspective. *Natural Hazards* 70, 607–627.

Liu, D.L., Timbal, B., Mo, J., & Fairweather, H. 2011. A GIS-Based Climate Change Adaptation Strategy Tool. *International Journal of Climate Change Strategies and Management* 3(2), 140–155.

Niemeijer, D. 2002. Developing Indicators for Environmental Policy: Data-Driven and Theory-Driven Approaches Examined by Example. *Environmental Science and Policy* 5, 91–103.

Organisation for Economic Co-operation and Development (OECD). 2008. *Handbook on Constructing Composite Indicators: Methodology and User Guide*. Paris: Organisation for Economic Co-operation and Development. Retrieved from www.oecd.org/std/42495745.pdf.

Peacock, W.G., Brody, S.D., Seitz, W.A., Merrell, W.J., Vedlitz, A., Zahran, S., Harriss, R.C., & Stickney, R. 2010. *Advancing Resilience of Coastal Localities: Developing, Implementing, and Sustaining the Use of Coastal Resilience Indicators: A Final Report*. Hazard Reduction and Recovery Center. Retrieved from http://hrrc.arch.tamu.edu/_common/documents/10-02R.pdf.

Siebeneck, L., Arlikatti, S., & Andrew, S.A. 2015. Using Provincial Baseline Indicators to Model Geographic Variations of Disaster Resilience in Thailand. *Natural Hazards* 79(2), 955–975.

Simpson, D.M. 2006. *Indicator Issues and Proposed Framework for a Disaster Preparedness Index (Dpi)*. Fritz Institute Disaster Assessment Project. San Francisco, CA: The Fritz Institute.

Sniffer. 2012. *Climate Change Adaptation-Related Indicators*. Retrieved from www.sniffer.org.uk/sniffer-er23-phase-1-final-report-pdf (accessed July 2017).

Summers, J.K., Smith, L.M., Harwell, L.C., & Buck, K.D. 2017. Conceptualizing Holistic Community Resilience to Climate Events: Foundation for a Climate Resilience Screening Index. *GeoHealth* 1. DOI: 10.1002/2016GH000047.

Sustainable Cities International. 2012. *Indicators for Sustainability: How Cities Are Monitoring and Evaluating Their Success*. Retrieved from www.sustainable-measures.com/indicators.

Tate, E. 2013. Uncertainty Analysis for a Social Vulnerability Index. *Annals of the Association of American Geographers* 103(3), 526–543.

Thatcher, C.A., Brock, J.C., & Pendleton, E.A. 2013. Economic Vulnerability to Sea-Level Rise Along the Northern US Gulf Coast. *Journal of Coastal Research* 63(sp1), 234–243.

United States Agency for International Development (USAID). 2009. *Glossary of Evaluation Terms*. Retrieved from http://pdf.usaid.gov/pdf_docs/pnado820.pdf.

Wisner, B. 2016. Vulnerability as Concept, Model, Metric and Tool. *Oxford Research Encyclopedia—Natural Hazards Science*. Retrieved from http://naturalhazard science.oxfordre.com/view/10.1093/acrefore/9780199389407.001.0001/ acrefore-9780199389407-e-25.

World Economic Forum (WEF). 2013. *Global Risks 2013* (8th edition). Geneva, Switzerland: World Economic Forum. Retrieved from www3.weforum.org/ docs/WEF_GlobalRisks_Report_2013.pdf.

Young, B.E., Byers, E., Hammerson, G., Frances, A., Oliver, L., & Treher, A. 2015. *Guidelines for Using the NatureServe Climate Change Vulnerability Index Release 3.0*. Arlington, VA: NatureServe.

8 Applications of Remote Sensing and GIS in Climate Change Assessments

Chapter Objectives

This chapter provides an overview of satellite remote sensing applications in climate adaptation planning seeking to answer the following questions:

- What is the role of satellite remote sensing (SRS) in climate system observations and vulnerability assessments?
- What are some oceanographic and climatological applications of SRS in climate change assessments?
- How have advances in remote sensing improved the monitoring of ice sheet, mountain glaciers, and snow depth dynamics?
- What are some applications of SRS in sea level change assessments and coastal vulnerability?
- How are SRS data used in the assessment of ecosystem vulnerability to climate change?
- What is the role of SRS in drought monitoring and assessment of changes in urban microclimate?

Climate System Observations and Vulnerability Assessments Using Satellite Remote Sensing

Data collected through satellite remote sensing (SRS) provides critical information about the warming of the climate system, including changes in global average air and ocean temperatures; mass loss in polar ice sheets, glaciers, and ice caps; and rising global average sea level (Yang et al., 2013). Satellite remote sensing provides observational data to investigate both the mechanisms and causes of the changing ice mass balance (Prichard, Luthcke & Fleming, 2010). The accumulation and ablation of ice at the ocean surface and on land is a key indicator of both short-term and long-term climatic processes (Jacob, Wahr, Pfeiffer & Swensen, 2012; Yang et al., 2013; Yi, Wang, Chang & Sun, 2016; Rabatel et al., 2017; Pimentel, Herrero & Polo, 2017). Changes in glacial mass in the Himalayas, Tibetan Plateau, and Tian Shan mountains can potentially affect the water supply for billions of people and cause dangerous glacial lakes outbursts (Yi et al., 2016; Nie et al., 2017). Furthermore, melting glaciers, ice caps, and ice sheets accelerate sea level rise (Jacob et al., 2012; Shepherd et al., 2012).

Monitoring and evaluating biophysical parameters indicative of the changes occurring in global terrestrial, estuarine, and marine ecosystems are key to understanding the ecological responses to global environmental change. In many cases, conventional field-based methods of ecological research do not readily apply to regional and global contexts (Devi et al., 2018). Optical remote sensing in the visible and thermal wavelengths and radar instruments sensing in the microwave wavelengths have been widely used for monitoring water availability, plant growth, and soil moisture conditions and conducting risk assessments of drought severity (Zhang et al., 2013). Satellite remote sensing plays an increasingly important role in monitoring and evaluating changes in urban microclimate.

Oceanographic and Climatological Applications of Satellite Remote Sensing

Monitoring of land surface temperatures (LST), sea surface temperatures (SST), and climatological variables, including atmospheric temperature, humidity, and precipitation, provides key input variables for weather and climate analysis and forecasting (Yang et al., 2013; Kuleshov et al., 2016). In fact, time series analysis of the near-surface air temperature over land based on observational data from *in situ* weather stations provided the initial evidence of the global climate warming trends over the past 100 years (Intergovernmental Panel on Climate Change [IPCC], 2007). Remote sensing data from active and passive sensors have shown immense potential for observing and detecting changes in global temperature trends at the ocean surface, on land, and in the atmosphere (Yang et al., 2013).

Landsat imagery has been providing moderate resolution multispectral data of the Earth's surface on a global basis, ensuring data continuity since the early 1970s (Irons, Dwyer & Barsi, 2012; United States Geological Survey (USGS), 2016). The newest **Landsat 8 Observatory,** launched in 2013, supports nine spectral bands with varying levels of spatial resolution: 30 meters for bands 1 through 7 and band 9, 15 meters for Band 8 (panchromatic), and 100 meters for bands 10 and 11 (thermal) (USGS, 2016). The spectral bands of the **Operational Land Imager (OLI)** sensor adds two additional spectral bands: a deep blue visible channel (Band 1) specifically designed for water resources, aerosol analysis, and coastal zone studies, and a new short-wave infrared channel (Band 9) for cirrus cloud detection (Gao & Kaufman, 1995; USGS, 2016). The **Thermal Infrared Sensor (TIRS)** enhances the L8 data acquisition (compared to prior Landsat instruments) by adding two thermal bands—Band 10 (10.60–11.19μm) and Band 11 (11.50–12.51μm)—which are widely used for computation of land surface temperatures (Barsi, Lee, Kvaran, Markham & Pedelty, 2014; Wenny et al., 2015; USGS, 2016).

Multichannel sea surface temperature products have been developed from the Advanced Very High Resolution Radiometer (AVHRR) by the National Oceanographic and Atmospheric Administration (NOAA)'s National Environmental Satellite, Data, and Information Service (NESDIS) since the late 1970s (Emery, Baldwin, Schluessel & Reynolds, 2001; Yang et al., 2013; NASA, 2018). The AVHRR-Pathfinder SST time series contains the most extensive continual global ocean observational data from space (NASA, 2018). A significant amount of data about the Earth's water cycle, including SST, ocean water evaporation, atmospheric water vapor content, cloud type and distribution, rainfall, soil moisture, ice, and snow cover is collected by NASA's Earth Science satellite missions including *Aqua*, *Terra*, and *Aura* (NASA, 2018). *Aqua*, launched in 2002, carries the Moderate Resolution Imaging Spectroradiometer (MODIS), Atmospheric Infrared Sounder (AIRS), Advanced Microwave Sounding Unit (AMSU-A), and Advanced Microwave Scanning Radiometer for EOS (AMSR-E) delivering high temporal resolution observational data on sea surface and atmospheric temperature, cloud temperature and altitude, and aerosol content on a global scale (NASA, 2018). Figure 8.1 displays the 2016 February mean SST anomaly based on 1981–2009 climate data generated by the Ocean Reanalysis System 4 (ORAS4) of the European Centre for Medium-Range Weather Forecasts.

A new generation of global, near-real-time SST products has been developed by the Group for High-Resolution Sea Surface Temperature (GHRSST), an international consortium established to address the emerging needs for multi-sensor, high-resolution products to support analysis and forecasting at short, medium and decadal/climate timescales (NASA, 2018). The Suomi National Polar-orbiting Operational Environmental Satellite System (NPOESS) Preparatory Project, or S-NPP, launched in 2011, is a pilot mission for the Joint Polar Satellite System (JPSS). S-NPP is the next generation polar-orbiting operational environmental satellite system for the United States carrying five state-of-the-art instruments/sensors including the Advanced Technology Microwave Sounder (ATMS), Visible Infrared Imaging Radiometer Suite (VIIRS), Cross-track Infrared Sounder (CrIS), Ozone Mapping Profiler Suite (OMPS), and Clouds and the Earth's Radiant Energy System (CERES). Its primary objective is to advance short-term weather forecasts and support long-term climate modeling by providing multi-sensor data for research and operational applications (NASA, 2018).

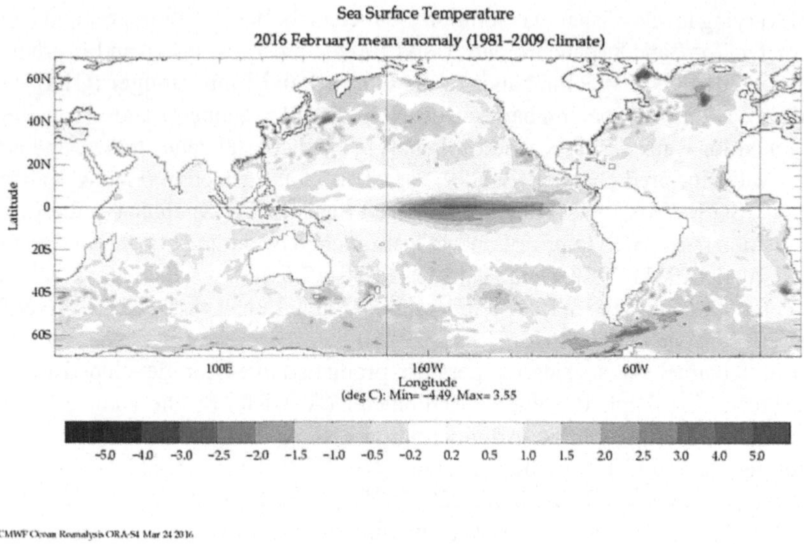

Figure 8.1 A map of the average SST anomaly for February 2016 based on 1981–2009 climate data generated by the Ocean Reanalysis System 4 (ORAS4).

Source: European Centre for Medium-Range Weather Forecasts (http://www.ecmwf.int/en/forecasts/charts/oras4/oras4_xymaps_1m?time=2016020100,0,2016020100&AnomalyMode=Anomaly&Field=Sea%20Surface%20Temperature)

Remote Sensing Applications in Reanalysis

Remote sensing data have been used for quality control and validation of atmospheric reanalysis datasets. In reanalysis, numerical climate modeling is calibrated, validated, and updated through data assimilation and real-time meteorological inputs from a variety of sources, including ground stations, satellites, vessels, airplanes, and RAOB[1] (Christensen & Lettenmaier, 2006; Maurer & Hidalgo, 2007; Bader et al., 2008). Reanalysis datasets for North America are available through NOAA's Earth System Research Laboratory–Physical Sciences Division, National Centers for Environmental Prediction (NCEP), and the National Center for Atmospheric Research (NCAR). Several atmospheric reanalysis datasets are available through the European Center for Medium-Range Weather Forecasts (ECMWF, 2015). They include ERA-Interim/Land (1979-2010), ERA-20CM (1900-2010), ERA-40 (1957-2002), ORAS4 (1958-2015), and ORAP5 (Ocean Reanalysis Pilot 5; 1979-2013) (Tietsche, Balmaseda, Zuo & Mogensen, 2017; Zuo, Balmaseda & Mogensen, 2017). The Ocean Reanalysis System 4 (ORAS4) and its real-time extension Ocean Real-Time Analysis System 4 (ORTA4) have tracked ocean circulation and surface conditions since 1957 to the present (ECMWF, 2015). Figure 8.2 displays raster layers of sea surface temperature and combined SST/land surface temperature from the ECMWF 40 Years Reanalysis provided through the ECMWF's ORAS4.

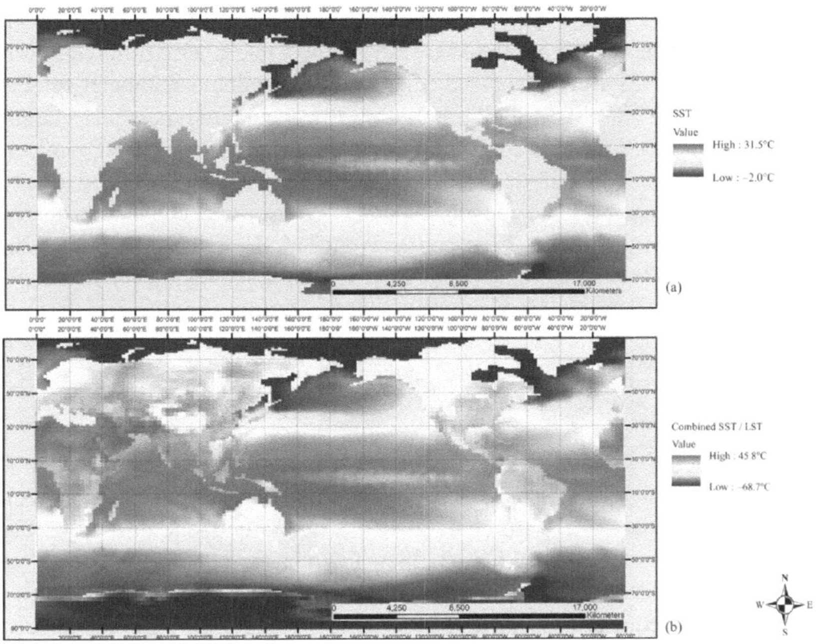

Figure 8.2 Raster layers of sea surface temperatures (SST) and combined SST/land surface temperatures created in ArcMap using the *Make NetCDF Raster Layer* tool and the following NetCDF datasets: (a) sea surface temperature collected by PCMDI (Program for Climate Model Diagnosis and Intercomparison) for use by the IPCC (2001–2002), (b) SST and 2-meter land surface temperature data for 2002 from the ECMWF 40 Years Reanalysis, daily time-step.

Data Source: ECMWF ERA-40 dataset, retrieved from http://data.ecmwf.int/data/

Although the reanalysis datasets provide the best available four-dimensional data for analyzing changes in climatic variables (Wang, Goldberg, Liu & Zhou, 2010), there is a growing scientific literature on objective quantification and validation of reanalysis parameters using remote sensing (Koster et al., 2006; Ferguson & Wood, 2011; Tang et al., 2017). Wang, Goldberg, Liu, and Zhou (2010) derived spectral signatures for stratosphere temperature and upper troposphere water vapor from data acquired from the Atmospheric Infrared Sounder operated by the National Aeronautics and Space Administration (NASA). The study evaluated the data quality of several reanalysis datasets, including NASA's Modern Era Retrospective Analysis for Research and Applications (MERRA) (Rienecker et al., 2008), European Centre for Medium-Range Weather Forecast's ERA-Interim Reanalysis, and the Japanese 25-year Reanalysis (JRA-25). Ferguson and Wood (2011) used satellite remote sensing data records for 2002–2009 to classify feedbacks between atmospheric controls and soil moisture–rainfall conditions employing the convective triggering potential (CTP)–humidity index (HI) framework. The study systematically assessed the

NASA Modern Era Retrospective Analysis for Research and Applications against satellite-based observations of HI and CTP from the Atmospheric Infrared Sounder and soil moisture from the Advanced Microwave Scanning Radiometer for Earth Observing System (EOS-AMSR-E). Based on the evaluation, the study classified the global land area into four convective regimes: (1) areas with deep convection over wet soils, (2) areas with deep convection over dry soils, (3) areas with atmospheric conditions suppressing convection, and (4) areas where convection develops over any type of land surface (Ferguson & Wood, 2011). The modified classification is applied to 20 climatic regions and three Global Land–Atmosphere Coupling Experiment (GLACE) hotspots to improve seasonal weather predictions.

Monitoring of Ice Sheets, Mountain Glaciers, Ice Caps, and Snow Depth

Satellite altimetry, gravimetry, and interferometry are routinely employed in scientific studies to investigate changes in ice mass and snow depth (Prichard, Luthcke & Fleming, 2010; Shepherd et al., 2012). Spaceborne radar altimetry sensors onboard satellites such as TOPEX, ICESat, CryoSat-2, ERS-1, ERS-2, Envisat, GFO, and Jason-1 collect observations on the Earth's surface "by transmitting a series of radio-frequency pulses and recording their echoes" (Quartly & Chen, 2006, p. 616). Satellite radar altimetry mapping provides reliable quantitative measurements of the overall growth or reduction of the ice sheet shape and volume (Rémy & Parouty, 2009). The interferometric synthetic aperture radar (SAR) data further facilitate the computation of the ice sheet mass budget. Two new satellites: Ice, Cloud, and land Elevation Satellite (ICESat) and Gravity Recovery and Climate Experiment (GRACE), launched in 2003, enable combined altimetric and gravimetric measurements of ice sheet mass, thereby advancing the methods for mass balance determination, volume change, and glacier ice flow mapping (Prichard, Luthcke & Fleming, 2010).

Adolph, Albert, and Hall (2018) combined MODIS-derived surface temperatures with *in situ* measurements during a field campaign near **Summit Station in Greenland**. The findings indicated that the 2-meter air temperature was significantly higher than the temperature of the actual snow surface measured *in situ*, suggesting the presence of a near-surface summer temperature inversion. Interferometric Synthetic Aperture Radar (InSAR) provides robust measurements of surface deformations and is widely used in reconstructing ice sheet dynamics from estimates of ice flow velocity. Zhou et al. (2014) used ERS-1/2, Envisat, and PALSAR data with D-InSAR and intensity tracking to retrieve height and rate of displacement of the **Polar Record Glacier, East Antarctica**. The study found that the interactions between the glacier tongue, iceberg blocking, and sea ice presence affected ice flow patterns and velocities (Zhou et al., 2014). The height and displacement of the glacier were also subject to localized seasonal variations (Zhou et al., 2014).

Simões, da Rosa, Czapela, Vieira, and Simões (2015) processed an extended record of remotely sensed data (1983–2006) and *in situ* measurements and analyzed changes in the ice front position of **Collins Glacier, King George Island, South Shetland Islands, Antarctic Peninsula**, a region particularly vulnerable to climate change due to significant temperature increases over the past few decades. The results indicated that during the study period the retreat was uneven, with higher rates in the northern sector. The rate of retreat was found to correlate with the mean monthly temperature and annual variations in the number of melting-degree days (Simões, da Rosa, Czapela, Vieira & Simões, 2015). Suet al. (2018) evaluated the interannual anomalies of mass change and elevation change over the **Antarctic ice sheet** from 2003–2009 using gravimetric data from the Gravity Recovery and Climate Experiment and Envisat altimetry data. The study found significant variability of the density of snow/ice associated with excess snow accumulation and the accelerated ice discharge occurring in the **Amundsen Sea** sector of the Antarctic ice sheet. Sea ice thickness and snow depth provide critical information to both climate change studies and navigation forecasts in polar regions. Xu, Zhou, Liu, Lu, and Wang (2017) measured sea ice thickness and snow depth using a combined retrieval of active laser altimetry data and the L-band passive microwave data, illustrating the usefulness of the combined approach in analyzing sea ice cover. Osipov and Osipova (2018) investigated seven small (< 1.5 kilometers squared) glaciers located in the northern slope of the **East Sayan Range, southeast Siberia.** The analysis of Landsat TM/ETM+ scenes acquired between 1986 and 2010 revealed accelerated mass loss between 1991 and 2001 consistent with the regional climatic trends such as winter precipitation and summer temperatures (Osipov & Osipova, 2018).

Veettil, Wang, Florêncio de Souza, Bremer, and Simões (2017) discussed glacier monitoring and mapping in the **tropical Andes** using remote sensing data. The study focused on observed glacial changes in **Venezuela**, **Colombia**, **Ecuador**, **Peru**, and **Bolivia**. The authors hypothesized that climate change would precipitate rapid glacier mass loss in the inner tropics and the southern wet outer tropics, leading to their potential disappearance. The study indicated that the cold phases of El Niño Southern Oscillation (ENSO) and Pacific Decadal Oscillation (PDO) had a positive effect on glaciers in the northern wet outer tropics and dry outer tropics, which showed lower rates of retreat. Mountain glaciers in Ecuador located above 5,750 meters were found to be relatively stable while the glaciers in Colombia and Venezuela exhibited accelerated mass loss. According to Veettil, Wang, Florêncio de Souza, Bremer, and Simões (2017), smaller glaciers at lower altitudes near the Amazon Basin are the most vulnerable to climate change in the tropical Andes.

The high Asian mountains, also known as the "Asian water tower" (Yi, Wang, Chang & Sun, 2016, p. 1), supply glacial meltwater to major rivers in **Southeast Asia** on which billions of people depend for sustaining their livelihoods. Yi, Wang, Chang, and Sun (2016) examined changes in glacier mass and glacial lake levels in the **Tian Shan mountains** using GRACE, ICESat, and MODIS satellite data. The spaceborne gravimetry analysis indicated that from 2003–2014, glaciers in the central Tian Shan have been losing mass at a rate of -4.0 ± 0.7 gigaton/year^{-1}, which is in agreement with the results from the laser altimetry (-3.4 ± 0.8 gigaton/year^{-1}). The study also found that rising temperatures have temporarily increased the supply of meltwater to **Bosten Lake, China**, a short-term benefit that can be compromised in the long-term as glaciers continue to lose mass. Ran, Li, and Cheng (2018) and Zhang, Wang, Zhang, Tang, and Liu (2018) used high-resolution SAR data to examine the permafrost thermal degradation of the **Qinghai-Tibet Plateau**.

Mishra, Babel, and Tripathi (2014, p. 685) used a variety of snow products obtained from MODIS and the Advanced Spaceborne Thermal Emission and Reflection Radiometer (ASTER) to estimate changes in perennial snow cover, seasonal snow cover, forest snow cover, and north- and south-faced snow cover in the **Kaligandaki River Basin, Himalaya, Nepal**, which generates nearly 20% of the national hydroelectric power. The study found greater warming trends at higher altitudes and decreasing snow cover trends in all parts of the study area, highlighting potential future impacts of diminishing snow depth/snow accumulation on water availability. Mukherjee, Bhattacharya, Pieczonka, Ghosh, and Bolch (2018) examined glacier mass loss in **Lahaul-Spiti, Western Himalaya, India**. The findings indicated that glacier mass loss has accelerated since the year 2000 (-0.30 ± 0.1 m w.e. year^{-1}). Analysis of climatic data provided evidence that the rate of acceleration was correlated with an increase in average annual temperature since 1995 and decrease in winter precipitation since 2000.

Using satellite remote sensing data, Banerjee and Shankar (2013) evaluated the response of the glaciers of the **Himalayas** to climate change, indicating that the observed changes are not uniform but exhibit both spatial and seasonal variability. Processing Landsat scenes from 1976 to 2016, Bandyopadhyay, Singh, and Birajdar (2018) evaluated changes in the **Chorabari glacier in Central Himalaya, India**. The study found that the Chorabari glacier area is decreasing by 0.8% per year (Bandyopadhyay, Singh & Birajdar, 2018). The authors noted that increased meltwater supply from Chorabari to the **Mandakini River** resulted in massive flash floods in **Uttarakhand, India**, in 2013, highlighting the increasing hazard exposure of local communities. Using over 300 Landsat images at 30-meter resolution, Nie et al. (2017) conducted a regional-scale assessment of changes occurring in nearly 5,000 **Himalayan glacial lakes** between 1990 and 2015. The results indicated that Himalayan glacial lakes increased in extent by approximately 14.1%. The formation of emergent glacial lakes has also been observed, especially in areas of glacier retreat. Himalayan atmospheric warming is considered a primary cause for the observed changes. The study identified 118 glacial lakes at high risk of glacial lake outburst floods, which are known for causing loss of life and catastrophic damage to communities.

Assessment of Sea Level Change and Coastal Vulnerability

The IPCC Fifth Assessment Report (AR5) identified ocean thermal expansion and glacier melting as the dominant factors explaining 20th-century global mean sea level rise (Church et al., 2013, p. 1139). Based on simulations of the Coupled Model Intercomparison Project phase 5 (CMIP5) Atmosphere-Ocean General Circulation Models (AOGCMs), the report estimated (with a high level of confidence) that the sum of thermal expansion and glacier mass loss could explain "65% of the observed global mean sea level rise for 1901–1990 and 90% for 1971–2010 and 1993–2010" (Church et al., 2013, p. 1139).

Contributions to sea level rise of the **Greenland** and **Antarctic ice sheets** have been shown to increase, particularly in the past three decades (Prichard, Luthcke &

Fleming, 2010; Rignot, Velicogna, van den Broeke, Monaghan & Lenaerts, 2011; Shepherd et al., 2012; Horwath, Legresy, Remy, Blarel & Lemoine, 2012). Processing GRACE satellite data, Prichard, Luthcke, and Fleming (2010, p. 1151) estimated the net mass loss from the Greenland ice sheet at 195 ± 30 gigatons per year. The study highlighted the effect of an intense summer melt on the **Antarctic Peninsula**, which resulted in an ice-shelf collapse and rapid dynamic thinning of tributary glaciers at up to 70 meters per year. Similar trends are observed in the **Gulf of Alaska** (Prichard, Luthcke & Fleming, 2010, p. 1151). Shepherd et al. (2012) found that, between 1992 and 2011, the ice sheets of Greenland, East Antarctica, West Antarctica, and the Antarctic Peninsula lost mass contributing, on average, 0.59 ± 0.20 mm $year^{-1}$ to the rate of global sea level rise.

Similarly, Jacob, Wahr, Pfeiffer, and Swensen (2012) calculated mass change from monthly GRACE-derived gravity fields over ice-covered regions, excluding Greenland and Antarctic peripheral glaciers and ice caps. The results indicated ice mass loss at a rate of 148 ± 30 gigaton/yr^{-1} from 2003 to 2010, contributing 0.41 ± 0.08 mm yr^{-1} to sea level rise (Jacob, Wahr, Pfeiffer & Swensen, 2012). Comparing GRACE satellite gravimetry with independent altimetric estimates from the ICESat from 2003 to 2010, Gardner et al. (2013, p. 852) developed mass-budget estimates for **Arctic Canada**, **Alaska**, **coastal Greenland**, **the southern Andes**, **high-mountain Asia**, and **Antarctica**. The study suggested that all glacierized regions (except Antarctica) were losing mass at an increasing rate, contributing 13% of the observed sea level rise. The study found some discrepancies with existing glaciological records indicating that "global glacier mass wastage is much less than previously thought" (Gardner et al., 2013, p. 852), possibly due to higher rates of thinning in peripheral areas than the respective regional mean.

Studies have also shown how sea level changes are expected to impact coastal settlements. Neumann, Vafeidis, Zimmermann, and Nicholls (2015) estimated that by 2030 between 879 million (low estimates) and 949 million people (high estimates) could reside in the low-lying coastal zone. By 2060, over 400 million people could be at risk from coastal flooding, with 30% inhabiting the 100-year floodplain (Neumann, Vafeidis, Zimmermann & Nicholls, 2015). Sea level change is not likely to occur uniformly throughout the globe (Edwards, 2007). Based on a report by Climate Central on the potential impacts of sea level rise, the **Netherlands**, the **United Kingdom**, **Ireland**, **France**, and **Denmark** in **Europe** and **China**, **Vietnam**, **Thailand**, **Japan**, **Myanmar**, **Bangladesh**, and **Indonesia** in **Asia** are among the countries most at risk (Strauss & Kulp, 2014). The east coast of the **United States** and the **Caribbean** are also considered highly vulnerable areas (Strauss & Kulp, 2014). In addition to ocean circulation factors such as increasing intensity of the subtropical gyres due to thermal expansion of the ocean, other factors such as geologic processes (e.g., post-glacial rebound, settling of sediment deposits) and anthropogenic activities (e.g., oil extraction and groundwater withdrawals) can cause changes in shoreline elevations and affect sea level.

In the **United States,** 272 coastal communities could be effectively inundated by 2030 (Dahl, Spanger-Siegfried, Caldas & Udvardy, 2017). It is expected that the number could double by 2100 (Dahl, Spanger-Siegfried, Caldas & Udvardy, 2017). Local trends in sea level change may differ considerably from the patterns observed on a global scale. High-resolution elevation datasets derived from LiDAR (Light Detection and Ranging) and InSAR have improved the accuracy of the sea level rise vulnerability assessments (Wang et al., 2012; Oliver-Cabrera & Wdowinski, 2016). It was found, for example, that the rates of sea level rise in the northern parts of the **Gulf of Mexico** are much higher (10 mm/yr) than those in other parts of the world due to high subsidence rates caused by settling of young sediments in the **Mississippi delta** (González & Törnqvist, 2006; Oliver-Cabrera & Wdowinski, 2016). The observed trends in areas dominated by isostatic rebound such as the southeastern portion of **Alaska** indicate a decrease in relative sea level change (Shugar et al., 2014). Using high-resolution elevation data, Strauss, Ziemlinski, Weiss, and Overpeck (2012, p. 4) estimated the extent of the low-lying areas in the contiguous United States (i.e., "land less than 1 m above local Mean High Water"), ranking **Florida** as the top vulnerable state due to the number of housing units (nearly 900,000) and population (over 1.6 million) exposed to sea level rise. Lateral inundation of coastal areas due to sea level rise is just one of many areas of concern for Floridians who already experience flooding associated with high tides and seasonally high water table levels.

For those who live close to the water, some of the effects of sea level rise are already visible. Many coastal communities have experienced widespread coastal flooding during high tides, including **Jamaica Bay, New York**; **Norfolk, Virginia**; **South Jersey Shore, New Jersey**; **Outer Banks, North Carolina**; **Annapolis, Maryland**; and **Savannah** and **Tybee Island, Georgia** (Spanger-Siegfried, Fitzpatrick & Dahl, 2014). **South Florida** and the **Florida Keys** have been particularly affected, with many communities experiencing unusual spring tides[2] at levels high enough to disrupt daily life.

Concerns have also been raised regarding the combined effects of sea level rise and other coastal hazards (e.g., storm surge and spring tides) and the implications of more frequent and intense coastal flooding for flood insurance coverage and flood rate mapping (Brody, Highfield & Kang, 2011; AECOM, 2013). Pinter, Santos, Hui, and Schaefer (2016) used satellite images from the European Space Agency's Sentinel 2A satellite to measure the inundation extent of the flooding resulting from a 1,000-year rainfall event which led to displacement of nearly 100,000 residents in **Baton Rouge, Louisiana**, in the summer of 2016 (Pinter, Santos, Hui & Schaefer, 2016). Pre- and post-flood maps were compared based on the *Normalized Difference Water Index* (NDWI) to determine the inundation extent (Pinter, Santos, Hui & Schaefer, 2016). The findings indicated that nearly 30% of the flooded areas occurred outside the 100-year flood zone designated by the Federal Emergency Management Agency (FEMA) (Pinter, Santos, Hui & Schaefer, 2016). Figure 8.3 shows flooded areas in comparison to FEMA's flood zones.

Assessment of Ecosystem Vulnerability to Climate Change

Vegetation supports a range of ecosystem services and plays an essential role in preserving the global ecological balance (Wu et al., 2014). Satellite remote sensing provides data sources and analytical techniques to address the challenges of

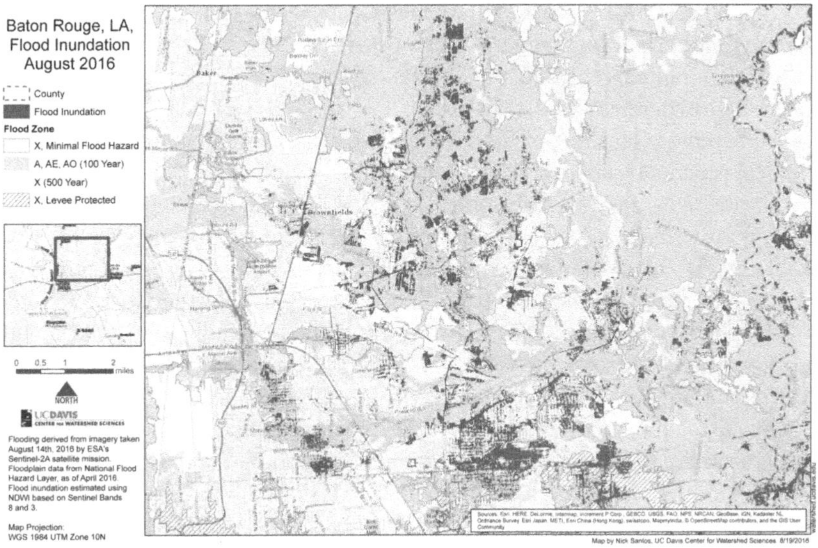

Figure 8.3 Flooded areas in the Baton Rouge area, Louisiana, in August 2016.

Source: This map was made by staff in the Natural Hazards Research and Mitigation Group at UC Davis (Prof. N. Pinter), in association with the UC Davis Center for Watershed Sciences. The map was made by Nick Santos, with consultation from Nicholas Pinter, Quinn Hart, and Rui Hui, available at https://watershed.ucdavis.edu/experiments/baton_rouge_2016/map/, used with permission.

assessing both short-term and long-term ecosystem vulnerability to climatic variability and change. Muraoka and Koizumi (2009) introduced an interdisciplinary approach named Satellite Ecology (SATECO) with the objective to link ecology, remote sensing, and micrometeorology in a coherent framework for the study of the impacts of climate change and land cover alterations on terrestrial ecosystems. Vegetation indices derived from NOAA/AVHRR (Gutman & Ignatov, 1998) and MODIS (Fensholt & Proud, 2012) have long been used to map vegetative cover, monitor vegetation conditions and health, and examine environmental changes. Wu et al. (2014) used the Normalized Difference Vegetation Index (NDVI) datasets from the Global Inventory Monitoring and Modeling Systems (GIMMS) to investigate the spatiotemporal variability of the global Fractional Vegetation Cover[3] (FVC) from 1981 to 2011. The study covered six continents and four temperature zones and indicated an increase in the average global biomass production by 12%. The highest annual maximum FVC growth rate of 25.17% was observed in the **North Frigid Zone**, potentially linked to the global rise in average temperatures, Arctic sea-ice loss, and growing season alterations (Wu et al., 2014). Using the Normalized Difference Vegetation Index derived from satellite remote sensing and climatological observations from 1982–2013, Wei et al. (2018) examined the effects of warming hiatuses on vegetation greening in the **Northern Hemisphere**. Increasing temperatures and rainfall were found to be positively correlated with vegetation greenness in **Mongolia** and **central China**. In regions where warming hiatuses did not occur, including **northern Russia** and **northern central Asia**, over 28% of the vegetation growth was inhibited, revealing the differential impact of meteorological factors on vegetation growth rates (Wei et al., 2018).

Satellite remote sensing provides critical information for mapping and analyzing ecological processes in forest ecosystems to inform the development of best management practices. Bokhorst, Tømmervik, Callaghan, Phoenix, and Bjerke (2012) examined the effects of extreme winter warming events in the **sub-Arctic region**. Changes in NDVI indicated substantial vegetation loss and damage due to rapid temperature changes and thinning snow cover. Although the sub-Arctic forest recovered relatively quickly from the recent warming event, the study highlighted the potential for recurrent losses associated with an increase in extreme weather frequency in the Arctic region due to climate change. Wylie et al. (2008) proposed a methodology for analyzing disturbances in ecosystem performance of high-latitude boreal forests which are exposed to stronger warming effects than mid-latitude or tropical forests. The study examined ecological changes in the **Alaskan boreal forests**, distinguishing climate-driven anomalies from other disturbances. The findings suggested that approximately 10% of the Alaskan boreal forest suffered ecological disturbances associated with the effects of recent fires and soil desiccation resulting from permafrost degradation (Wylie et al., 2008). Bright et al. (2014) investigated the climate effects of alternative forest management strategies within a boreal ecosystem of **eastern Norway**, employing satellite remote sensing data, field measurements, and empirical modeling. Accounting for changes in daily, seasonal, and annual mean surface temperatures and albedo, Bright et al. (2014) showed that several forest management practices could result in direct cooling benefits despite predicted long-term regional warming. Urbazaev et al. (2018) mapped the spatial distribution of the aboveground biomass of the **Mexican national forest** using airborne LiDAR, SAR, and optical satellite data in combination with field measurements and machine learning techniques. The study highlighted the importance of remote sensing data for mapping vegetative cover in areas where the national forest inventory data were unavailable.

Studies employing satellite remote sensing have also examined the effects of climate change on global grassland biomes. Li, Zhang, Chen, and Feng (2018) found a steady increase in grassland growth rate from 1982 through 2013 in most regions globally. However, grassland growth rates appeared to be more sensitive to climate variability in the steppes in the **Northern Hemisphere** compared to the savanna biomes in the **Southern Hemisphere**. Anjos and de Toledo (2018) used high-resolution remote sensing data to explore climate-related changes in three critical ecosystems in **South America**: forest, savanna, and grassland. The findings suggested that forests were more vulnerable to climate change than savannas or grasslands, raising concerns about biodiversity conservation in South America. Goldblatt, Ballesteros, and Burney (2017) conducted an ecological assessment of the semi-arid **Brazilian Sertão** using high-resolution optical and spectral imagery. Using Landsat remote sensing imagery, Park, Hooper, Flegal, and Jenerette (2018) evaluated the ecological transitions throughout the chaparral-dominated communities in the **Angeles National Forest, California**, suggesting that soil moisture variability, fire-induced disturbances, anthropogenic activity, and topography played a role in the observed increases in invasive cover. Fluctuations in soil moisture conditions explained 47% of observed variation, indicating that "chaparral vulnerability to the invasion in southern California may increase in the next century due to reduced precipitation associated with projected climate change" (Park, Hooper, Flegal & Jenerette, 2018, p. 497). Zhou et al. (2017) estimated the extent of grassland degradation in **China** from 1982 to 2010. The study found a 22% reduction in net primary productivity (NPP), indicating that the impacts of climate change had a stronger effect on NPP losses than anthropogenic activities (Zhou et al., 2017).

Wetland remote sensing is an emerging field of research providing a wealth of information related to the evolution and spatial distribution of freshwater wetlands and coastal marshes. Guo, Li, Sheng, Xu, and Wu (2017) conducted a detailed overview of the wetland remote sensing methods including optical imagery at low, medium and high resolution, synthetic interferometric aperture radar (InSAR), hyperspectral imagery, and Light Detection and Ranging (LiDAR). The study discussed specific applications of each sensor for wetland classification, assessment of biodiversity and habitat quality, biomass estimation, water quality assessment, mangrove distribution analysis, and the potential effects of sea level rise (Guo, Li, Sheng, Xu & Wu, 2017). Processing multi-year Linear Imaging Self Scanning images (LISS) II (IRS-1B) and III (IRS P6/RESOURCESAT-1) obtained from the Indian Remote Sensing (IRS) archive, Srivastava, Mehta, Gupta, Singh, and Islam (2015) examined changes in the **Mundra mangrove forest** on the western coast of **Gujarat, India**, from 1994 to 2010. The analysis, which involved phenology classification, climatology (e.g., temperature, precipitation, and wind), hydrology (including sea level rise), and population growth analysis, revealed that the Mundra mangrove forest, famous for its unique biodiversity, may experience considerable habitat loss in the coming decades (Srivastava, Mehta, Gupta, Singh & Islam, 2015).

Using the northeastern region of **Puerto Rico** as a case study, Comarazamy, González, Luvall, Rickman, and Bornstein (2013) tested the combined effects of land use/ and cover change and global warming on tropical islands. Based on scenario simulations derived from an integrated mesoscale atmospheric modeling approach and remote sensing, the study projected considerable temperature increase along the tropical coastal plains and inland lowlands (Comarazamy, González, Luvall, Rickman & Bornstein, 2013). Brown, Pearce, Leon, Sidle, and Wilson (2018) combined historical and current satellite imagery with participatory mapping to assess mangrove change on the **Maroochy River, Queensland, Australia**. The results indicated that mangrove forests in the lower Maroochy River were declining due to past and present stressors such as land clearing for agricultural activities, population growth, and pollutant discharges (Brown, Pearce, Leon, Sidle & Wilson, 2018).

Monitoring of Drought Conditions

Droughts are complex phenomena most commonly associated with less than average annual precipitation for an extended period ranging from weeks to years. Meteorological drought caused by decreased rainfall often translates into a hydrological drought with considerable streamflow and water table decline, often resulting in lower water levels in lakes and reservoirs. Persisting dry weather and reduced water supply often lead to agricultural droughts affecting millions of people every year on a global scale. The EM-DAT database suggests that droughts have affected 345 million people in **Asia** and over 40 million people in **Africa** in 2016 (Guha-Sapir, Hoyois, Wallemacq & Below, 2017). Drought conditions were reported in nearly 70% of the conterminous **United States** in 2012–2013 (Guha-Sapir, Hoyois, Wallemacq & Below, 2017). The amount of precipitation on local, regional, and global scales is closely linked to atmospheric circulation, barometric pressure, ocean evaporation, and surface wind speed (Wei, Jin, Yang & Dirmeyer, 2016). The negative teleconnections between droughts in the tropical regions and the El Niño Southern Oscillation phenomenon (e.g., above average SST in the tropical regions of the central and eastern Pacific Ocean) have been well-documented in the scientific literature (Hoell, Hoerling, Eischeid, Quan & Liebmann, 2017). However, linkages between droughts, El Niño, climate change, and anthropogenic influences are less well understood (Funk et al., 2018).

Early satellite remote sensing tools for drought detection and tracking include the AVHRR-based Temperature Condition Index (TCI) and Vegetation Condition Index (VCI) developed by NOAA in the 1990s (Kogan, 1995). The indices provided useful information on temperature-related vegetation stress as well as the time of onset, intensity, and duration of both widespread, intensive drought events and localized, short-term droughts (Kogan, 1995). Rhee, Im, and Carbone (2010) developed a Scaled Drought Condition Index (SDCI) integrating precipitation data from Tropical Rainfall Measuring Mission (TRMM), vegetation data based on the NDVI, and land surface temperature data from MODIS. The index performance was tested in both arid (**Arizona** and **New Mexico**) and humid (**North Carolina** and **South Carolina**) regions over a 10-year period (2000–2009). The index was validated against the United States Drought Monitor (USDM) maps, indicating a good agreement with existing observational data. Mu, Zhao, Kimball, McDowell, and Running (2013) proposed a near-real-time remotely sensed Drought Severity Index (DSI) with a high spatial (1 kilometer) and temporal (8-day intervals) resolution using satellite-based terrestrial evapotranspiration (ET) and NDVI to monitor and detect drought onset on a global scale. When tested with historical data, the index identified observed regional droughts over the last decade, including the drought events in **Europe** in 2003, **Russia** in 2010, and **the Amazon** in 2005 and 2010 (Mu, Zhao, Kimball, McDowell & Running, 2013). Zhang et al. (2013) provided a detailed overview of recent application and contributions of remote sensing and model-data fusion techniques to monitoring and assessing the impact of drought conditions on forest productivity.

Droughts are particularly devastating in **Africa**, where the food security of the vast majority of the population depends on predominantly rainfed agriculture. Rojas, Vrieling, and Rembold (2011) introduced a method for estimating drought probability in agricultural areas using AVHRR-based Vegetation Health Index (VHI) and a phenological model based on NDVI. Derived probability values were highly correlated with the droughts in Africa during the 1981–2009 period (Rojas, Vrieling & Rembold, 2011). Agutu et al. (2017) used rainfall, VCI, and terrestrial water storage from satellite remote sensing in combination with atmospheric reanalysis datasets and soil moisture models to analyze the severity, duration, and spatial extent of the drought conditions in **East Africa** between 1983 and 2013. The findings highlighted the importance of remote sensing indices for rapid detection of agricultural drought conditions in Africa, particularly in the regions strongly dependent on rainfed agriculture (Agutu et al., 2017).

Early detection and monitoring of droughts play a vital role in water resources management. Satgé et al. (2017) used seven indices derived from satellite remote sensing data to analyze the causes of the **Poopó Lake** drought, **Bolivia**, in 2015, which prompted the Bolivian government to declare a state of emergency due to dwindling water resources. The study showed that the effect of regional annual rainfall anomalies between 2000 and 2014 are exacerbated by increased evapotranspiration over the agricultural areas and unsustainable water use. Ma, Wu, Wang, and Zhang (2017) investigated the spatial distribution of severe drought conditions in **Yunnan Province, China**, between 2002 and 2014 using the GRACE Total Water Storage Anomalies (TWSA) dataset, observed precipitation data, and the Climate Change Initiative Soil Moisture (CCI SM) dataset. The study found that data derived from CCI SM and TWSA provided a better representation of drought conditions compared to precipitation data, which was only indicative of meteorological drought. Seasonal trends in CCI SM and TWSA captured observed drought conditions in Yunnan from 2009–2010 (Ma, Wu, Wang & Zhang, 2017).

Remote Sensing of Urban Microclimate

Land surface temperature and NDVI derived from remote sensing data are increasingly used to evaluate changes in urban microclimate. Fan et al. (2017) investigated the spatial patterns and dynamics of the daytime surface urban heat island (SUHI) effect in five subtropical desert cities: **Las Vegas, Nevada, United States**; **Beer Sheva, Israel**; **Jodhpur, India**; **Kharga, Egypt**; and **Hotan, China**, computing LST and NDVI from Landsat imagery for the summer months of 1990, 2000, and 2010. The study considered city size, population growth, and the rate of urbanization, finding that the amplitude of the SUHI effect increased differentially with significant differences in the urban-rural NDVI. Hamdi (2010) analyzed the urban heat island effect in **Uccle (Brussels, Belgium)** using a 40-year (1960–1999) time series temperature record derived from both ground stations and remote sensing. The study found a warming trend and a stronger urban heat island effect over the past decade. Chen, Ding, Yang, Hu, and Qi (2018) developed a heat hazard and vulnerability index for the **Yangtze River delta, China**, using demographic and socioeconomic census data and land surface temperature derived from satellite remote sensing. The study found that the population in highly urbanized areas is the most vulnerable to adverse health effects due to both high human exposure and high heat hazard (Chen, Ding, Yang, Hu & Qi, 2018).

Méndez-Lázaro, Muller-Karger, Otis, McCarthy, and Rodríguez (2018) developed the Heat Vulnerability Index (HVI) combining remote sensing and socioeconomic data for **San Juan, Puerto Rico**. The land surface temperature was derived from the Landsat 8 TIRS data. The LST layers were overlaid with socioeconomic variables such as age, unemployment, education, and health insurance coverage to determine the level of heat-related health vulnerability. High-density residential areas (especially those with a higher percentage of people with disabilities, people without health insurance, and people above 65 and living alone) were ranked as the most vulnerable to heat stress. Areas with more green space and near urban water bodies were found to have a lower LST and lower exposure to heat stress. Shih (2017) examined the effect of green space on mitigating land surface temperature in **Taipei, Taiwan**, noting that the presence of vegetated landscapes is not sufficient to enhance the cooling effect of urban greenery. The study recommendations highlighted the need for careful consideration of green space location, size, and connectivity.

Data and Methodological Challenges

Advances in satellite remote sensing have exponentially increased the range of applications in climatological, coastal, environmental, health, and urban planning studies. The main advantage of satellite remote sensing is its global coverage and ability to collect information in complex, dynamic, and difficult-to-access environments at a relatively low cost. Remotely sensed products are often delivered in near-real time, enhancing the monitoring capabilities needed to assist decision-making and response and mitigation efforts (Mu, Zhao, Kimball, McDowell & Running, 2013). Despite these advances, several critical research gaps and challenges remain. The accuracy of the products derived from remote sensing data often depends on the sensor properties and retrieval algorithms (Yang et al., 2013; Zhang et al., 2013). Both can introduce biases and uncertainties that can "affect the magnitude of detected trends and even change their direction" (Yang et al., 2013, p. 881). Additional challenges arise from the sensors' spectral resolution, which can introduce a bias in the estimation of indices and produce different values for the observed variable based on the data derived from various sensors (Yang et al., 2013; Zhang et al., 2013).

In some cases, it is necessary to extensively validate remote sensing records with ground measurements to improve the product accuracy. For example, the initial retrieval algorithms of satellite sea surface temperatures used in the 1970s were found to produce some artificial SST fields and were routinely regressed against the buoy SST, which was considered more reliable and less noisy than the satellite data (Emery, Baldwin, Schluessel & Reynolds, 2001). Also, optical and thermal sensors are often affected by climatological conditions such as cloud cover, smoke, or haze. Other limitations are related to (1) short time frames of satellite datasets; (2) uncorrected background atmospheric effects; and (3) gaps in knowledge linking meteorological variables with other remotely sensed times series (Yang et al., 2013; Devi et al., 2018). Improvements in instrumentation and retrieval algorithms could resolve some of the persistent biases (Yang et al., 2013, p. 881). Additional approaches have also been used to address these challenges, including (1) model-data fusion, which enhances the accuracy of the remotely sensed data through integration with empirical or process-based models (Zhang et al., 2013), and (2) integrating remote sensing with participatory mapping approaches, particularly where the availability of ground-based observational data is limited (Brown, Pearce, Leon, Sidle & Wilson, 2018).

Notes

1 RAOB stands for RAdiosonde OBservation program, the world's most advanced sounding software (www.RAOB.com) that works with a variety of formats to display over 200 atmospheric parameters.

2 Spring tides are higher than normal tides which occur as a result of the combined gravitational pull of the sun and moon. In Florida, the spring tides occur between September and November with the highest tides recorded in mid-October.

3 FVC is a biophysical parameter defined as a ratio of the estimated biomass production (e.g., stalks, leaves, and branches) to the total vegetation area (Wu et al., 2014, p. 4218).

References

Adolph, A.C., Albert, M.R., & Hall, D.K. 2018. Near-surface Temperature Inversion During Summer at Summit, Greenland, and Its Relation to MODIS-derived Surface Temperatures. *Cryosphere* 12(3), 907–920.

AECOM. 2013. *The Impact of Climate Change and Population Growth on the National Flood Insurance Program*. A Report Commissioned for the Federal Emergency Management Agency (FEMA). Retrieved from http://www.aecom.com/fema-climate-change-report/ (accessed 07/21/1917).

Agutu, N.O., Awange, J.L., Zerihun, A., Ndehedehe, C.E., Kuhn, M., & Fukuda, Y. 2017. Assessing Multi-satellite Remote Sensing, Reanalysis, and Land Surface Models' Products in Characterizing Agricultural Drought in East Africa. *Remote Sensing of Environment* 194(1), 287–302.

Anjos, L.J.S., & de Toledo, P.M. 2018. Measuring Resilience and Assessing Vulnerability of Terrestrial Ecosystems to Climate Change in South America. *PLoS ONE* 13(3), 1–15.

Bader, D.C., Covey, C., Gutowski, W.J., Held, I.M., Kunkel, K.E., Miller, R.L., Tokmakian, R.T., & Zhang, M.H. 2008. *Climate Models: An Assessment of Strengths and Limitations*. A Report by the U.S. Climate Change Science Program and the Subcommittee on Global Change Research, Department of Energy, Office of Biological and Environmental Research, Washington, DC.

Bandyopadhyay, D., Singh, G., & Birajdar, F. 2018. Remote Sensing Analysis of Changes in Chorabari Glacier, Central Himalaya, India. *Current Science* 114(4), 771–775.

Banerjee, A., & Shankar, R. 2013. On the Response of Himalayan Glaciers to Climate Change. *Journal of Glaciology* 59(215), 480–490.

Barsi, J.A., Lee, K., Kvaran, G., Markham, B.L., & Pedelty, J.A. 2014. The Spectral Response of the Landsat-8 Operational Land Imager. *Remote Sensing* 6, 10232–10251.

Bokhorst, S., Tømmervik, H., Callaghan, T.V., Phoenix, G.K., & Bjerke, J.W. 2012. Vegetation Recovery Following Extreme Winter Warming Events in the sub-Arctic Estimated Using NDVI from Remote Sensing and Handheld Passive Proximal Sensors. *Environmental & Experimental Botany* 81, 18–25.

Bright, R.M., Antón-Fernández, C., Astrup, R., Cherubini, F., Kvalevåg, M., & Strømman, A.H. 2014. Climate Change Implications of Shifting Forest Management Strategy in a Boreal Forest Ecosystem of Norway. *Global Change Biology* 20(2), 607–621.

Brody, S., Highfield, W., & Kang, J.E. 2011. *Rising Waters: Causes and Consequences of Flooding in the United States*. London, UK: Cambridge University Press.

Brown, M.I., Pearce, T., Leon, J., Sidle, R., & Wilson, R. 2018. Using Remote Sensing and Traditional Ecological Knowledge (TEK) to Understand Mangrove Change on the Maroochy River, Queensland, Australia. *Applied Geography* 94, 71–83.

Chen, Q., Ding, M., Yang, X., Hu, K., & Qi, J. 2018. Spatially Explicit Assessment of Heat Health Risk by Using Multi-sensor Remote Sensing Images and Socioeconomic Data in Yangtze River Delta, China. *International Journal of Health Geographics* 17(1), 1–14.

Christensen, N., & Lettenmaier, D.P. 2006. A Multimodel Ensemble Approach to Assessment of Climate Change Impacts on the Hydrology and Water Resources of the Colorado River Basin. *Hydrology and Earth System Sciences Discussions* 11(4), 1417–1434.

Church, J.A., Clark, P.U., Cazenave, A., Gregory, J.M., Jevrejeva, S., Levermann, A., Merrifield, M.A., Milne, G.A., Nerem, R.S., Nunn, P.D., Payne, A.J., Pfeffer, W.T., Stammer, D., & Unnikrishnan, A.S. 2013. Sea Level Change. In Stocker, T.F., Qin, D., Plattner, G.-K., Tignor, M., Allen, S.K., Boschung, J., Nauels, A., Xia, Y., Bex, V., & Midgley, P.M. (Eds.) *Climate Change 2013: The Physical Science Basis. Contribution of Working Group I to the Fifth Assessment Report of the Intergovernmental Panel on Climate Change*. Cambridge, UK/New York, NY: Cambridge University Press.

Comarazamy, D.E., González, J.E., Luvall, J.C., Rickman, D.L., & Bornstein, R.D. 2013. Climate Impacts of Land-Cover and Land-Use Changes in Tropical Islands under Conditions of Global Climate Change. *Journal of Climate* 26(5), 1535–1550.

Dahl, K., Spanger-Siegfried, E., Caldas, A., & Udvardy, S. 2017. Effective Inundation of Continental United States Communities with 21st-century Sea Level Rise. *Elementa Science of the Anthropocene* 5(37), 1–21.

Devi, R.M., Patasaraiya, M.K., Sinha, B., Saran, S., Dimri, A.P., & Jaiswal, R. 2018. Understanding the Linkages between Climate Change and Forest. *Current Science* 114(5), 987–996.

Edwards, R. 2007. Sea Levels: Resolution and Uncertainty. *Progress in Physical Geography* 31(16), 621–632.

Emery, W.J., Baldwin, D.J., Schluessel, P., & Reynolds, R.E. 2001. Accuracy of in Situ Sea Surface Temperatures Used to Calibrate Infrared Satellite Measurements. *Journal of Geophysical Research* 106(C2), 2387–2405.

European Centre for Medium-Range Weather Forecasts (ECMWF). 2015. *Data Assimilation*. Retrieved from http://www.ecmwf.int/en/research/data-assimilation.

Fan, C., Myint, S.W., Kaplan, S., Middel, A., Zheng, B., Rahman, A., Huang, H.-P., Brazel, A., & Blumberg, D.G. 2017. Understanding the Impact of Urbanization on Surface Urban Heat Islands--A Longitudinal Analysis of the Oasis Effect in Subtropical Desert Cities. *Remote Sensing* 9(7), 1–15.

Fensholt, R., & Proud, S.R. 2012. Evaluation of Earth Observation-based Global Long-term Vegetation Trends: Comparing GIMMS and MODIS Global NDVI Time Series. *Remote Sensing of the Environment* 119, 131–147.

Ferguson, C.R., & Wood, E.F. 2011. Observed Land–Atmosphere Coupling from Satellite Remote Sensing and Reanalysis. *Journal of Hydrometeorology* 12, 1221–1254. https://doi.org/10.1175/2011JHM1380.1.

Funk, C., Davenport, F., Harrison, L., Magadzire, T., Galu, G., Artan, G.A., Shukla, S., Korecha, D., Indeje, M., Pomposi, C., Macharia, D., Husak, G., & Nsadisa, F.D. 2018. Anthropogenic Enhancement of Moderate-to-strong El Niño Events Likely Contributed to Drought and Poor Harvests in Southern Africa During 2016. *Bulletin of the American Meteorological Society*. https://doi.org/10.1175/BAMS-D-17-0112.1.

Gao, B.C., & Kaufman, Y.J. 1995. Selection of the 1.375-micrometer MODIS Channel for Remote Sensing of Cirrus Clouds and Stratospheric Aerosols from Space. *Journal of the Atmospheric Sciences* 52(23), 4231–4237.

Gardner, A.S., Moholdt, G., Cogley, J.G., Wouters, B., Arendt, A.A., Wahr, J., Berthier, E., Hock, R., Pfeffer, W.T., Kaser, G., Ligtenberg, S.R.M., Bolch, T., Sharp, M.J., Hagen, J.O., van den Broeke, M.R., & Paul, F. 2013. A Reconciled

Estimate of Glacier Contributions to Sea Level Rise: 2003 to 2009. *Science* 340, 852–857.

Goldblatt, R., Ballesteros, A.R., & Burney, J. 2017. High Spatial Resolution Visual Band Imagery Outperforms Medium Resolution Spectral Imagery for Ecosystem Assessment in the Semi-arid Brazilian Sertão. *Remote Sensing* 9(12), 1–26.

González, J.L., & Törnqvist, T.E. 2006. Coastal Louisiana in Crisis: Subsidence or Sea Level Rise? *EOS Transactions of the American Geophysical Union* 87(45), 493–508.

Guha-Sapir, D., Hoyois, P., Wallemacq, P., & Below, R. 2017. *Annual Disaster Statistical Review 2016*. Brussels, Belgium: Centre for Research on the Epidemiology of Disasters (CRED), Institute of Health and Society (IRSS), Université Catholique de Louvain.

Guo, M., Li, J., Sheng, C., Xu, J., & Wu, L. 2017. A Review of Wetland Remote Sensing. *Sensors* 17(4), 1–36.

Gutman, G., & Ignatov, A. 1998. The Derivation of the Green Vegetation Fraction from NOAA/AVHRR Data for Use in Numerical Weather Prediction Models. *International Journal of Remote Sensing* 19, 1533–1543.

Hamdi, R. 2010. Estimating Urban Heat Island Effects on the Temperature Series of Uccle (Brussels, Belgium) Using Remote Sensing Data and a Land Surface Scheme. *Remote Sensing* 2(12), 2773–2784.

Hoell, A., Hoerling, M., Eischeid, J., Quan, X., & Liebmann, B. 2017. Reconciling Theories for Human and Natural Attribution of Recent East Africa Drying. *Journal of Climate* 30, 1939–1957.

Horwath, M., Legresy, B., Remy, F., Blarel, F., & Lemoine, J.-M. 2012. Consistent Patterns of Antarctic Ice Sheet Interannual Variations from ENVISAT Radar Altimetry and GRACE Satellite Gravimetry. *Geophysical Journal International Banner* 189, 863–876.

Intergovernmental Panel on Climate Change (IPCC). 2007. Climate Change 2007: Synthesis Report. In Core Writing Team, Pachauri, R.K., & Reisinger, A. (Eds.) *Contribution of Working Groups I, II and III to the Fourth Assessment Report of the Intergovernmental Panel on Climate Change*, 104. Geneva, Switzerland: IPCC.

Irons, J.R., Dwyer, J.L., & Barsi, J.A. 2012. The Next Landsat Satellite: The Landsat Data Continuity Mission. *Remote Sensing of Environment* 122, 11–21.

Jacob, T., Wahr, J., Pfeiffer, W.T., & Swensen, S. 2012. Recent Contributions of Glaciers and Ice Caps to Sea Level Rise. *Nature* 482, 514–518.

Kogan, F.N. 1995. Application of Vegetation Index and Brightness Temperature for Drought Detection. *Advances in Space Research* 15(11), 91–100.

Koster, R.D., Guo, Z., Dirmeyer, P.A., Bonan, G., Chan, E., Cox, P., Davies, H., Gordon, C.T., Kanae, S., Kowalczyk, E., Lawrence, D., Liu, P., Lu, C.-H., Malyshev, S., Mvavaney, B., Mitchell, K., Mockp, D., Oki, T., Oleson, K.W., Pitman, A.W., Sud, Y.C., Taylor, C.M., Verseghy, D., Vasic, R., & Yongkang, X. 2006. GLACE: The Global Land–Atmosphere Coupling Experiment. Part I: Overview. *Journal of Hydrometeorology* **7**, 590–610.

Kuleshov, Y., Choy, S., Fu, E.F., Chane-Ming, F., Liou, Y.-A., & Pavelyev, A.G. 2016. Analysis of Meteorological Variables in the Australasian Region Using Ground- and Space-based GPS Techniques. *Atmospheric Research* 176–177, 276–289.

Li, F., Zhang, X., Chen, J., & Feng, B. 2018. Increased Growth Rate (1982-2013) in Global Grasslands Biomes. *Remote Sensing Letters* 9(6), 550–558.

Ma, S., Wu, Q., Wang, J., & Zhang, S. 2017. Temporal Evolution of Regional Drought Detected from GRACE TWSA and CCI SM in Yunnan Province, China. *Remote Sensing* 9(11), 1–15.

Maurer, E.P., & Hidalgo, H.G. 2007. Utility of Daily vs. Monthly Large-scale Climate Data: An Intercomparison of Two Statistical Downscaling Methods. *Hydrology and Earth Systems Sciences Discussions* 12, 551–563.

Méndez-Lázaro, P., Muller-Karger, F.E., Otis, D., McCarthy, M.J., & Rodríguez, E. 2018. A Heat Vulnerability Index to Improve Urban Public Health Management in San Juan, Puerto Rico. *International Journal of Biometeorology* 62(5), 709–722.

Mishra, B., Babel, M.S., & Tripathi, N.K. 2014. Analysis if Climatic Variability and Snow Cover in the Kaligandaki River Basin, Himalaya, Nepal. *Theoretical and Applied Climatology* 116, 681–694.

Mu, Q., Zhao, M., Kimball, J.S., McDowell, N.G., & Running, S.W. 2013. A Remotely Sensed Global Terrestrial Drought Severity Index. *Bulletin of the American Meteorological Society* 94, 83–98.

Mukherjee, K., Bhattacharya, A., Pieczonka, T., Ghosh, S., & Bolch, T. 2018. Glacier Mass Budget and Climate Reanalysis Data Indicate a Climatic Shift Around 2000 in Lahaul-Spiti, Western Himalaya. *Climatic Change* 148(1/2), 219–233.

Muraoka, H., & Koizumi, H. 2009. Satellite Ecology (SATECO)—Linking Ecology, Remote Sensing and Micrometeorology, from Plot to Regional Scale, for the Study of Ecosystem Structure and Function. *Journal of Plant Research* 122(1), 3–20.

National Aeronautics and Space Administration (NASA). 2018. *Sea Surface Temperature*. Physical Oceanography Distributed Active Archive Center, NASA & Jet Propulsion Laboratory, California Institute of Technology. Retrieved from https://podaac.jpl.nasa.gov/SeaSurfaceTemperature.

Neumann, B., Vafeidis, A.T., Zimmermann, J., & Nicholls, R.J. 2015. Future Coastal Population Growth and Exposure to Sea-level Rise and Coastal Flooding – A Global Assessment. *PLoS ONE* 10(3), e0118571.

Nie, Y., Sheng, Y., Liu, Q., Liu, L., Liu, S., Zhang, Y., & Song, C. 2017. A Regional-scale Assessment of Himalayan Glacial Lake Changes Using Satellite Observations from 1990 to 2015. *Remote Sensing of Environment* 189, 1–13.

Oliver-Cabrera, T., & Wdowinski, S. 2016. InSAR-based Mapping of Tidal Inundation Extent and Amplitude in Louisiana Coastal Wetlands. *Remote Sensing* 8, 393. doi:10.3390/rs8050393.

Osipov, E.Y., & Osipova, O.P. 2018. Glacier Changes on the Pik Topografov Massif, East Sayan Range, Southeast Siberia, from Remote Sensing Data. *Geosciences* 8(5), 148, 1–12. doi:10.3390/geosciences8050148.

Park, I.W., Hooper, J., Flegal, J.M., & Jenerette, G.D. 2018. Impacts of Climate, Disturbance and Topography on Distribution of Herbaceous Cover in Southern California Chaparral: Insights from a Remote-sensing Method. *Diversity & Distributions* 24(4), 497–508.

Pimentel, R., Herrero, J., & Polo, M.-J. 2017. Quantifying Snow Cover Distribution in Semiarid Regions Combining Satellite and Terrestrial Imagery. *Remote Sensing* 9, 995. doi: 10.3390/rs9100995.

Pinter, N., Santos, N., Hui, R., & Schaefer, K. 2016. *New Baton Rouge Flood Map Show Limits of Current Risk and Planning Methods*. UC Davis Center for Watershed Sciences. Retrieved from https://californiawaterblog.com/2016/08/28/new-baton-rouge-flood-map-show-limits-of-current-risk-and-planning-methods/.

Prichard, H.D., Luthcke, S.B., & Fleming, A.H. 2010. Understanding Ice-sheet Mass Balance: Progress in Satellite Altimetry and Gravimetry. *Journal of Glaciology* 56(200), 1151–1161.

Quartly, G., & Chen, G. 2006. Satellite Altimetry: New Sensors and New Applications. *Sensors* 6, 616–619.

Rabatel, A., Sirguey, P., Drolon, V., Maisongrande, P., Arnaud, Y., Berhier, E., Davaze, L., Dedieu, J.-P., & Dumont, M. 2017. Annual and Seasonal Glacier-wide Surface Mass Balance Quantified from Changes in Glacier Surface State: A Review on Existing Methods Using Optical Satellite Imagery. *Remote Sensing* 9, 507. doi: 10.3390/rs9050507.

Ran, Y., Li, X., & Cheng, G. 2018. Climate Warming Over the Past Half-century has Led to Thermal Degradation of Permafrost on the Qinghai-Tibet Plateau. *Cryosphere* 12, 595–608.

Rémy, F., & Parouty, S. 2009. Antarctic Ice Sheet and Radar Altimetry: A Review. *Remote Sensing* 1, 1212–1239. doi:10.3390/rs1041212.

Rhee, J., Im, J., & Carbone, G.J. 2010. Monitoring Agricultural Drought for Arid and Humid Regions Using Multi-sensor Remote Sensing Data. *Remote Sensing of Environment* 114(12), 2875–2887.

Rienecker, M.M., Suarez, M.J., Todling, R., Bacmeister, J., Takacs, L., Liu, H.-C., Gu, W., Sienkiewicz, M., Koster, R.D., Gelaro, R., Stajner, I., and Nielsen, J.E. 2008. The GEOS-5 Data Assimilation System—Documentation of Versions 5.0.1, 5.1.0 and 5.2.0. NASA GSFC Tech. Rep. TM-2008-104606. Retrieved from https://ntrs.nasa.gov/archive/nasa/casi.ntrs.nasa.gov/20120011955.pdf.

Rignot, E., Velicogna, I., van den Broeke, M.R., Monaghan, A., & Lenaerts, J. 2011. Acceleration of the Contribution of the Greenland and Antarctic Ice Sheets to Sea Level Rise. *Geophysical Research Letters* 38, L05503–L05508.

Rojas, O., Vrieling, A., & Rembold, F. 2011. Assessing Drought Probability for Agricultural Areas in Africa with Coarse Resolution Remote Sensing Imagery. *Remote Sensing of Environment* 115(2), 343–352.

Satgé, F., Espinoza, R., Zolá, R.P., Roig, H., Timouk, F., Molina, J., Garnier, J., Calmant, S., Seyler, F., & Bonnet, M.-P. 2017. Role of Climate Variability and Human Activity on Poopó Lake Droughts between 1990 and 2015 Assessed Using Remote Sensing Data. *Remote Sensing* 9(3), 1–17.

Shepherd, A., Ivins, E.R., Geruo, A., Barletta, V.R., Bentley, M.J., Bettadpur, S., Briggs, K.H., Bromwich, D.H., Forsberg, R., Galin, N., Horwath, M., Jacobs, S., Joughin, I., King, M.A., Lenaerts, J.T.M., Li, J., et al. 2012. A Reconciled Estimate of Ice-sheet Mass Balance. *Science* 338(6111), 1183–1189.

Shih, W. 2017. Greenspace Patterns and the Mitigation of Land Surface Temperature in Taipei Metropolis. *Habitat International* 60, 69–80.

Shugar, D.H., Walker, I.J., Lian, O.B., Eamer, J.B.R., Neudorf, C., McLaren, D., & Fedje, D. 2014. Post-glacial Sea-level Change along the Pacific Coast of North America. *Quaternary Science Reviews* 97, 170–192.

Simões, C.L., da Rosa, K.K., Czapela, F.F., Vieira, R., & Simões, J.C. 2015. Collins Glacier Retreat Process and Regional Climatic Variations, King George Island, Antarctica. *Geographical Review* 105(4), 462–471.

Spanger-Siegfried, E., Fitzpatrick, M.F., & Dahl, K. 2014. *Encroaching Tides: How Sea Level Rise and Tidal Flooding Threaten U.S. East and Gulf Coast*

Communities Over the Next 30 Years. Cambridge, MA: Union of Concerned Scientists.

Srivastava, P., Mehta, A., Gupta, M., Singh, S., & Islam, T. 2015. Assessing Impact of Climate Change on Mundra Mangrove Forest Ecosystem, Gulf of Kutch, Western Coast of India: A Synergistic Evaluation Using Remote Sensing. *Theoretical & Applied Climatology* 120(3/4), 685–700.

Strauss, B.H., Ziemlinski, R., Weiss, J.L., & Overpeck, J.T. 2012. Tidally Adjusted Estimates of Topographic Vulnerability to Sea Level Rise and Flooding for the Contiguous United States. *Environmental Research Letters* 7, 014033. doi:10.1088/1748-9326/7/1/014033.

Strauss, B.H., & Kulp, S. 2014. *New Analysis Shows Global Exposure to Sea Level Rise, Research Report by Climate Central*. Retrieved from http://www.climatecentral.org/news/new-analysis-global-exposure-to-sea-level-rise-flooding-18066.

Su, X., Guo, J., Duan, J., Yi, Y., Shum, C.K., Howat, I.M., Jezek, K.C., & Kuo, C. 2018. High-Resolution Interannual Mass Anomalies of the Antarctic Ice Sheet by Combining GRACE Gravimetry and ENVISAT Altimetry. *IEEE Transactions on Geoscience & Remote Sensing* 56(1), 539–546.

Tang, G., Zeng, Z., Ma, M., Liu, R., Wen, Y., & Hong, Y. 2017. Can Near-Real-Time Satellite Precipitation Products Capture Rainstorms and Guide Flood Warning for the 2016 Summer in South China? *Geoscience and Remote Sensing Letters IEEE* 14(8), 1208–1212.

Tietsche, S., Balmaseda, M.A., Zuo, H., & Mogensen, K. 2017. Arctic Sea Ice in the Global Eddy-permitting Ocean Reanalysis ORAP5. *Climate Dynamics* 49(3), 775–789.

United States Geological Survey (USGS). 2016. *LANDSAT 8 (L8) Data Users Handbook, Version 2.0, LSDS-1574, Department of the Interior, U.S. Geological Survey Landsat Project Science Office*. Sioux Falls, SD: Earth Resources Observation and Science (EROS) Center. Retrieved from https://landsat.usgs.gov/sites/default/files/documents/LSDS-1574_L8_Data_Users_Handbook.pdf.

Urbazaev, M., Thiel, C., Cremer, F., Dubayah, R., Migliavacca, M., Reichstein, M., & Schmullius, C. 2018. Estimation of Forest Aboveground Biomass and Uncertainties by Integration of Field Measurements, Airborne LiDAR, and SAR and Optical Satellite Data in Mexico. *Carbon Balance & Management* 13(1), 1–20.

Veettil, B.K., Wang, S., Florêncio de Souza, S., Bremer, U.F., & Simões, J.C. 2017. Glacier Monitoring and Glacier-climate Interactions in the Tropical Andes: A Review. *Journal of South American Earth Sciences* 77, 218–246.

Wang, H., Wright, T.J., Yu, Y., Lin, H., Jiang, L., Li, C., & Qiu, G. 2012. InSAR Reveals Coastal Subsidence in the Pearl River Delta, China. *Geophysical Journal International* 191(3), 1119–1128.

Wang, L., Goldberg , M., , X., & Zhou, L. 2010. *Assessment of Reanalysis Datasets Using AIRS and IASI Hyperspectral Radiances*. Proceedings of the 2010 IEEE International Geoscience and Remote Sensing Symposium, 25-30 July 2010, Honolulu, HI, USA. doi: 10.1109/IGARSS.2010.5650744.

Wei, H., Zhao, X., Liang, S., Zhou, T., Wu, D., & Tang, B. 2018. Effects of Warming Hiatuses on Vegetation Growth in the Northern Hemisphere. *Remote Sensing* 10(5), 683. https://doi.org/10.3390/rs10050683.

Wei, J., Jin, Q., Yang, Z.-L., & Dirmeyer, P.A. 2016. Role of Ocean Evaporation in California Droughts and Floods. *Geophysical Research Letters* 43(12), 6554–6562.

Wenny, B.N., Helder, D., Hong, J., Leigh, L., Thome, K.J., & Reuter, D. 2015. Pre-and Post-launch Spatial Quality of the Landsat 8 Thermal Infrared Sensor. *Remote Sensing* 7(2), 1962–1980.

Wu, D., Wu, H., Zhao, X., Zhou, T., Tang, B., Zhao, W., & Jia, K. 2014. Evaluation of Spatiotemporal Variations of Global Fractional Vegetation Cover Based on GIMMS NDVI Data from 1982 to 2011. *Remote Sensing* 6(9), 4217–4239.

Wylie, B.K., Zhang, L., Bliss, N., Ji, L., Tieszen, L.L., & Jolly, W.M. 2008. Integrating Modeling and Remote Sensing to Identify Ecosystem Performance Anomalies in the Boreal Forest, Yukon River Basin, Alaska. *International Journal of Digital Earth* 1(2), 196–220.

Xu, S., Zhou, L., Liu, J., Lu, H., & Wang, B. 2017. Data Synergy between Altimetry and L-Band Passive Microwave Remote Sensing for the Retrieval of Sea Ice Parameters—A Theoretical Study of Methodology. *Remote Sensing* 9(10), 1–24.

Yang, J., Gong, P., Fu, R., Zhang, M., Chen, J., Liang, S., Xu, B., Shi, J., & Dickinson, R. 2013. The Role of Satellite Remote Sensing in Climate Change Studies. *Nature Climate Change* 3, 875–883.

Yi, S., Wang, Q., Chang, L., & Sun, W. 2016. Changes in Mountain Glaciers, Lake Levels, and Snow Coverage in the Tianshan Monitored by GRACE, ICESat, Altimetry, and MODIS. *Remote Sensing* 8, 798. doi: 10.390/rs8100798.

Zhang, Y., Peng, C., Li, W., Fang, X., Zhang, T., Zhu, Q., Chen, H., & Zhao, P. 2013. Monitoring and Estimating Drought-induced Impacts on Forest Structure, Growth, Function, and Ecosystem Services Using Remote-sensing Data: Recent Progress and Future Challenges. *Environmental Reviews* 21(2), 103–115.

Zhang, Z., Wang, C., Zhang, H., Tang, Y., & Liu, X. 2018. Analysis of Permafrost Region Coherence Variation in the Qinghai-Tibet Plateau with a High-resolution TerraSAR-X Image. *Remote Sensing* 10, 298. doi: 10.3390/rs10020298.

Zhou, C., Zhou, Y., Deng, F., Ai, S., Wang, Z., & Dongchen, E. 2014. Seasonal and Interannual Ice Velocity Changes of Polar Record Glacier, East Antarctica. *Annals of Glaciology* 55(66), 45–51.

Zhou, W., Yang, H., Huang, L., Chen, C., Lin, X., Hu, Z., & Li, J. 2017. Grassland Degradation Remote Sensing Monitoring and Driving Factors Quantitative Assessment in China from 1982 to 2010. *Ecological Indicators* 83, 303–313.

Zuo, H., Balmaseda, M.A., & Mogensen, K. 2017. The New Eddy-permitting ORAP5 Ocean Reanalysis: Description, Evaluation and Uncertainties in Climate Signals. *Climate Dynamics* 49(3), 791–811.

PART 3
GIS and Climate Vulnerability Assessments

9 Mapping and Assessing the Impacts of SLR on Coastal Regions

Chapter Objectives

This chapter answers the following questions:

- Why are rising seas a global concern?
- What are the impacts of sea level rise (SLR) on coastal communities?
- What type of data is needed to conduct sea level rise vulnerability assessments?
- What are the tools and types of spatial analyses that can be used to support adaptation planning?
- What are some commonly encountered methodological issues?

Physical and Socioeconomic Impact Assessments

Coastal communities and barrier islands around the globe are highly diverse in terms of population density, capital stock, socioeconomic vulnerability, and risk of floods, storm surge, hurricanes, and sea level rise. The U.S. coastlines, once known simply as tourist attractions and resort communities, have seen tremendous growth and development pressure partly as a result of increases in permanent population (Beach, 2002; Pielke et al., 2008). A report by the National Oceanic and Atmospheric

Administration in support of the 2013 National Climate Assessment (Burkett & Davidson, 2012, p. 4) estimated that over 50% of the U.S. population lived along the coast in 2010. Future U.S. coastal population is expected to increase by 50%, or 46 million, by 2100 under the assumptions of the Intergovernmental Panel on Climate Change (IPCC) Special Report on Emissions Scenarios (SRES) scenario B1 and 144%, or 132 million, under the assumptions of scenario A2 (Environmental Protection Agency [EPA], 2009, as cited in Burkett and Davidson, 2012, p. 5). In addition to increasing densities, Mileti (1999) characterized "capital stock"—lifeline infrastructure, such as transportation networks, electrical networks, water networks, and marinas, seaports, and airports and other critical facilities—as expanding and increasingly vulnerable to natural hazards and extreme events.

GIS (geographic information system)-based analyses capture a broad range of physical and socioeconomic vulnerability assessments in an attempt to answer questions such as: What are the major risks for people, infrastructure, and property associated with sea level rise in low-lying coastal areas? What is the economic value associated with these risks, and what are the costs of managing them? How can spatial analyses be used to inform choices, solutions, and activities? How can we reduce our vulnerability to sea level rise in the near and longer term? This section summarizes studies by various scholars around the globe.

In the mid-1990s, Tol (1995) estimated the number of people forced to migrate due to the rising seas as a function of the area loss and the average population density in the region. In follow-up research, estimates of the cost of population displacement were presented. In that later study, the value of land loss into income density loss per unit area per period of time ($/square kilometer) was calculated (Tol, 1999, 2002a, 2002b). Under the assumption of linear distribution of income, the model yielded an average of four million US dollars per square kilometer for the Organisation for Economic Co-operation and Development (OECD) countries (Tol, 2002b). The average cost of displacement of one person was found to be three times the regional per capita income (Tol, 1995). Additionally, Tol (2002b) performed a cost-benefit analysis of shoreline protection, assuming constant annual costs. Different levels of protection were incorporated into the model, and the trade-offs between wetland protection and building long-lived structures along the coastline were evaluated against estimates of wetland loss (Tol, 2002a).

Simpson et al. (2010) evaluated the risks of sea level rise for the nations of the **Caribbean**. The study utilized the Global Digital Elevation Model (GDEM) dataset derived from the Advanced Spaceborne Thermal Emission and Reflection Radiometer (ASTER) at 30-meter spatial resolution and other publicly available geospatial datasets to examine the exposure of total land area, population, urbanized areas, infrastructure, tourism and recreational assets, and natural habitats to 1 meter and 2 meters of sea level rise (Simpson et al., 2010). The findings indicated that some island nations would face dramatic coastline changes resulting in loss of land and infrastructure. The study found that the low-lying coastal plains of **Belize**, **Guyana**, and **Suriname**, and the low-lying islands of the **Bahamas**, the **Grenadines**, and **Barbuda**, would suffer the worst impacts, becoming increasingly vulnerable to storm surge and coastal flooding. Other potential impacts such as accelerated rates of erosion, degradation of mangrove ecosystems, and saltwater intrusion were also highlighted (Simpson et al., 2010). Similar impacts, although more limited in scope, were expected to occur in **Haiti**, **Barbados**, **Antigua**, **Jamaica**, and **Trinidad and Tobago** (Simpson et al., 2010). Although the impacts would not be as widespread as those in the previous group of island states due to varied topography and the presence of cliffs and mountainous areas, most of the beaches and coastal wetlands would be degraded, affecting primary sources of revenue such as tourism and fishing (Simpson et al., 2010). According to the analysis, the volcanic islands of **St. Lucia**, **St. Vincent**, **Dominica**, **Grenada**, and **St. Kitts and Nevis** world be the least affected. Nevertheless, they would most likely face loss of low-lying beach areas and, potentially, severe landslides (Simpson et al., 2010).

Dutta, Babel, and Gupta (2005) conducted flood risk analysis using socioeconomic and elevation data and hydrological models; population, land use, and buildings data; and road and rail networks data. The analysis was conducted for the years 2005, 2025, 2050, 2075, and 2100 for low-lying coastal areas in **Bangladesh (Barisal and Patuakhali cities)**, **India (Bhubaneswar-Cuttack-Puri)**, **Pakistan (Karachi)**, **Sri Lanka (Matara)**, **Thailand (Bangkok)**, and **Vietnam (Hue City)**. Overall, each of the six case study areas became more susceptible to flooding over time, albeit at different points in time. The regions of **Barisal** and **Patuakhali** in **Bangladesh** were the most affected relative to all impact measures (population, buildings, and transportation network). It is noteworthy that at the time of the study, **Bangladesh** was the only country among the nations included in the study that had implemented relevant climate change policies. Marfai and King (2008) examined various sea level rise scenarios against the backdrop of accelerated land subsidence in the central part of the island of Java in **Indonesia**. Marfai (2014) investigated the impact of sea level rise on mangrove forests, aquaculture, agriculture, and urban infrastructure on the northern coast of Java.

Some vulnerable areas are economically vital to their regions. For example, the **Red River delta in Vietnam** is not only agriculturally fertile but is also home to **Hanoi** and **Ho Chi Minh City**—these are Vietnam's largest cities. In the study by Boateng (2012), GIS was used to map flood risk zones in population centers. Three GIS layers were generated corresponding to sea level rise increments of 1 meter, 2 meters, and 5 meters, respectively. Large sections of the **Vietnam** coastline were found to be at severe risk of flooding from hurricanes, increased monsoonal rainfall, storm surges, and flash floods (Boateng, 2012).

Vulnerability to sea level rise on a global scale has been analyzed using the Shuttle Radar Topography Mission (SRTM) elevation data, mean sea surface data based on satellite altimetry from the TOPEX/Poseidon mission, and population data from LandScan (Beckley, Lemoinse, Lutchke, Ray & Zelensky, 2007; Merrifield, Genz, Kontoes & Marra, 2013; Strauss & Kulp, 2014). Using topography and hydrography data, Lichter and Felsenstein (2012) used GIS to analyze the impact of 0 to 2 meters of sea level rise on floodplains and potential damage to buildings and capital stock in the coastal cities in **Israel**. The entire range of flood impacts was examined by using a combination of sea level rise (under various scenarios), a 1-meter-high tide, and a 4-meter tsunami. In the **Haifa** and **Tel Aviv** case studies, findings indicated that 6.44% of **Haifa** would be inundated under a 0.5-meter SLR scenario. Lichter and

Felsenstein (2012) found that **Tel Aviv** would experience minor impacts with only 1.5% of the city land potentially at risk. A 4-meter tsunami with a 0.5-meter SLR is expected to flood 16.42% of **Haifa** and 5.64% of **Tel Aviv**.

The city of **Mombasa** in **Kenya** is a center for regional growth and home to the largest international seaport in East Africa. **Mombasa** is at risk of the potential impact of other natural hazards, and sea level rise is expected to exacerbate damage from storm surges and rain events. Kebede, Nicholls, Hanson, and Mokrech (2012) used elevation data, projected population, and a global estimated annual sea level rise of 3.1 millimeters per year to evaluate the exposure of coastal areas of Mombasa to flooding as a result of sea level rise. The findings indicate that 195,000 people lived in areas vulnerable to sea level rise in 2012 with potential losses of nearly $470 million from a 100-year flood event. With the projected rate of sea level rise, the number of people in risk areas is expected to increase to 392,000 with US $16 billion in assets at risk by 2080. Findings suggest that the results are more sensitive to changes in socioeconomic scenarios than changes in climate scenarios. Rising sea levels will not only exacerbate the 100-year flood events but also decrease drainage in low-lying areas, thereby expanding the extent of the flooded areas during rain events.

A study by Wilson, Trenholm, Bornemann, and Lieske (2012) evaluated the effect of sea level rise for the **Tantramar** region in southeast **New Brunswick**, one of **Canada's** 10 provinces, and found that in 100 years, the economic vulnerability due to flooding will increase by 22%. The analysis focuses on flood impacts using prior flood damages on various land use types—rural, agricultural, mixed-use, commercial institutional, industrial, and residential.

Hallegatte et al. (2011) drew our attention to the low-lying **coastal city of Copenhagen**, Denmark's capital and the most populous city with vital infrastructure and capital stock at risk. The study analyzed the direct and indirect economic losses stemming from sea level rise of 0 to 1.25 meters above the current levels. Exposure to flood risk of the population and city assets were modeled using geoprocessing algorithms. A digital terrain model (DTM) was derived based on elevation data from the Shuttle Radar Topography Mission. The terrain analysis found that 2% of the population of **Copenhagen** lives below 1 meter above sea level and 4% below 2 meters above sea level. Overall, 13% of the population lives below 5 meters above sea level. **Copenhagen** has made considerable investments in coastal flooding control measures, but in the absence of increased protection, economic losses would be expected to magnify over time. Economic losses were calculated based on the loss of capital assets, loss of production, and net loss of jobs (jobs lost minus jobs created as part of reconstruction). At the time of the study, a storm surge of 1.5 meters above mean sea level (a 100-year flood in this scenario) in the city of **Copenhagen** was estimated to produce EUR 3 billion (US$3.4 billion) of economic losses (Hallegatte et al., 2011). A 0.25-meter rise in mean sea level (in the absence of new city protections) would increase the economic losses stemming from a 150-centimeter storm surge to EUR 4 billion. A sea level rise of 1 meter above the current level would be expected to produce EUR 8 billion (US$9.1 billion) in losses. The study by Hallegatte et al. (2011) also suggested that the costs of building city protection from SLR in the form of dikes and other measures would be less costly than the economic loss from a potential disaster.

Delineating and Assessing Erosion Risk

Worldwide, coastal erosion is recognized as a significant threat for beach environments and remains a concern for scientists, policy-makers, and the general public (Gornitz & White, 1992; Koukoulas et al., 2006; Greenlaw, Roff, Redden & Allard, 2011; Saravanan, Chandrasekar, Rajamanickam, Hentry & Joevivek, 2014; Glick, Kostyack, Pittman, Briceno & Wahlund, 2014). This continuous process is accelerated by storms, sea level rise, and tidal action and is a major underlying cause of coastal infrastructure and property damage. Geo-scientists typically document erosion rates by observing the changing beach profile, by determining the location of the seasonal mean high water line, and by calculating differences in the sand volume (Esnard & Puszkin-Chevlin, 2010). GIS is a key tool for delineating and assessing the impact of erosion risk on coastal areas.

Multi-temporal images are commonly used to assess the degree of shoreline change, coastal erosion, and sediment dynamics, and remote sensing techniques using multi-temporal Landsat images have been used to visualize coastline evolution. Using these and other relevant data sources, Kuleli (2009) estimated erosion rates of more than 24.5 meters per year in the **southeast coasts of the Mediterranean Sea in Turkey** over a 30-year period (1972–2002). Shoreline change analyses of **South India** have been conducted by Saravanan, Chandrasekar, Rajamanickam, Hentry, and Joevivek (2014), who sought to categorize the 24-kilometer coastal stretch between Kallar and Vembar into five different zones: very high, high, medium, low, and very low erosion hazard. Other flexible process-based models such as SCAPE (Soft Cliff and Platform Erosion) are incorporated in SCAPEGIS to explore sea level rise and wave climate scenarios, to visualize data of past trends and future scenarios of shoreline evolution, and to support shoreline management and cliff-top land use planning for the **coastline of England and Wales** (Koukoulas et al., 2006).

The United States Geological Survey (USGS) has developed multiple tools to help identify and manage erosion risk and prevent economic losses along the nation's shorelines. The agency has also recognized the need for consistent and standardized methods for mapping and analyzing shoreline movement and understanding how past changes will affect the vulnerability of open-ocean sandy shores of the conterminous United States and parts of Alaska and Hawaii to future hazards. The **USGS Coastal and Marine Geology Program** (http://marine.usgs.gov/coastalchangehazards) features a range of research initiatives, including a coastal change hazards portal, which provides access to long- and short-term rates of shoreline change and probabilities of erosion for various storm surge and SLR scenarios. See Chapter 15 for more details.

Ecological and Ecosystem Assessments

GIS can be used to map multiple land characteristics, land development practices, environmental hazards, and regulatory boundaries in order to identify and protect environmentally sensitive areas requiring appropriate mitigation, conservation, or management strategies. As part of an effort to develop tailored recommendations and coastal management strategies for the mangrove ecosystem at **Bragança** in the northern areas of **Brazil**, Krause, Bock, Weiers, and Braun (2004) integrated remote sensing data, aerial photographs, and scientific field work to assess spatiotemporal changes of the mangrove peninsula, evaluating the socioeconomic impacts of mangrove degradation on the adjacent rural areas.

Coastal development often results in loss of natural ecosystems and ecosystem services, declining natural coastal defenses, and lost scenic values and recreational opportunities. Using the analytical capabilities of GIS, Greenlaw, Roff, Redden, and Allard (2011) examined the characteristics of coastal inlet area waters to determine their classification as various types of bays, estuaries, and coves, their functions, and prioritization for protection as biologically and ecologically significant areas. A study by Mendoza-Gonzalez, Martinez, Lithgow, Perez-Maqueo, and Simonin (2012) analyzed land use changes and calculated the value of these changes in terms of lost ecosystem services using the benefit-transfer method for three study sites in the **central region of the Gulf of Mexico**. For the period from 1995–2006, the authors found that mangroves, grasslands, croplands, and beach areas were impacted by urban sprawl, and all study sites experienced a net loss over time. As part of a different study on the valuation of coastal recreation, Ghermandi and Nunes (2013) computed recreation values of coastlines using GIS techniques to combine geologic, climatic, urban, and natural ecosystems to generate a global recreation value map and to prioritize coastal conservation priorities from a socioeconomic perspective.

Mitsova and Esnard (2012, p. 10) highlighted the value of using Light Detection and Ranging (LiDAR) to characterize shore zone elevations and identify natural areas where beaches can naturally migrate with SLR. These types of simple but important assessments can inform and prioritize appropriate adaptation responses and regulatory strategies. Mitsova et al. (2018) developed an analytical framework to assess the suitability and feasibility of nature-based shoreline stabilization options to protect the shoreline and respond to coastal hazards. The generic model was applied to densely populated areas in Southeast Florida, one of the most vulnerable coastal regions in the world. The investigators identified a set of criteria for locating shoreline segments suitable for "living shorelines" and conducted an expert opinion survey to elicit weights for the decision parameters. The survey responses were summarized using the Analytic Hierarchy Process (AHP) in which a reciprocal matrix of pairwise comparisons was constructed. The weighted parameters were then mathematically aggregated to derive a composite score of suitability for various types of generic living shoreline projects. Despite Florida's history of disturbance, the results reveal multiple opportunities to enhance shoreline protection using soft and hybrid technique (Mitsova et al., 2018).

Groundwater and aquifers in coastal areas can be particularly susceptible to saltwater intrusion and other sources of contamination. In a study by Kattaa, Al-Fares, and Al Charideh (2010), five layers (bedrock, infiltration rates, soils, karst geology, and vegetation) were indexed and combined into a RISKE model for the **Banyas catchment of the Syrian coastal area**. RISKE is an index model designed to produce vulnerability scores for different locations by combining several thematic layers (Kattaa, Al-Fares & Al Charideh, 2010, p. 1103) The RISKE index values range between 0.2 (low vulnerability) and 2.2 (high vulnerability); 52% of the study area was found to be under moderate groundwater vulnerability and 38% under high groundwater vulnerability.

Economic Impact Assessment

The economic costs of coastal flood hazards to communities can be categorized into three main components: the value of the land that can be potentially inundated, the value of the structures at risk, and the range of potential investments in protection measures (Harrington & Walton, 2007; Hurlimann, Barnett, Fincher, Osbaldiston, Mortreux & Graham, 2014). **U.S.** national assessments of sea level rise of 0.5 meters found that approximately 9,000 square miles of **U.S.** coastal areas could be inundated by 2100 (Neumann, Yohe, Nicholls & Manion, 2000; Neumann, Vafeidis, Zimmermann & Nicholls, 2015). According to Yohe and Schlesinger (1998), land and structures should be considered separately in any economic analysis of the potential impacts of sea level rise. The assumption is that the price gradient, as a result of property values decreasing from the shoreline to the interior, would simply be transferred inland as the sea advances (Yohe & Schlesinger, 1998). Because of depreciation over time, the economic loss due to structural damage or abandonment can be minimized if planned with sufficient foresight, taking into account the likelihood of imminent retreat and abandonment (Yohe & Schlesinger, 1998; Füssel & Klein, 2006; Jacob, Gornitz & Rozenzweig, 2007; Franck, 2009).

Yohe and Schlesinger (1998) conducted a cost-benefit analysis considering three scenarios of adaptive response to sea level rise. The scenarios include: (i) "perfect foresight," which is expected to minimize damage costs; (ii) "incomplete reaction" because of uncertainty and deficient information, which will slow down the decision-making process and undermine the cost-effectiveness of retreat; and (iii) "disbelief," which commands no action and opportunities for the market to adjust, resulting in the maximum possible loss of economic value (Yohe & Schlesinger, 1998). The proposed scenarios were based on cost-benefit analysis taking into consideration the present value of the net benefit of shoreline protection measures, the property value per unit area over the projected period of time, and the increments of land threatened from inundation or erosion due to sea level rise.

Thatcher, Brock, and Pendleton (2013) developed a Coastal Economic Vulnerability Index (CEVI), with the U.S. Gulf Coast as the case study. The index depicts the vulnerability of coastal areas to sea level rise at a 1-kilometer resolution. The CEVI considers both physical coastline attributes such as coastal slope, wave height, and tide range and critical infrastructure such as utility facilities, schools, and hospitals located at elevations below 1 meter. The study utilized population density from the U.S. Census Bureau and information on potential replacement costs of disaster-damaged residential and commercial buildings and infrastructure (such as oil refineries, natural gas facilities, power plants, hospitals, and water treatment plants) from the Federal Emergency Management Agency (FEMA)'s Hazus software. Dasymetric mapping was used to standardize census tract data to 30-meter raster cells using land use cover to redistribute census tract data. GIS layers were created for each variable and used to create the index.

The benefits of an immediate large-scale investment in shoreline protection may not justify the costs because (i) the level of protection required is unknown due to uncertainties in sea level rise projections and (ii) currently built long-lived structures may not be adequate for future conditions as structural reliability diminishes over time due to aging and capacity degradation. Similarly, early retreat from areas that are potentially threatened by lasting inundation may reduce tax revenues and thus hinder the ability of local communities to respond to the threats posed by the rising seas. Alternatively, a delay in incorporating sea level rise projections in future land use plans, new development, and infrastructure projects, especially in low-lying coastal areas, may increase the impending societal costs of adaptation. None of these policy decisions can be made without sufficient information, clearly defined priorities, and reasonable timelines. Possibly, the greatest obstacle to successful implementation of adaptation strategies is not lack of options, science, or certainty, but rather, the time and ability of coastal planning agencies and resource managers to incorporate climate change concepts into their day-to-day practice.

Land use plans, revision of zoning ordinances, and building codes are examples of decisions that are worthwhile to prioritize as they cost little and can potentially offset some of the risks associated with sea level rise. This might also create opportunity costs for developers or investors if there is a large percentage of developable land in these low-lying areas. However, since sea level rise proceeds slowly, dramatic measures of prohibiting development or imposing expensive preventive costs on developers or property owners might not be realistic. Land use, zoning, and building codes should continue to focus on traditional disaster mitigation perspectives associated with storm surge and flooding. Stricter and stronger building codes will be more appropriate while local communities closely monitor the impacts of sea level rises and educate citizens about these impacts.

Sea level rise assessments require a thorough understanding of uncertainty. Placing projections and decisions within "a probabilistic framework" is one such approach (Nordhaus, 2007). Such an approach will be particularly useful in floodplain management including floodplain mapping. Currently, there is no sufficient knowledge on the surge potential of sea level rise and how it will affect FEMA flood designations and evacuation routes. As indicated earlier, regulatory measures and insurance policies can gradually incorporate the probability of sea level rises into the insurance pricing and property taxation model. However, given the controversial and hypothetical nature of scientific projections, it might be difficult to convince residents and developers to follow probability-based pricing.

Case Study: Using Future Land Use and Dasymetric Mapping to Assess Potential Population Exposure to Sea Level Rise

Study Objective

The primary objective of this case study is to illustrate the use of spatially disaggregated population projections using ancillary data to assess exposure of future urban development to sea level rise. Thematic mapping is often used in spatial representations of population data using flat boundary files (e.g., census tracts or census block groups). A common assumption in thematic mapping is that the population counts or density within the boundary of the areal unit are homogeneously distributed (Mennis, 2003; Thurstain-Goodwin, 2003; Mennis & Hultgren, 2006; Mennis, 2009). Zandbergen and Ignizio (2010), Zandbergen (2011), and Mitsova, Esnard, and Li (2012) provide an overview of the suite of techniques commonly known as dasymetric mapping. Broadly, dasymetric mapping is an interpolation procedure in which population data at the areal unit level is disaggregated using ancillary data (e.g., land use).

Researchers distinguish between "source" areas (i.e., the flat boundary files containing the demographic data) and "target" areas (i.e., the ancillary data files used in the interpolation). Dasymetric approaches can range from recalculating population density using residential area only (that is, removing the uninhabited areas from the calculation) (Wright, 1936) to more complex methods where population counts are redistributed between areas of low, medium, and high density (Mennis, 2003, 2009; Zandbergen & Ignizio, 2010; Zandbergen, 2011; Mitsova, Esnard & Li, 2012; Nagle, Buttenfield, Leyk & Speilman, 2014). Research shows that there are several ancillary data options that can yield reasonably accurate interpolations including USGS National Land Cover Dataset (NLCD) (Eicher & Brewer, 2001; Zandbergen & Ignizio, 2010), parcel data (Xie, 2006; Maantay, Maroko & Herrmann, 2007; Maantay, Maroko & Porter-Morgan, 2008; Mitsova, Esnard & Li, 2012), and road network density (Brinegar & Popick. 2010). This case study is based on the enhanced dasymetric mapping approach developed by Mitsova, Esnard, and Li (2012).

Planning and Policy Applications

- Assist local governments in addressing the challenges posed by growing coastal populations and the demand for new development in areas vulnerable to future inundation from extreme events and sea level rise.
- Improve the accuracy of the spatial representation of population data.
- Evaluate the potential impacts of future land use planning on population exposure and vulnerability to sea level rise.

Data and Methods

- 2010 census data at the block group level for the coastal Miami-Dade County obtained from PCensus[1]
- Ancillary datasets for disaggregating the population counts at the block group level: (1) 2009 tax appraiser's parcel data for Miami-Dade County; (2) existing 2010 land use dataset; and (3) the future 2015–2025 land use dataset obtained from the county
- Population projections at 5-year increments from 2010 to 2030 by minor statistical area (MnSA) provided in the Comprehensive Emergency Management Plan of Miami-Dade County (Miami-Dade County, 2013)

Summary of Analysis

The potential extent of the projected sea level rise inundation is estimated using the methodology developed by the National Oceanic and Atmospheric Administration (NOAA, 2010). The approach considers the relationship between the tidal and orthometric datums to render a more realistic representation of the inundation extent. Vertical errors in the elevation dataset and measurement errors in the tidal data are used to account for uncertainty. The method employs a probabilistic approach in which the probability of a specified inundation level is derived from a z-scores surface (NOAA, 2010). The following equation is used to perform the Z-score calculation (NOAA, 2010):

$$ProbZ_a = \frac{Elevation_{x,y} - Inundation\_Increment_{MHHW+SLR}}{RMSE} \quad (1)$$

where $ProbZ_a$ is the calculated z-score, $Elevation_{x,y}$ stands for ground elevation referenced to NAVD88, $Inundation\_Increment_{MHHW+SLR}$ is the water level height, which takes into account the Mean Higher High Water (MHHW) level and the increment of sea level rise, and *RMSE* is the cumulative root mean square error computed as follows (NOAA, 2010):

$$RMSE = \sqrt{RMSE_E^2 + TidalSurfaceError^2} \quad (2)$$

The estimated RMSE, which accounts for the combined effect of the root mean square errors (RMSEs) in both the LiDAR data and the tidal grid, serves as a denominator for Equations (1) and (3). The calculations are performed using *Raster Calculator* (*Spatial Analyst Tools* > *Map Algebra*) using the following equation:

$$z - score = \frac{(SLR + MHHW) - LiDAR\_DEM\_tile}{RMSE} \quad (3)$$

where SLR denotes the chosen height of floodwaters (e.g., 0.5, 1, or 2 meters), MHHW is the tidal datum, and LiDAR_DEM_tile is the elevation raster dataset. Expression (3) creates a z-score surface from which the probability of inundation for each raster cell can be computed.

In order to disaggregate block group population counts (P_i) to a parcel level, we employ the *limiting variable* method where the number of dwelling units (DU_i) per tax lot and vacancy rates (V_i) are used as weighting factors. Population counts at the parcel lot level are estimated from the 2010 block group population multiplied by the ratio of the number of dwelling units within the tax lot to the total number of units within the block group using Equation (4) with an adjustment for vacancy rates (Mitsova, Esnard, & Li, 2012, p. 363):

$$P_{i=}P_{BG}\frac{DU_iV_i}{DU_{BG}} \quad (4)$$

where P_i provides an estimate of the population count at the parcel level, P_{BG} is the total block group population, DU_i is the number of residential units at the parcel level, DU_{BG} is the number of dwelling units within each block group, and V_i is a proxy of the vacancy rate at the lot level (Mitsova, Esnard & Li, 2012, p. 363). Figure 9.1 displays a thematic map of population counts at the block group level. Future population projections were disaggregated to a parcel level using a similar approach. Table 9.1 provides a summary of the procedure for disaggregating future population projections to a parcel level.

Key Findings and Sample Output

The 2030 population projections were disaggregated taking into account changes in residential land use and vacant land as described in the 2015–2025 future land use map. Between 6.5% (under the low-growth scenario) and 8% (under the rapid-growth scenario) of the projected population increase was expected to be accommodated by the existing vacant land. The calculation accounted for the minimum and maximum allowable housing units, household size, and estimated vacancy rates (Mitsova, Esnard & Li, 2012). The future land use map also indicated which tax lots are expected to be rezoned from low and medium density to high residential density through up-zoning. The process of up-zoning calculation accommodated roughly 36% (under the low-growth scenario) to 57% (under the rapid-growth scenario) of the future population growth in the study area (Mitsova, Esnard & Li, 2012). The results indicated that in the absence of further planning measures to divert growth from highly vulnerable areas, the population at risk in the coastal zone might increase by as much as 33% (Mitsova, Esnard & Li, 2012). Figure 9.2(a)

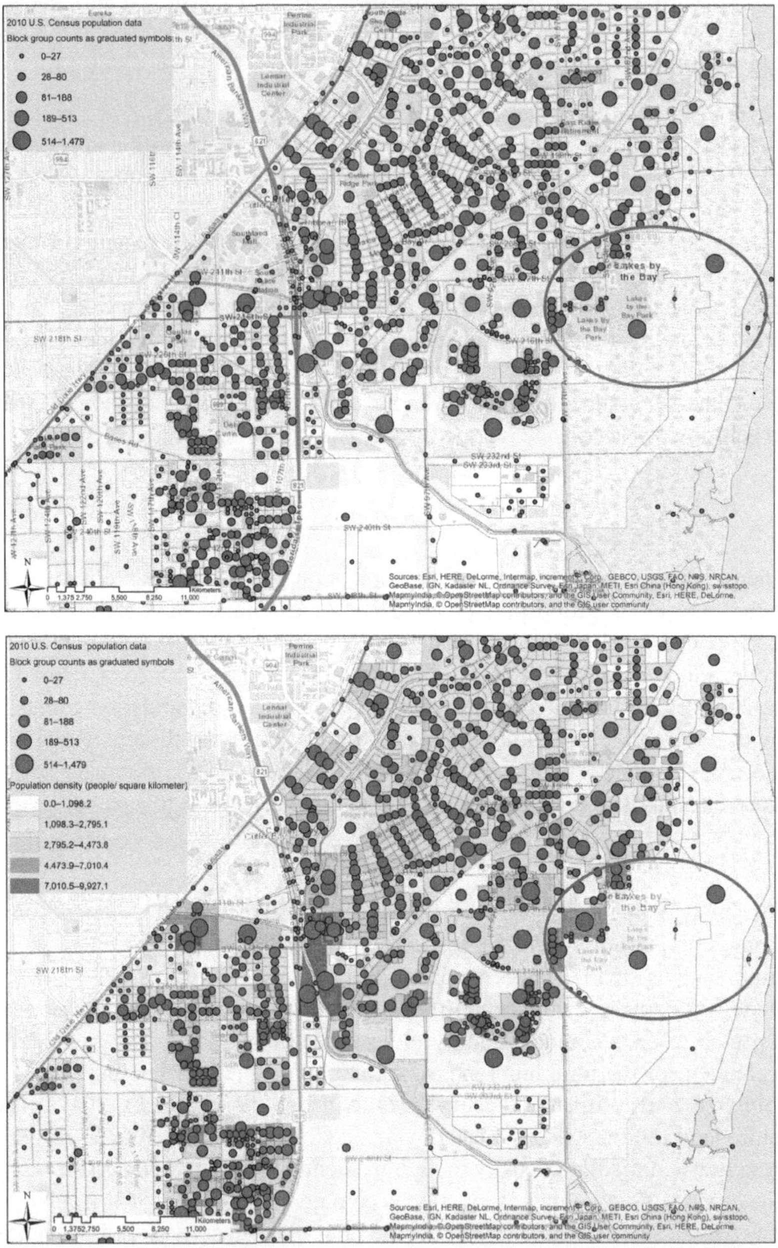

Figure 9.1 Both the graduated symbol map (a) and the thematic map (b) show high population counts in an area predominantly covered by wetlands in south Miami-Dade County, Florida, U.S.

Source: PCensus 2010

MnSA/Census Tract/Parcel	Pop in 2010	Total # of Units	Total # of Future Units	Total # of Vacant Units	Average Household Size	Projected Pop in 2030	Allowable Vacancy Rate	Projected Pop in 2030 (Adjusted)
Minor Statistical Area (MnSA)	P_1	H_1		V_1		P_2		
Census Tract					S		V_2	
Parcels								
Residential parcels with up-zoning	P_{33}	H_{33}	H_{331}(min) – H_{332} (max)		S	H_{331}*S (low) H_{332}*S (high)	V_2	H_{331}*S*(1 – V_2) H_{332}*S*(1 – V_2)
Vacant parcels			H_{441}(min) – H_{442}(max)		S	H_{441}*S (low) H_{442} *S (high)	V_2	H_{441}*S*(1 – V_2) H_{442}*S*(1 – V_2)
Low occupancy residential parcels without up-zoning	P_{11}	H_{11}		V_{11}	S	P_{11}+ (V_{11}/V_1)* (P_2 – P_1 – H_{331} *S-H_{441}*S), or P_{11}+ (V_{11}/V_1)* (P_2 – P_1 – H_{332}* S-H_{442}*S)	V_2	[P_{11}+ (V_{11}/V_1) *(P_2 – P_1 – H_{331} *S-H_{441}*S)]* (1 – V_2), or [P_{11}+ (V_{11}/ V_1)*(P_2 – P_1 – H_{332} *S-H_{442}*S)]*(1 – V_2)
High occupancy residential parcels without up-zoning	P_{22}	H_{22}		V_{22}	S	No change	V_2	No change

Table 9.1 Methods for disaggregating future population projections to a parcel level

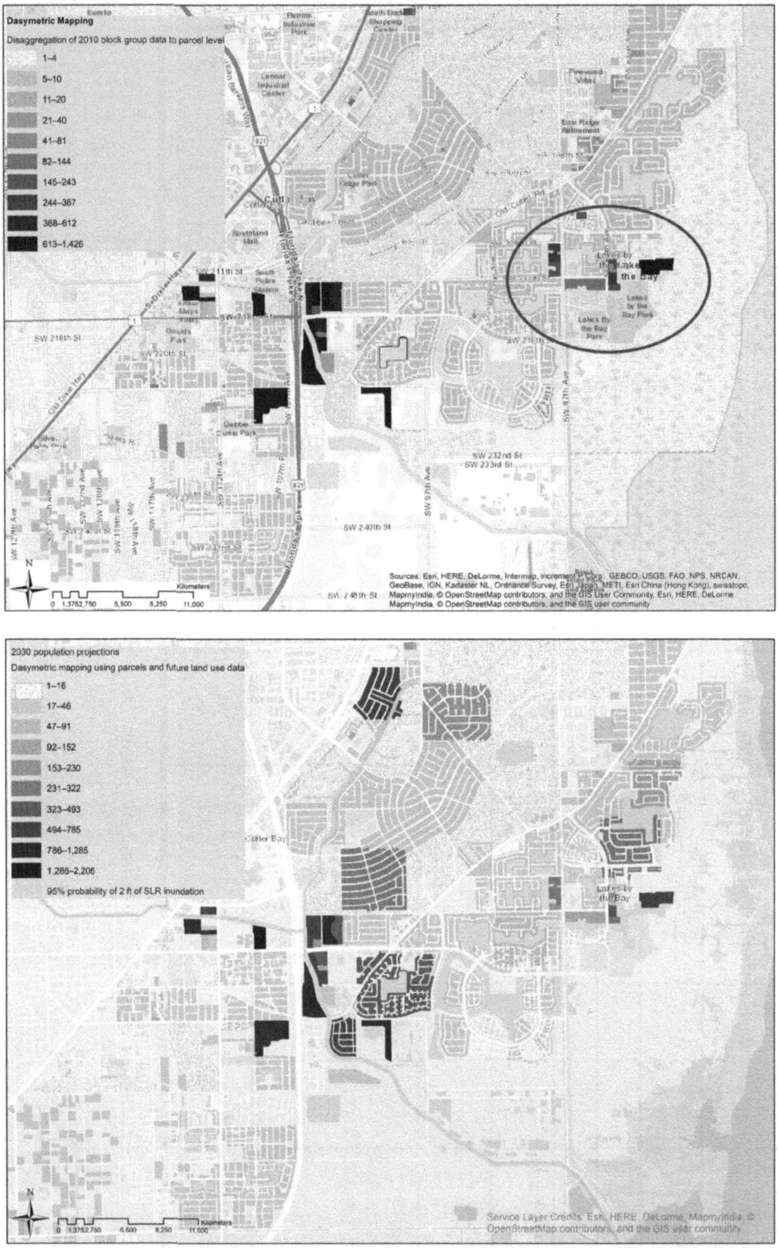

Figure 9.2 Dasymetric maps obtained from disaggregating the 2010 census block population counts to a tax lot level (a); dasymetric mapping from disaggregating the 2030 Miami-Dade County Comprehesive Emergency Management Plan (CEMP) population projections to a tax lot level and the potential impact of projected 2 feet of sea level rise (b)

displays the result of the dasymetric interpolation and the location of the residential units. Figure 9.2(b) illustrates the potential impact of population growth and planned development without consideration of the potential impact of 2 feet of sea level rise.

Miami-Dade County and the Southeast Florida Regional Climate Change Compact (SFRCCC) continue to advocate policy measures to address these potential impacts. A key achievement of the Compact is the optional designation of Adaptation Action Areas (AAAs) as part of the coastal management element of the local comprehensive plans (South Florida Regional Planning Council [SFRPC], 2013; Vella, Butler, Sipe, Chapin, & Murley, 2016). The adoption of Adaptation Action Areas, as defined by the 2011 Florida Community Planning Act (FCPA) [§163.3164(1) & §163.3177(6)(g)(10), Florida Statutes] allows local governments to consider policies within the coastal management element to reduce vulnerability to sea level rise and allocate funds for infrastructure improvements. AAAs are typically low-lying areas (below or near MHHW) which are hydrologically connected to coastal waters and within the evacuation zones for storm surge [§163.3177(6)(g)(10), Florida Statutes].

Uncertainty as a Methodological and Policy Issue

Some of the most common methodological issues related to sea level rise vulnerability assessments include the use of relatively coarse elevation data, inconsistency between vertical accuracy and increments of sea level rise, delineation of inundation zones based on increments that are within the margins of statistical uncertainty, reporting socioeconomic impacts without measures of uncertainty, and ignoring the differences between the geodetic reference system used for land elevation data and the tidal datums for measuring local mean sea level (Climate Change Science Program [CCSP], 2009; CaribSave, 2009; Kopp et al., 2014). Additionally, smaller-scale maps covering large areas and statistical information on impacts derived from these maps are not particularly useful for decision support and policy-making.

Understanding of uncertainty is inherent to effective long-term planning and risk management in coastal communities. Uncertainty in risk assessment is associated with two factors: uncertainty in the model results and uncertainty in the interpretation and application of results. Purvis, Bates, and Hayes (2008) point out that uncertainty has to be evaluated to avoid potential bias in the estimation models. Moser (2005) argues that "human-factors" can significantly influence the processes of assessment and policy responses to the potential impacts of sea level rise. Human values, interests, and preferences; technology choices; the degree to which decision-makers are comfortable with risk exposure and vulnerability; and the perception of responsibility, immediacy, interdependency, and scale, as well as the political process itself, affect the policy outcomes of climate adaptation planning (Moser, 2005). Furthermore, human factors affect the assessment of potential risks and the choice of policy options (Moser, 2005). The assessment approach determines how the problem is framed, what the priorities are, what conditions are considered "normal," how impacts are measured in terms of monetary and non-monetary values, what actions are considered politically feasible, what funding is available, and how one deals with uncertainty (Moser, 2005). As perceptions and values influence any decision-making process, risk perception within the affected constituencies is an essential component of the policy framework (Moser, 2005). According to Moser (2005), uncertainty and perceptions should not be an excuse to postpone action; refusing to act because of uncertainty could become a significant drawback, especially when adaptation requires early action.

Mitigation and Adaptation Strategies

The unavoidable impacts of sea level rise have spawned a broad spectrum of climate change adaptation strategies. Several countries and institutions have invested in climate adaptation planning and risk reduction initiatives (NRC, 2010; Burkett & Davidson, 2012; Boateng, 2012; SFRPC, 2013; European Union, 2013; Strauss, Kulp & Levermann, 2015; Bloetscher et al., 2016). As discussed in Chapter 3 and Chapter 4, capacity building and increased public awareness are key to the implementation of local adaptation projects, especially in rural communities and small island nations. This section reviews some commonly adopted approaches for sea level rise adaptation planning.

Marine Protected Areas

Marine Protected Areas (MPAs) are legislated geographic areas that impose restrictions on human activities within the specified boundaries. MPAs are protected areas of seas, oceans, and large lakes and are vital components of integrated management of coastal zones (covering both land and water) with the goals of sustainable development and conserving biodiversity (Kelleher & Kenchington, 1992; Gubbay, 1995; Salm, Clark & Siirila, 2000; Hoyt, 2012). Marine Protected Areas are central to common marine conservation program in many countries.

Protect, Adapt, or Retreat?

Three types of adaptive responses are typically considered in relation to the impact of sea level rise on coastal communities: planned retreat (zoning away from coastal areas), accommodation (retrofitting existing structures), or protection (constructing levees and seawalls) (Nicholls, Hoozemans & Marchand, 1999; Klein, Nicholls, Ragoonaden, Capabianco, Aston & Buckley, 2001; Deyle, Bailey & Matheny, 2007; Hurlimann, Barnett, Fincher, Osbaldiston, Mortreux & Graham, 2014). Smit, Burton, Klein, and Wandel (2000) point out that adaptation requires knowledge of what changes are most likely to occur, what response strategies should be envisioned, and how adaptation plans should be evaluated and adjusted. Adaptation also requires the involvement of various institutions, reliance on economic and legal instruments, and incorporation of advanced technological solutions (Smit, Burton, Klein & Wandel, 2000). Furthermore, the range of environmental and socioeconomic factors used to weigh possible alternatives and response strategies is immense because of the inherent variability and uncertainty associated with each of these factors (Klein, Nicholls, Ragoonaden, Capabianco, Aston & Buckley, 2001; Moser, 2005; Hurlimann, Barnett, Fincher, Osbaldiston, Mortreux & Graham, 2014). This further confounds risk assessment and response formulation.

Protection, adaptation, and retreat strategies have been widely reported in the literature (Mimura, 1999; Mimura & Harasawa, 2000; IPCC, 2001; Klein, Nicholls, Ragoonaden, Capabianco, Aston & Buckley, 2001; Smit, Burton, Klein & Wandel, 2000; Kleinosky, Yarnal & Fischer, 2007). Measures to protect the coast are commonly grouped into three categories: hard structures, soft measures, and hybrid approaches. Hard structures (such as seawalls, dikes, levees, groins, and breakwaters) are typically built parallel to the shoreline to act as a physical barrier. Major concerns with the structural measures for coastal protection are cost of construction and maintenance and disruption of natural sediment processes in the coastal zone (Wamsley, Cialone, Smith, Atkinson & Rosati, 2010; Glick, Kostyack, Pittman, Briceno & Wahlund, 2014). Such measures often reduce the recreational potential of adjoining areas. Soft measures include filling beaches with sand, dredging, and stabilizing sand dunes. Combined methods include submerged breakwaters, perched beaches that retain sand above the original level, and artificial headlands along dunes that allow the natural process of accretion to supply sand to the beach areas (Wamsley, Cialone, Smith, Atkinson & Rosati, 2010; Walker, Bendell & Wallendorf, 2011; United States Army Corps of Engineers [USACE], 2016).

Rolling Easements

In the United States, Titus (1998; 2011) has examined a variety of strategies for abandonment of coastal lowlands. By analyzing the costs and benefits, as well as the legal framework necessary for each strategy, the reviewed studies contend that abandonment options can be weighed and feasibility determined (Titus et al., 1991; Titus, 1998; Titus et al., 2009). Equity, uncertainty, constitutionality, and ecosystem protection are used to assess each strategy's likelihood for success (Titus et al., 1991; Titus et al., 2009). Adaptation strategies, explored by Titus (1998), range from purchasing land or prohibiting development on vacant land, to removing government subsidies for coastal protection infrastructure and incorporating rolling easements or conditional leases into coastal zoning laws. The last strategy is notable because it preserves the current land uses, allows for migration of wetlands in the future, and creates an actionable plan taking into account our current knowledge (Titus, 1998). When sea levels rise to the point of complete inundation of a particular property, rolling easements give the government the right to take the land for conservation at about 1% of its present value. Titus (1998) suggests that rolling easements have to be enacted at least 50 to 100 years before the predicted abandonment in order to be effective. In some situations, other strategies may be more feasible. Titus (1998) reiterates that the success of each strategy depends on the amount of time for its implementation and that long-term planning with regard to sea level rise is essential. A rolling easement is essentially a regulatory tool that allows property owners to place building structures near vulnerable areas but prohibits armoring of the shoreline and other structural protection measures, thus preventing protection from rising sea levels (Titus et al., 2009; Grannis, Wyman, Singer, Shoaf & Lynch, 2012).

Protection of Shorelines and Ecological Systems

Sea level rise will most likely amplify the shoreline management challenges faced by residents, planners, coastal managers, and decision-makers. Restoration and shoreline enhancement projects provide both ecosystem services and potential for shoreline protection against the threats of sea level rise (Dugan, Airoldi, Chapman, Walker & Schlacher, 2011; Arkema et al., 2013; Arkema et al., 2015). The shore zone, an attractive place to live and recreate, is affected by multiple natural and anthropogenic stressors. Natural shorelines such as mangrove swamps and sandy beaches provide numerous ecosystem services, including wave dissipation, habitat, and natural sediment buildup. These services are oftentimes degraded as a result of urban development, including dredging and filling, vegetation removal, shoreline armoring, and discharges of stormwater containing nutrients and contaminants (Bilkovic & Mitchell, 2013).

Nature-based shoreline stabilization options have been shown to provide "solutions of comparable engineering efficiency, with considerable economic savings and with the maintenance of collateral ecosystem services and functions" (Dugan, Airoldi, Chapman, Walker & Schlacher, 2011, p. 36). This finding is corroborated by a report of the National Wildlife Federation which emphasizes the benefits of preserving and restoring natural infrastructure for reducing the vulnerability of local communities to flooding and hurricane risks, in addition to numerous ecological and economic benefits (Glick, Kostyack, Pittman, Briceno & Wahlund, 2014). The report highlights the important role that healthy natural ecosystems can play in building resilient communities and provides several examples of successful implementation of nature-based strategies to counteract the damaging effects of extreme events. In **Jamaican Bay, New York**, for example, 150 acres of restored wetlands helped mitigate the destructive backwash waves brought about by Superstorm Sandy (Glick, Kostyack, Pittman, Briceno & Wahlund, 2014). Artificial reefs and seagrass can dampen the effects of waves, as do mangrove forests (Masria et al., 2015).

"Living" shorelines, which use a combination of sediment stabilization materials and habitat restoration techniques to provide shoreline stabilization and erosion control, are another alternative explored in the **United States.** (Currin, Delano & Valdes-Weaver, 2008; Currin, Chappell & Deaton, 2010; Mitsova & Esnard, 2012; Gittman, Popowich, Bruno, & Peterson, 2014; Mitsova et al., 2018).

Regulatory Tools

Given the uncertainty of sea level rise, timing is a key to policy initiatives. Acting too early by directing development away from the vulnerable areas may create potential losses to property owners and developers, raise legal disputes, and reduce the fiscal revenues of local governments. However, if action is postponed, losses associated with sea level rise may become multidimensional. For example, land use planning can incorporate sea level rise projections and thus direct development away from vulnerable areas.

The linkage between coastal zone management policy formulation processes and decision support systems remains a challenge that requires further exploration (Van Kouwen, Dieperink, Schot & Wassen, 2008). A study by Puszkin-Chevlin and Esnard (2009a, 2009b) examined how a 2006 legislative change to Florida's Coastal

High Hazard Area (CHHA) definition and compliance requirements *impacts* asset vulnerability in three **Florida Treasure Coast counties**. GIS was effective in evaluating the age of the building inventory in conjunction with land use in order to assess susceptibility to redevelopment and up-zoning pressure. Findings from the research showed that the new CHHA dropped a large inventory of the structures built in the 1970s and 1980s, as well as several mobile home parks, which could, in turn, allow additional asset accumulation in areas vulnerable to flood and high winds (Puszkin-Chevlin & Esnard, 2009a, 2009b). Although Puszkin-Chevlin and Esnard (2009b) focused on Florida, the lessons are transferable to policy-makers who seek a balance among economic development, property rights, and environmental risk.

Regulatory tools and requirements can be used in new development or redevelopment in vulnerable areas. In addition, land use planning has long-term consequences as what we built today cannot be relocated without significant costs. Shoreline protection measures and insurance rates can be adjusted as the sea advances, and therefore such decisions should not be necessarily taken in the immediate future. In addition, measures to control shoreline erosion and flooding are already in place, and therefore adjustment can gradually take place over time. Protecting or planning to relocate critical infrastructure requires long-term planning and consistency among strategic initiatives at all levels of government. The strategic plans should be the blueprint for future planning responses and should involve close cooperation with other professionals in disaster management, transportation, public works and utilities, insurance, and homeland security through inter-agency agreements. Partnerships involving various stakeholders with a strong leading agency can increase the efficiency of responses to sea level rise. GIS plays an integral role in support of policy formulation and is used by a broad range of professionals (i.e., planners, environmental advocates, regulators, insurers, attorneys, appraisers, lenders, (re)developers, and real estate professionals) involved in coastal land (re)development and land regulation.

Note

1 PCensus is an advance demographic package developed by Tetrad Computer Applications Inc. (www.tetrad.com/software/pcensus/pcensus-desktop.html).

References

Arkema, K.K., Guannel, G., Verutes, G.M., Wood, S.A., Guerry, A.D., Ruckelshaus, M., Kareiva, P., Lacayo, M., & Silver, J.M. 2013. Coastal Habitats Shield People and Property from Sea-Level Rise and Storms. *Nature Climate Change* 3, 913–918. DOI: 10.1038/NCLIMATE1944.

Arkema, K.K., Verutes, G.M., Wood, S.A., Clarke-Samuels, C., Rosado, S., Canto, M., Rosenthal, A., Ruckelshaus, M., Guannel, G., Toft, J., Faries, J., Silver, J.M., Griffin, R., & Guerry, A.D. 2015. Embedding Ecosystem Services in Coastal Planning Leads to Better Outcomes for People and Nature. *PNAS* 112(24), 7390–7395.

Beach, D. 2002. *Coastal Sprawl: The Effects of Urban Design on Aquatic Ecosystems in the United States*. Pew Oceans Commission. Retrieved from www.pewtrusts.org/en/research-and-analysis/reports/2002/04/08/coastal-sprawl-the-effects-of-urban-design-on-aquatic-ecosystems-in-the-united-states.

Beckley, B.D., Lemoinse, F.G., Lutchke, S.B., Ray, R.D., & Zelensky, N.P. 2007. A Reassessment of Global and Regional Mean Sea Level Trends from TOPEX and Jason-1 Altimetry Based on Revised Reference Frame and Orbits. *Geophysical Research Letters* 34, L14608.

Bilkovic, D.M., & Mitchell, M.M. 2013. Ecological Tradeoffs of Stabilized Salt Marshes as a Shoreline Protection Strategy: Effects of Artificial Structures on Macrobenthic Assemblages. *Ecological Engineering* 61, 469–481.

Bloetscher, F., Polsky, C., Bolter, K., Mitsova, D., Palbicke Garces, K., King, R., Cosio Carballo, I., & Hamilton, K. 2016. Assessing Potential Impacts of Sea Level Rise on Public Health and Vulnerable Populations in Southeast Florida and Providing a Framework to Improve Outcomes. *Sustainability 8*(4), 315.

Boateng, I. 2012. GIS Assessment of Coastal Vulnerability to Climate Change and Coastal Adaptation Planning in Vietnam. *Journal of Coastal Conservation* 16(1), 25–36.

Brinegar, S.J., & Popick, S.J. 2010. A Comparative Analysis of Small Area Population Estimation Methods. *Cartography and Geographic Information Science* 37(4), 273–284.

Burkett, V.R., & Davidson, M.A. (Eds.). 2012. *Coastal Impacts, Adaptation and Vulnerability: A Technical Input to the 2012 National Climate Assessment*. Cooperative Report to the 2013 National Climate Assessment, 150.

CaribSave. 2009. *Caribbean Climate Change, Tourism & Livelihoods: A Sectoral Approach to Vulnerability and Resilience. Water, Energy, Agriculture, Human Health, Biodiversity, Infrastructure and Settlement, Comprehensive Disaster Management*. Retrieved from http://research3.fit.edu/sealevelriselibrary/documents/doc_mgr/473/Negril__CC_Vulnerability_&_Adaptive_Capacity_-Caribsave_&_CCCCC.pdf.

Climate Change Science Program (CCSP). 2009. *Coastal Sensitivity to Sea-Level Rise: A Focus on the Mid-Atlantic Region*. Final Report: Synthesis and Assessment Product 4.1, Washington, DC. Retrieved from https://downloads.globalchange.gov/sap/sap4-1/sap4-1-final-report-all.pdf.

Currin, C.A., Chappell, W.S., & Deaton, A. 2010. Developing Alternative Shoreline Armoring Strategies: The Living Shoreline Approach in North Carolina. In: Shipman, H., Dethier, M.N., Gelfenbaum, G., Fresh, K.L., & Dinicola, R.S. (Eds.) *Puget Sound Shorelines and the Impacts of Armoring—Proceedings of a State of the Science Workshop, May 2009*. U.S. Geological Survey Scientific Investigations Report 2010–5254, 91–102.

Currin, C.A., Delano, P.C., & Valdes-Weaver, L.M. 2008. Utilization of a Citizen Monitoring Protocol to Assess the Structure and Function of Natural and Stabilized Unmodified Salt Marshes in North Carolina. *Wetlands Ecological Management* 16, 97–118.

Deyle, R.E., Bailey, K.C., & Matheny, A. 2007. *Adaptive Response Planning to Sea-level Rise in Florida and Implications for Comprehensive and Public-Facilities Planning*. Florida Planning and Development Lab, Department of Urban and Regional Planning, Florida State University (FSU).

Dugan, J.E., Airoldi, L., Chapman, M.G., Walker, S.J., & Schlacher, T. 2011. Estuarine and Coastal Structures: Environmental Effects, a Focus on Shore and Nearshore Structures. In: Wolanski, E., & McLusky, D. (Eds.) *Treatise on Estuarine and Coastal Science*. Waltham, MA: Academic Press.

Dutta, D., Babel, M.S., & Gupta, A.D. 2005. *An Assessment of the Socio-economic Impacts of Floods in Large Coastal Areas*. Final Report for APN CAPaBLE Project, Asian Institute of Technology (AIT), Thailand.

Eicher, C.L., & Brewer, C.A. 2001. Dasymetric Mapping and Areal Interpolation: Implementation and Evaluation. *Cartography and Geographic Information Science* 28(2), 125–138.

Environmental Protection Agency (EPA). 2009.

Esnard, A.-M., & Puszkin-Chevlin, A. 2010. Coastal Property Analysis Through a Hazard Exposure Lens. In: Linne, M., & Thompson, M.M. (Eds.) *Visual Valuation: Implementing Valuation Modeling and Geographic Information Solutions*. Chicago, IL: The Appraisal Institute.

European Union. 2013. *Adaptation to Climate Change*. European Comminssion, Brussels, Belgium. URL: https://ec.europa.eu/clima/policies/adaptation_en.

Franck, T. 2009. Coastal Adaptation and Economic Tipping Points. *Management of Environmental Quality* 20(4), 434–450.

Füssel, H., & Klein, R.J. 2006. Climate Change Vulnerability Assessments: An Evolution of Conceptual Thinking. *Climate Change* 75, 301–329.

Ghermandi, A., & Nunes, P.A. 2013. A Global Map of Coastal Recreation Values: Results from a Spatially Explicit Meta-Analysis. *Ecological Economics* 86, 1–15.

Gittman, R.K., Popowich, A.M., Bruno, J.F., & Peterson, C.H. 2014. Marshes with and Without Sills Protect Estuarine Shorelines from Erosion Better Than Bulkheads During a Category 1 Hurricane. *Ocean and Coastal Management* 102, 94–102.

Glick, P., Kostyack, J., Pittman, J., Briceno, T., & Wahlund, N. 2014. *Natural Defenses from Hurricanes and Floods: Protecting America's Communities and Ecosystems in an Era of Extreme Weather*. National Wildlife Federation & Earth Economics. Retrieved from www.nwf.org/news-and-magazines/media-center/

news-by-topic/global-warming/2014/10-20-14-new-report-action-needed-now-to-fix-climate-change.aspx.

Gornitz, V., & White, T.W. 1992. *A Coastal Hazards Data Base for the U.S. East Coast, CDIAC NDP-043A*. Retrieved from http://cdiac.ornl.gov/ndps/ndp043a.html (accessed 07.10/2015).

Grannis, J., Wyman, J., Singer, M., Shoaf, J., & Lynch, C. 2012. Coastal Management in the Face of Rising Seas: Legal Strategies for Connecticut. *Sea Grant Law and Policy Journal* 5(1), 59–88.

Greenlaw, M.E., Roff, J.C., Redden, A.M., & Allard, K.A. 2011. Coastal Zone Planning: A Geophysical Classification of Inlets to Define Ecological Representation. *Aquatic Conservation: Marine and Freshwater Ecosystems* 21(5), 448–461.

Gubbay, S. (Ed.) 1995. *Marine Protected Areas: Principles and Techniques for Management*. Dordrecht, NL: Springer Science.

Hallegatte, S., Ranger, N., Mestre, O., Dumas, P., Corfee-Morlot, J., Herweijer, C., & Wood, R.M. 2011. Assessing Climate Change Impacts, Sea-Level Rise and Storm Surge Risk in Port Cities: A Case Study on Copenhagen. *Climatic Change* 104(1), 113–137.

Harrington, J., & Walton, T.L. 2007. *Climate Change in Coastal Areas in Florida: Sea-Level Rise Estimation and Economic Analysis to Year 2080*. Center for Economic Forecasting and Analysis, Florida State University (FSU). Retrieved from www.cefa.fsu.edu/sites/g/files/imported/storage/original/application/2dec40fac4c55a937b4c2497db652815.pdf.

Hoyt, E. 2012. *Marine Protected Areas for Whales, Dolphins, and Porpoises: A World Handbook*. London, UK: Earthscan.

Hurlimann, A., Barnett, J., Fincher, R., Osbaldiston, N., Mortreux, C., & Graham, S. 2014. Urban Planning and Sustainable Adaptation to Sea-Level Rise. *Landscape and Urban Planning* 126, 84–93.

Intergovernmental Panel on Climate Change (IPCC). 2001. *Climate Change 2001: Synthesis Report. A Contribution of Working Groups I, II, and III to the Third Assessment Report of the Intergovernmental Panel on Climate Change*. Cambridge University Press, Cambridge.

Jacob, K., Gornitz, V., & Rozenzweig, C. 2007. Vulnerability of New York City Metropolitan Area to Coastal Hazards, Including Sea-Level Rise: Inferences for Urban Coastal Risk Management and Adaptation Policies. In: McFadden, Loraine, Robert Nicholls and Edmund Penning-Rowsell (Eds.) *Managing Coastal Vulnerability*, 141–158. Oxford, UK: Elsevier.

Kattaa, B., Al-Fares, W., & Al Charideh, A.R. 2010. Groundwater Vulnerability Assessment for the Banyas Catchment of the Syrian Coastal Area Using GIS and the RISKE Method. *Journal of Environmental Management* 91(5), 1103–1110.

Kebede, A.S., Nicholls, R.J., Hanson, S., & Mokrech, M. 2012. Impacts of Climate Change and Sea-Level Rise: A Preliminary Case Study of Mombasa, Kenya. *Coastal Research* 28(1A), 8–19.

Kelleher, G., & Kenchington, R.A. 1992. *Guidelines for Establishing Marine Protected Areas: A Marine Conservation and Development Report, The IUCN Conservation Library*. Charlottesville, VA: IUCN, University of Virginia.

Klein, R.J.T., Nicholls, R.J., Ragoonaden, S., Capabianco, M., Aston, J., & Buckley, E.N. 2001. Technological Options for Adaptation to Climate Change in Coastal Zones. *Journal of Coastal Research* 17(3), 531–543.

Kleinosky, L.R., Yarnal, B., & Fischer, A. 2007. Vulnerability of Hampton Roads, Virginia to Storm-Surge Flooding and Sea-Level Rise. *Natural Hazards* 40, 43–70.

Kopp, R. E., Horton, R.M., Little, C.M., Mitrovica, J.X., Oppenheimer, M., Rasmussen, D.J., Strauss, B.H., & Tebaldi, C. 2014. Probabilistic 21st and 22nd Century Sea-Level Projections at a Global Network of Tide-Gauge Sites. *Earth's Future* 2, 383–406.

Koukoulas, S., Nicholls, R., Dickson, M., Walkden, M., Hall, J., Pearson, S., Mokrech, M., & Richards, J. 2006. *A GIS Tool for Analysis and Interpretation of Coastal Erosion Model Outputs (SCAPEGIS)*. Fifth International Coastal Dynamics Conference, April 4–8, 2005, Barcelona, Spain.

Krause, G., Bock, M., Weiers, S., & Braun, G. 2004. Mapping Land-Cover and Mangrove Structures with Remote Sensing Techniques: A Contribution to a Synoptic GIS in Support of Coastal Management in North Brazil. *Environmental Management* 34(3), 429–440.

Kuleli, T. 2009. Quantitative Analysis of Shoreline Changes at the Mediterranean Coast in Turkey. *Environmental Monitoring and Assessment* 167(1–4), 387–397.

Lichter, M., & Felsenstein, D. 2012. Assessing the Costs of Sea-Level Rise and Extreme Flooding at the Local Level: A GIS-Based Approach. *Ocean & Coastal Management* 59, 47–62.

Maantay, J.A., Maroko, A.R., & Herrmann, C. 2007. Mapping Population Distribution in the Urban Environment: The Cadastral-Based Expert Dasymetric System (CEDS). *Cartography and Geographic Information Science* 34(2), 77–102.

Maantay, J.A., Maroko, A.R., & Porter-Morgan, H. 2008. Research Note—a New Method for Mapping Population and Understanding the Spatial Dynamics of Disease in Urban Areas: Asthma in the Bronx, New York. *Urban Geography* 29(7), 724–738.

Marfai, M.A., & King, L., 2008. Tidal Inundation Mapping under Enhanced Land Subsidence in Semarang, Central Java, Indonesia. *Natural Hazards* 44, 93–109.

Marfai, M.A. 2014. Impact of Sea-Level Rise to Coastal Ecology: A Case Study on the Northern Part of Java Island, Indonesia. *Quaestiones Geographicae* 33(1), 107–114.

Masria, A., Nadaoka, K., Negm, A., & Iskander, M. 2015. Detection of Shoreline and Land Cover Changes around Rosetta Promontory, Egypt, Based on Remote Sensing Analysis. *Land* 4(1), 216–230.

Mendoza-Gonzalez, G., Martinez, M.L., Lithgow, D., Perez-Maqueo, O., & Simonin, P. 2012. Land Use Change and Its Effects on the Value of Ecosystem Services Along the Coast of the Gulf of Mexico. *Ecological Economics* 82, 23–32.

Mennis, J. 2003. Generating Surface Models of Population Using Dasymetric Mapping. *The Professional Geographer* 55, 31–42.

Mennis, J. 2009. Dasymetric Mapping for Estimating Population in Small Areas. *Geography Compass* 3, 727–745.

Mennis, J., & Hultgren, T. 2006. Intelligent Dasymetric Mapping and Its Application to Areal Interpolation. *Cartography and Geographic Information Science* 33, 179–194.

Merrifield, M.A., Genz, A.S., Kontoes, C.P., & Marra, J.J. 2013. Annual Maximum Water Levels from Tide Gauges: Contributing Factors and Geographic Patterns. *Journal of Geophysical Research: Oceans* 118, 2535–2546.

Miami-Dade County. 2013. *Comprehesive Emergency Management Plan (CEMP).* Retrieved from www.miamidade.gov/fire/library/OEM/CEMP.pdf (accessed 09/02/17).

Mileti, D.S. 1999. *Disasters by Design: A Reassessment of Natural Hazards in the United States.* Washington, DC: Joseph Henry Press.

Mimura, N. 1999. Vulnerability of Island Countries in the South Pacific to Sea Level Rise and Climate Change. *Climate Research* 12, 137–143.

Mimura, N., & Harasawa, H. 2000. *Data Book of Sea-Level Rise 2000.* Centre for Global Environmental Research, National Institute for Environmental Studies, Ibaraki, Japan.

Mitsova, D., Bergh, C., Guannel, G., Lustic, C., Renda, M., Byrne, J., Graves, A., Cresswell, K., Alhawiti, R., Goldberg, A., & Reed, S. 2018. Spatial Decision Support for Nature-Based Shoreline Stabilization Options in Subtropical Estuarine Environments. *Journal of Environmental Planning and Management*, doi: 10.1080/09640568.2017.1398637.

Mitsova, D., & Esnard, A-M. 2012. Holding Back the Sea: An Overview of Shore Zone Planning and Management. *Journal of Planning Literature* 27(4), 446–459.

Mitsova, D., Esnard, A-M., & Li, Y. 2012. Using Enhanced Dasymetric Mapping Techniques to Improve the Spatial Accuracy of Sea Level Rise Vulnerability Assessments. *Journal of Coastal Conservation* 16, 355–372.

Moser, S.C. 2005. Impact Assessments and Policy Responses to Sea-Level Rise in Three US States: An Exploration of Human-Dimension Uncertainties. *Global Environmental Change* 15, 353–369.

Nagle, N.N., Buttenfield, B.P., Leyk, S., & Speilman, S. 2014. Dasymetric Modeling and Uncertainty. *Annals of the American Association of Geographers* 104(1), 80–95.

National Oceanic and Atmospheric Administration (NOAA). 2010. *Mapping Inundation Uncertainty*. Charleston, SC: NOAA Coastal Services Center. Retrieved from https://coast.noaa.gov/data/digitalcoast/pdf/mapping-inundation-uncertainty.pdf.

National Research Council (NRC). 2010. *Adapting to the impacts of climate change*. Washington, D.C.: The National Academies Press.

Neumann, J.E., Yohe, G., Nicholls, R., & Manion, M. 2000. *Sea-Level Rise and Global Climate Change: A Review of Impacts to U.S. Coasts*. Pew Center on Global Climate Change. Retrieved from www.pewclimate.org/docUploads/env_sealevel.pdf (accessed 08/15/2017).

Neumann, B., Vafeidis, A.T., Zimmermann, J., & Nicholls, R.J. 2015. Future Coastal Population Growth and Exposure to Sea-Level Rise and Coastal Flooding—a Global Assessment. *PLoS ONE* 10(3), e0118571.

Nicholls, R.J., Hoozemans, F.M.J., & Marchand, M. 1999. Increasing Flood Risk and Wetland Loss Due to Global Sea-Level Rise: Regional and Global Analyses. *Global Environmental Change* 9, S69–S87.

Nordhaus, W. 2007. *Key Potential Improvements in Statistics and Data for Policies Governing Global Warming: The Role of Federal Statistical Agencies*. National Research Council Committee on National Statistics, Yale University.

Pielke Jr., R.A., Gratz, J., Landsea, C.W., Collins, D., Saunders, M.A., & Musulin, R. 2008. Normalized Hurricane Damage in the United States: 1900–2005. *Natural Hazards Review* 9(1), 29–42.

Purvis, M.J., Bates, P.D., & Hayes, C.M. 2008. A Probabilistic Methodology to Estimate Future Coastal Flood Risk Due to Sea-Level Rise. *Coastal Engineering* 55(12), 1062–1073.

Puszkin-Chevlin, A., & Esnard, A.-M. 2009a. Evaluating Spatial Impacts of Change to Coastal Hazard Policy Language. *URISA* 21(1), 41–51.

Puszkin-Chevlin, A., & Esnard, A-M. 2009b. Incremental Evolution and Devolution of Florida's Coastal High Hazard Area Policy. *Journal of Environmental Planning and Management* 52(3), 1–17.

Salm, R.V., Clark, J.R., & Siirila, E. 2000. *Marine and Coastal Protected Areas: A Guide for Planners and Managers*. Gland, Switzerland and Cambridge, UK: IUCN.

Saravanan, S., Chandrasekar, N., Rajamanickam, M., Hentry, C., & Joevivek, V. 2014. Management of Coastal Erosion Using Remote Sensing and GIS Techniques (SE India). *International Journal of Ocean and Climate Systems* 5(4), 211–222.

Simpson, M.C., Scott, D., Harrison, M., Sim, R., Silver, N., O'Keeffe, E., Harrison, S., Taylor, M., Lizcano, G., Rutty, M., Stager, H., Oldham, J., Wilson, M., New, M., Clarke, J., Day, O.J., Fields, N., Georges, J., Waithe, R., & McSharry, P. 2010. *Quantification and Magnitude of Losses and Damages Resulting from the Impacts of Climate Change: Modelling the Transformational Impacts and Costs of Sea Level Rise in the Caribbean*. United Nations Development Programme (UNDP), Barbados, West Indies. Retrieved from www.bb.undp.org/content/barbados/en/home/library/environment_energy/modelling-the-impacts-and-costs-of-slr-in-the-cbean.html (accessed 06/13/17).

Smit, B., Burton, I., Klein, R.J., & Wandel, J. 2000. An Anatomy of Adaptation to Climate Change and Variability. *Climatic Change* 45, 223–251.

South Florida Regional Planning Council (SFRPC). 2013. *Adaptation Action Areas: Policy Options for Adaptive Planning for Rising Sea Levels*. URL: http://www.southeastfloridaclimatecompact.org/wp-content/uploads/2014/09/final-report-aaa.pdf.

Strauss, B.H., & Kulp, S. 2014. New Analysis Shows Global Exposure to Sea Level Rise, Research Report. *Climate Central*. Retrieved from www.climatecentral.org/news/new-analysis-global-exposure-to-sea-level-rise-flooding-18066 (accessed 04/15/2017).

Strauss, B.H., Kulp, S., & Levermann, A. 2015. Carbon Choices Determine US Cities Committed to Futures Below Sea Level. *PNAS* 112(44), 13508–13513.

Thatcher, C.A., Brock, J.C., & Pendleton, E.A. 2013. Economic Vulnerability to Sea-Level Rise Along the Northern U.S. Gulf Coast. In: Brock, J.C.; Barras, J.A., and Williams, S.J. (Eds.) *Understanding and Predicting Change in the Coastal Ecosystems of the Northern Gulf of Mexico, Journal of Coastal Research*, Special Issue 63, 234–243.

Thurstain-Goodwin, M. 2003. Data Surfaces for a New Policy Geography. In: Longley P.A., & Batty M. (Eds.) *Advanced Spatial Analysis. The CASA Book of GIS*, 145–170. Redlands, CA: ESRI Press.

Titus, J.G. 1991. Greenhouse Effect and Coastal Wetland Policy: How Americans Could Abandon an Area the Size of Massachusetts at Minimum Cost. *Journal of Environmental Management* 15(1): 39–58.

Titus, J.G., Park, R.A., Leatherman, S.P., Weggel, J.R., Greene, M.S., Mausel, P.W., Brown, S., Gaunt, G., Trehan, M., and Yohe, G. 1991. Greenhouse effect and 11 sea level rise: The coast of holding back the sea. *Coastal Management*, 19, 171- 204.

Titus, J.G. 1998. Rising Seas, Coastal Erosion, and the Takings Clause: How to Save Wetlands and Beaches Without Hurting Property Owners. *Maryland Law Review* 57(4), 1279–1399.

Titus, J.G., Hudgens, D.E., Trescott, D.L., Craghan, M., Nuckols, W.H., Hershner, C.H., Kassakian, J.M., Linn, C.J., Merritt, P.G., McCue, T.M., O'Connell, J.H., Tanski, J., & Wang. J. 2009. State and Local Governments Plan for Development of Most Land Vulnerable to Rising Sea Level Along the US Atlantic Coast. *Environmental Research Letters* 4, 044008.

Titus, J.G. 2011. *Rolling Easements*. Climate Ready Estuaries Program, United States Environmental Protection Agency, Division of Water. Retrieved from http://water.epa.gov/type/oceb/cre/upload/rollingeasementsprimer.pdf.

Tol, R.S.J. 1995. The Damage Cost of Climate Change: Towards More Comprehensive Calculations. *Environmental and Resource Economics* 5, 353–374.

Tol, R.S.J. 1999. Spatial and Temporal Efficiency in Climate Change: Applications of FUND. *Environmental and Resource Economics* 14(1), 33–49.

Tol, R.S.J. 2002a. Estimates of the Damage Cost of Climate Change, Part 1: Benchmark Estimates. *Environmental and Resource Economics* 21(1), 47–73.

Tol, R.S.J. 2002b. Estimates of the Damage Cost of Climate Change, Part 1: Dynamic Estimates. *Environmental and Resource Economics* 21(1), 135–160.

United States Army Corps of Engineers (USACE). 2016. *Federal Register / 81(105), Proposed Rules—Nationwide Permit B: Living Shorelines, 35205–7*. Retrieved from www.gpo.gov/fdsys/pkg/FR-2016-06-01/pdf/2016-12083.pdf#page=21 (accessed 12/31/2016).

United States Environmental Protection Agency (EPA). 2009. Land-Use Scenarios: National-Scale Housing Density Scenarios Consistent with Climate Change Storylines. U.S. EPA/600/R-08/076F. Washington, D.C.: Environmental Protection Agency.

Van Kouwen, F., Dieperink, C., Schot, P., & Wassen, M. 2008. Applicability of Decision Support Systems for Integrated Coastal Zone Management. *Coastal Management* 36(1), 19–34.

Vella, K., Butler, W.H., Sipe, N., Chapin, T., & Murley, J. 2016. Voluntary Collaboration for Adaptive Governance: The Southeast Florida Regional Climate Change Compact. *Journal of Planning Education and Research*, 36(3), 363-376.

Walker, R., Bendell, B., & Wallendorf, L. 2011. Defining Engineering Guidelines for Living Shoreline Projects. *Coastal Engineering Practice*, ASCE: 1064–1077.

Wamsley, T.V., Cialone, M.A., Smith, J.M., Atkinson, J.H., & Rosati, J.D. 2010. The Potential of Wetlands in Reducing Storm Surge. *Ocean Engineering* 37, 59–68.

Wilson, J., Trenholm, R., Bornemann, J., & Lieske, D. 2012. *Forecasting Economic Damages from Storm Surge Flooding: A Case Study in the Tantramar Region of New Brunswick*. Atlantic Climate Adaption Solutions Association.

Wright, J.K. 1936. A Method of Mapping Densities of Population. *The Geographical Review*, 103–110.

Xie, Z. 2006. A Framework for Interpolating the Population Surface at the Residential-Housing Unit Level. *GIScience & Remote Sensing* 43(3), 233–251.

Yohe, G.W., & Schlesinger, M.E. 1998. Sea-Level Change: The Expected Economic Cost of Protection Or Abandonment in the United States. *Climatic Change* 38(4), 447–472.

Zandbergen, P.A. 2011. Dasymetric Mapping Using High Resolution Address Point Datasets. *Transactions in GIS* 15, 5–27.

Zandbergen, P.A., & Ignizio, D.A. 2010. Comparison of Dasymetric Mapping Techniques for Small-Area Population Estimates. *Cartography and Geographic Information Science* 7(3), 199–214.

10 Vulnerability of Critical Infrastructure to Climate Impacts

Chapter Objectives

This chapter focuses on answering the following questions:

- What are examples of critical infrastructure and the various interdependencies between them?
- How do climate-related hazards impact infrastructure?
- How is GIS (geographic information system) used by scholars across the globe to assess exposure and vulnerability of critical infrastructure?
- What are common conceptual and methodological issues encountered?
- What are some important considerations to ensure that mitigation and adaptation strategies lead to reduction in vulnerability and risk of critical infrastructure to climate-related hazards?

Critical Infrastructure: Types and Interdependencies

> *The interrelated nature of these systems is particularly significant when we consider the problem of infrastructure protection.*
>
> —Austin, DiSera, and Brooks (2015), p. 5

Critical infrastructure is a term that refers to the systems that are necessary for the day-to-day functioning of society (Oh, Deshmukh & Hastak, 2010; Bach, Gupta,

Nair & Birkmann, 2013). This infrastructure provides "lifeline services" that are vital to the successful operation of civil society (Austin, DiSera & Brooks, 2015) and to national security (Bigger, Willingham, Krimgold & Mili, 2009; Deshmukh, Ho Oh & Hastak, 2011). Critical infrastructure can be subdivided into two main types: physical and socioeconomic. Physical infrastructure systems include electricity generation and transmission, oil and gas supplies, landlines, cell phone and radio transmission, telecommunications, water/sewer, wastewater treatment, and transportation (air, roads, bridges, rail, water). These are themselves broad categories—comprising a hierarchy of systems, individual infrastructures, individual systems, and technical components (Cavelty, 2005). Socioeconomic systems are composed of hospitals, schools, public administration, fire, police, and emergency management (Bach, Gupta, Nair & Birkmann, 2013). Other authors have made a case for expanding the list to include the chemical sector, the defense industrial base, banking and finance, agriculture, food, and postal and shipping (Auerswald, Branscomb, La Porte & Michel-Kerjan, 2006), as well as key industry sites, national monuments, and large gathering venues (Parfomak, 2005).

Intra- and interdependencies between critical infrastructure are gaining attention from practitioners and researchers. For example, electrical infrastructure can affect water and sewer systems, transportation, banking, and communication. But the delivery of electrical energy cannot occur without the transportation of fuel, management of water resources, and economic maintenance of the transmission grid and energy-generation operations (Little, 2003; Bigger, Willingham, Krimgold & Mili, 2009; Ouyang, 2014). Bigger, Willingham, Krimgold, and Mili (2009, p. 208) provide several examples of infrastructure dependencies and interdependencies including those between energy and transportation given that loss of electricity can result in loss of traffic signals and can have adverse impacts on traffic flow and overall community safety. Closure of streets, highways, and bridges can also impede emergency vehicle and service access. The ripple effects throughout the national economy after the 2005 failure of the levee system caused by Hurricane Katrina highlight the cascading impacts of infrastructure disruptions on large geographic regions (Austin, DiSera & Brooks, 2015). Cascading impacts were even more evident after the Great East Japan Earthquake (GEJE) when more than 4 million people were affected by power outages, lack of potable water, and damage to roads, bridges, dikes, and railways (Austin, DiSera & Brooks, 2015).

Mileti (1999) in his book *Disasters by Design* characterized critical infrastructure as "capital stock" and as expanding and increasingly exposed to hazards in urbanized areas. Austin, DiSera, and Brooks (2015, p. 192) warn that "cascading infrastructure failures are not always the result of connectivity or intra- or inter-system relationships . . . the spatial arrangement of seemingly disparate systems may often have unintended consequences." Critical infrastructure tends to be geographically concentrated. A single hazard can, therefore, affect many otherwise independent systems. For example, the **Texas** and **Louisiana** coastal areas are home to 43% of U.S. national oil refining capacity (Parfomak, 2005). In 2005, Hurricane Katrina affected national refining capacity due to damage to pipelines and oil rig platforms (Herman, 2006). This co-location of infrastructure raises the question of geographic interdependencies and geospatial interdependencies. The locational/geographic attributes of critical infrastructure and their networks make GIS technology useful for assessing vulnerability and risks at multiple scales and for informing adaptation planning priorities.

*Definition: geographic interdependencies (*Austin, DiSera & Brooks, 2015*, p. 26)*

- Geographic interdependencies arise when infrastructure components (e.g., electric transmission lines, water pipelines, gas pipelines, and telecommunications cables) share common corridors, such as public rights-of-way and railway lines. This proximity increases the vulnerabilities to, and consequences from, disasters in the same geographic area.

Definition: geospatial interdependencies (Austin, DiSera & Brooks, 2015*, p. 27)*

- Geospatial interdependencies involve the physical proximity of one infrastructure to another. An event such as an explosion of a gas main in an urban area could create correlated disruptions with other infrastructures, such as water and electric services to a community.

Climate Hazards and Their Impacts on Infrastructure

> *Historical events that have compromised a city's infrastructure under conditions similar to those associated with climate change also provide information about what a city might expect in the way of consequences from a future of increased temperatures, precipitation, and sea level rise.*
>
> —Zimmerman and Farris (2010), p. 63

Overall, climate-related hazards can affect infrastructure through several mechanisms. The wide-ranging negative effects of changing climate on energy infrastructure are noteworthy. One study of Russian infrastructure from 1992 to 2010 reported that 70% of all electrical blackouts and 20% of water/sewer and heat problems were caused by the adverse impact of natural hazards on infrastructure (Petrova, 2011). When analyzed by types of natural hazards, the study showed that 37% of the damage was caused by windstorms and hurricanes, 20% by snowfall, 16% by rainfall, and 12% by cold and ice (Petrova, 2011).

Floods can cause severe damage and lead to infrastructure failures, such as breached and broken levees, road closures and failed bridges, and closure of power and wastewater facilities. Hurricane Katrina and the flooding in **New Orleans** resulted in the failure of levees and canals, which in turn caused other infrastructure to fail (Deshmukh, Ho Oh & Hastak, 2011). In **Louisiana**, **Mississippi**, and **Alabama**, flooding from Hurricane Katrina left 1,220 water systems inoperable because electrical power was shut down and water treatment plants were inoperable (Bach, Gupta, Nair & Birkmann, 2013, p. 14). Further, drinking water that was able to be pumped was contaminated by flooding. Flooding can also increase silt levels in the water, which in turn can cause hydroelectric plants to shut down (Bach, Gupta, Nair & Birkmann, 2013). For example, flash floods in July 2012 in the Himalayan region triggered the shutdown of several hydroelectric plants, causing a shortage of 18–20 million megawatts per day (Bach, Gupta, Nair & Birkmann, 2013).

Permafrost covers 20% of the planet, and thawing produces flooding, landslides, and erosion. Pipelines, power lines, railroads, and other linear installations are particularly subject to failure in such conditions. In **Tibet**, the railroad is built upon permafrost that is subject to thawing, and some sections have already sunk. In **Alaska**, thawing ice roads can slow construction by restricting the movement of heavy equipment for longer portions of the year.

Given the density of urban coastal areas, any impact on power or water infrastructure can affect a large population and various economic sectors. Zimmerman and Farris (2010) provide a thorough review of **New York City's** dense infrastructure networks (i.e., energy, transportation, water supply, wastewater, and communications) and the impacts of climate hazards (i.e., temperature increases, precipitation, and sea level rise [SLR]). For example, more frequent precipitation events and higher storm surge from SLR can lead to declines in levels of service (LOS) due to flooded transportation and transit infrastructure.

A study to assess the damage from three hurricanes that crossed **Florida** in 2004 identified interdependencies in the overall system. Railroads in central Florida were inoperable due to electrical signal failures, and offshore shipments were prevented from docking (due to an impending storm in **Tampa** and silting in **Port Canaveral**), thereby preventing any fuel from being delivered to the region (Bigger, Willingham, Krimgold & Mili, 2009). Gas power plants in southern Florida were unable to receive shipments because gas production facilities were shut down in anticipation of another hurricane. Given that on-site oil storage was limited, the gas power plants could not run at full capacity (Bigger, Willingham, Krimgold & Mili, 2009). A higher incidence of storms can also affect offshore drilling, delivery, and coastal refineries and other installations (Paskal, 2010). In **Leningradskaya Oblast, Russia**, in July of 2010, a hurricane disconnected five high-voltage transmission lines, causing power outages to 250,000 households (Petrova, 2011).

Nuclear and fossil fuel plants utilize large volumes of water in energy production. Therefore, drought and heat waves heavily impact energy infrastructure. Additionally, heat waves increase the demand for electricity to cool buildings and facilities. The production of electricity can, however, be affected by drought given the reduction in the availability of cooling water and the reduction in fuel delivery capacity when river levels are low. During heat waves in France in 2003, 17 nuclear plants had to be shut down due to lack of water, while the extreme heat increased demand for power for cooling. The heat wave impacted electricity and rail and road infrastructure networks. The 2003 European heat wave caused rivers to run low and impacted hydroelectric, coal, and nuclear electricity production.

Further, wastewater temperature regulations are often relaxed during heat waves, leading to potential environmental damage. Water is used for cooling nuclear and coal-fired power plants, but the discharged water has a warming effect on the ambient water temperature and becomes a threat to river ecosystems. A reduction in the flow of water and high ambient air temperatures forced a reduction in the amount of legal discharge ranging from 20% to 100% and led to a reduction in electricity production. Furthermore, the low water level of rivers caused ships to have to reduce their loads to 20% to 30% of ship capacity, leading to a reduction in the supply of coal (Bach, Gupta, Nair & Birkmann, 2013).

Wildfires can pose a danger to the transmission grid, and the heat from these fires makes transmission less efficient and more prone to failure. In January 2008, **Melbourne, Australia**, experienced rolling blackouts (McEvoy, Ahmed & Mullett, 2010). Demand for electricity reached 7% over the previous peak demand in Victoria, and supply was hampered by heat-induced capacity reductions. Rail transport was impacted; metal expansion caused the rails to buckle in hot conditions, as well as the failure of train car air conditioners. On the third and hottest day of the heat wave, an explosion at a transformer station cut off power to signals and electrical overhead lines. According to McEvoy, Ahmed, and Mullett (2010), 750 of the 2,400 rail services were canceled over the heat wave period.

The electric transmission grid is also at risk during heavy snowfall events. The 2005 heavy snowfall in **Munsterland, Germany**, affected electric service to 250,000 households for up to 5 days. As reported by Bach, Gupta, Nair, and Birkmann (2013), generators had to be airlifted to retirement homes and hospitals. In **Canada** and the northeastern **United States**, a major ice storm knocked out power in 1998, resulting in rolling blackouts and communication outages (Parfomak, 2005). In cold climates, melting permafrost is problematic given the instability of pipeline and energy transmission and road, port, and airport infrastructure. For example, water infrastructure can be vulnerable due to bursting of pipelines (Laucelli, Rajani, Kleiner & Giustolisi, 2014).

Assessing Exposure, Vulnerability, and Risk of Critical Infrastructure to Climate Hazards

In the **United States**, the vast majority of the communities that are highly exposed to wildfires are located in **Southern California** and the **western region of the United States.** Population increases in these communities have placed a strain on the road network and have increased the minimum evacuation times. Using a GIS-based methodology, Cova, Theobald, Norman, and Siebeneck (2013) identified and ranked communities for the most impaired road infrastructure networks for wildfire evacuations in the west. LANDFIRE dataset, a 30-meter resolution fire hazard layer, was combined with road network and housing density data layers. A model was created to calculate the number of households utilizing the local road infrastructure to evacuate. The calculations were based on the number of egress points from a community (Cova, Theobald, Norman & Siebeneck, 2013).

Permafrost makes up about 24% of the land mass in the northern hemisphere, which creates a problem since the melting of permafrost reduces the load-carrying capacity of the ground. GIS can be used to map areas subject to thaw-induced instability. In particular, the **Siberian region of Russia** contains energy infrastructure, such as pipelines and transmission lines, as well as roads and railroads. Streletskiy, Shiklomanov, and Nelson (2012) estimated the near-surface effect on permafrost parameters, utilizing GIS to map ground temperature and air temperature over time and to calculate the effect on permafrost under current and predicted scenarios in **Russia** and **Alaska**. The analysis took into account climate, snow cover, vegetation, and soil properties. The largest decrease, 40%, of weight bearing capacity was found in **Nadym, Russia** (Streletskiy, Shiklomanov & Nelson, 2012).

Crete, the largest Greek island and the fifth largest island in the Mediterranean Sea, is an important agricultural area and producer of fruit crops. The island has a mix of topographic features, including river basins and mountainous areas, as well as vegetation, rainfall, and geology. The island is also prone to flooding. A study by Kourgialas and Karatzas (2016) used ArcGIS to produce a flood map for Crete based on a weighted average of six thematic maps: flow accumulation, rainfall intensity, elevation, geology, land use, and slope. Flood risk was also examined for two climate change/future climate condition scenarios to better understand which fruit trees are most tolerant to flood conditions and to inform flood risk mitigation planning (Kourgialas & Karatzas, 2016).

Jalayer et al. (2014) utilized a topographic wetness index (a method for identifying flood-prone areas by identifying a location's capability to accumulate water), urban morphology, and population layers in a GIS framework. Flood risk "hotspots" were identified in **Addis Ababa, Ethiopia**. The risk to urban structures and major roads was taken into account, and Bayesian analysis was used to characterize uncertainties in potentially flood-prone areas. According to Jalayer et al. (2014), half of the residential buildings in flood-prone areas are made of mud and wood, and 3.4% to 24.4% of major urban roads are expected to be impacted by flooding.

As noted in an earlier chapter, flooding (both permanent and episodic) can exacerbate erosion and land loss, degrade ecosystems and habitats, impede land drainage, and accelerate saline intrusion into rivers, estuaries, and coastal aquifers. The study by Tate and Frazier (2013) assessed the susceptibility of wells to inundation from each level of hurricane strength for **Sarasota County, Florida**. Using GIS, the study combined LiDAR (Light Detection and Ranging) data and SLOSH (Sea, Lake, and Overland Surge from Hurricanes) model output, in addition to wellhead and substation point data and elevation data. Findings show no damage from a Category 1 hurricane, but 50% or more of the infrastructure is expected to be flooded in a Category 5 hurricane (Tate & Frazier, 2013).

Molarius, Könönen, Leviäkangas, Rönty, Hietajärvi, and Oiva (2014) operationalized an extreme weather risk indicator (EWRI) for the transport system in **Europe** utilizing hazard and vulnerability factors. Vulnerability was defined as a function of risk exposure, susceptibility, and coping capacity. This study identified six weather phenomena (i.e., wind, snowfall, blizzards, heavy precipitation, cold spells, and heat waves) as having a negative effect on transportation infrastructure and transit modes: road, railway, aviation, short sea shipping, and inland water transport (Molarius, Könönen, Leviäkangas, Rönty, Hietajärvi & Oiva, 2014). The severity, extent, and type of extreme weather events are tied to the local climate—for example, mountainous and northern European regions are more susceptible to cold weather events, while southern European zones have a higher frequency of heat-related events.

The study by Molarius, Könönen, Leviäkangas, Rönty, Hietajärvi, and Oiva (2014) divided Europe into regions: Northern, Temperate Central European, Temperate Eastern European, Oceanic, Mediterranean, and Mountainous. Each country was placed into a region for climatic effects but had its own vulnerability score. Overall, road and rail transport were found to be the most vulnerable transport modes, as they are susceptible in some way to all hazards.

Sea level rise studies employ GIS techniques to document vulnerable transportation systems and critical infrastructure. A study by Heberger, Cooley, Herrera, Gleick, and Moore (2009) focused on the adverse economic impacts of sea level rise on the **California** coast and on key critical infrastructure, including the **San Francisco** and **Oakland (U.S.)** airports. The **Philadelphia County (U.S.)** transit system provides rail and bus passenger services and lies along the Delaware River. Oswald and Treat (2013) reported on a Transit Inundation Modeling Method (TIMM) that was developed to identify transit networks that are at risk to sea level rise inundation levels of 1 to 5 meters. Their study showed that southern Philadelphia is most at risk to SLR inundation.

Scarborough, Maine (U.S.), was used as a case study in an assessment of rising sea levels on coastal infrastructure. Three conditions were tested to create a vulnerability score: Highest Astronomical Tide (HAT), HAT plus 2 feet, and highest historical level plus 2 feet. Four flood impacts were considered (economic impacts, social impacts, health and safety impacts, and environmental impacts) and each rated on a four-level scale from low impact to severe impact to create a consequence score (Johnston, Slovinsky & Yates, 2014). GIS was used to locate and display areas of infrastructure vulnerability based on elevation and past flooding. The overall vulnerability impact was calculated from a vulnerability score and a consequence score. The overall vulnerability score was used to identify systemic weaknesses for planning purposes. That study found that high-traffic roads and utility corridors were the most vulnerable infrastructure (Johnston, Slovinsky & Yates, 2014). More details and examples of SLR assessments are covered in Chapter 9.

Case Study: Impact of SLR on Land Use and Public Infrastructure in Broward County, Florida

Source: Mitsova, D., Li, Y., & Vos, J. 2011. *Planning for Sea Level Rise in the Face of Uncertainty: Implications for Mitigating the Increased Risk of Flooding in Coastal Communities*, Working Paper, School of Urban and Regional Planning, Florida Atlantic University, Fort Lauderdale, FL.

Study Objective

In vulnerable low-lying areas such as south **Florida**, public infrastructure and utilities that lie at elevations of 1 to 3 feet above present sea level are at risk of flooding. This case study seeks to understand the extent to which changes in the mean sea level expose both inland and coastal infrastructure and utilities to more frequent flooding. Incorporating sea level rise projections in future land use maps and infrastructure projects has been recognized as one of the most cost-effective adaptive responses to the impacts of climate change.

Planning and Policy Applications

- Identify land use and public infrastructure vulnerable to potential inundation from sea level rise.
- Inform planning strategies to mitigate the impacts of sea level rise.
- Provide land use planners and local decision-makers with data and maps to revisit and adjust land use designations in order to minimize the costs of potential future relocations.

Data and Methods

High-resolution (5-foot) digital elevation data generated from the airborne LiDAR system was used to develop sea level rise scenarios for **Broward County, Florida**. The data were obtained from the International Hurricane Research Center at the Laboratory for Coastal Research at the Florida International University. Airborne LiDAR is a remote sensing technology based on 1.1-meter infrared laser pulses that go back and forth below the flight path scanning the ground (Flood, 2001). Figure 10.1 indicates the extent of the study area.

Data sources included:

- Sea level trends, maximum tidal range, maximum significant wave tide, and mean long-term erosion trend from the *Coastal Hazards Data Base for the U.S. East Coast* (Gornitz, White & Daniels, 1992)
- Output from the SLOSH model of the National Hurricane Center to map storm surge risk zones for Broward County from the Florida Division of Emergency Management (FDEM, 2017).
- Data on insurance claims from the hurricane summary datasheets of the Florida Office of Insurance Regulation (FOIR, 2006).
- 2004 land use dataset from the South Florida Water Management District in addition to the future land use dataset developed by the Broward County Planning Council.

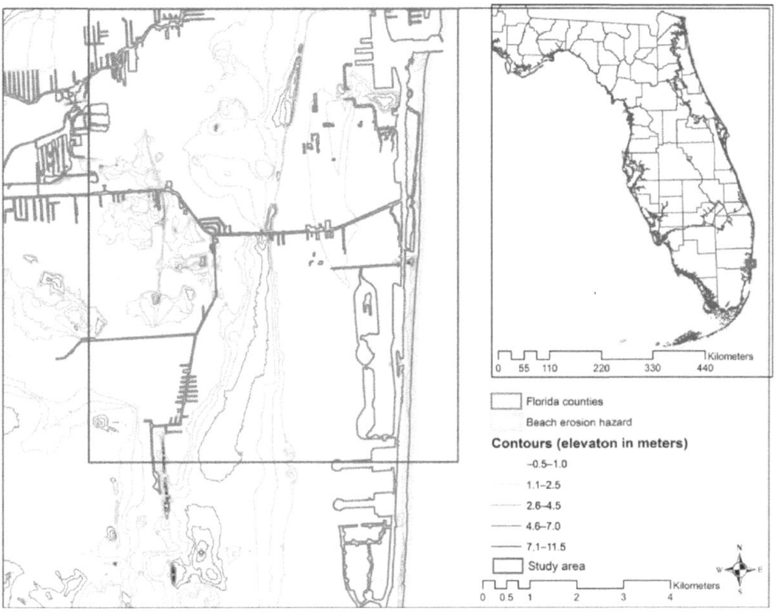

Figure 10.1 Study area

- Dataset with the 2007 parcel tax appraisals from the Broward County Property Appraiser (BCPA).
- Additional datasets including cost and extent of beach nourishment projects, number of permits for seawalls and other hard structures for shoreline protection, economic impact of beach-related tourism, and impact of coastal areas on property values and tax base obtained from various reports, studies, and personal communication

Summary of Analysis

Areas vulnerable to potential flood inundation from sea level rise were delineated from LiDAR data by using a series of boolean algebraic operations. The LiDAR dataset used in this study has a root mean square error (RMSE)[1] of 0.41 feet (12 cm) consistent with a vertical accuracy of 0.8 feet (24 cm) at the 95% confidence level. The vertical accuracy information is used to derive minimum increments of sea level rise. The linear error (LE) at the 95% confidence level shown in Figure 10.2 is an estimate of the average digital elevation model (DEM) accuracy. It accounts for random error when point measurements are linearly interpolated using the Triangulated Irregular Networks (TIN) method (Zhu et al., 2005).

Inundation scenarios of 1.5 and 3 feet were selected as they reflect both the midrange and the high end of published estimates of potential sea level rise and are consistent with the vertical accuracy of the data. For each increment of sea level rise, additional areas were delineated to visualize the spatial representation of uncertainty inherent to the elevation data (Figure 10.2). The impact assessment of sea level rise was conducted in two steps: (1) selecting appropriate sea level rise scenarios; and (2) overlaying potential inundation scenarios under various increments of sea level rise (considering also the output of the SLOSH model of the National Hurricane Center) and assessing the impact on land use and public infrastructure. Table 10.1 provides a summary of the potential SLR impacts on infrastructure assets.

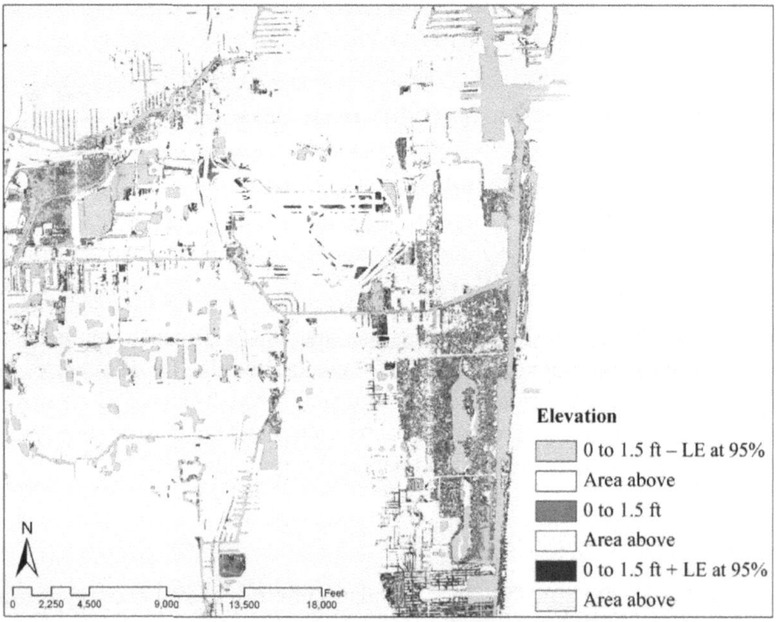

Figure 10.2 Delineation of areas at risk of inundation from sea level rise (portion of the coastline of Broward County, Florida, where LE denotes linear error as an estimate of the average DEM accuracy): left image: areas vulnerable to 1.5 feet of sea level rise developed from LiDAR-derived DEM with spatial representation of uncertainty; right image: areas vulnerable to 3 feet of sea level rise developed from LiDAR-derived DEM with spatial representation of uncertainty

Source: Created by authors

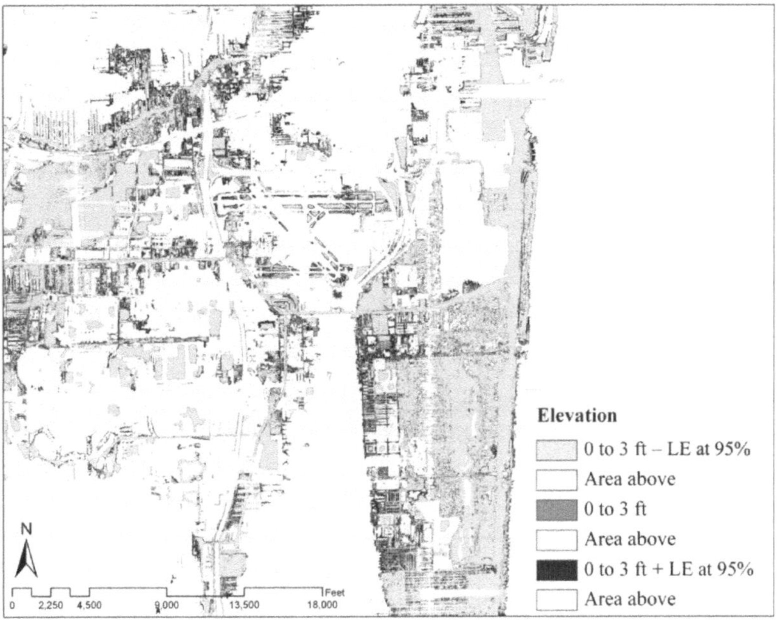

Figure 10.2 (Continued)

Key Findings

The results indicate that a number of utilities and commercial and residential areas lie at elevations of 1 to 3 feet above present sea level. Due to shallow water table and direct interactions between rising sea level and surface and subsurface flows, even slight changes in the mean sea level would expose both inland and coastal land uses to more frequent flooding. Depending on the timing and rate of sea level rise, between 4.8% and 11.4% of the study area is at risk of inundation, which may result in 15 square miles (under the 1.5 feet SLR scenario) to 36 (under the 3 feet SLR scenario) square miles of land loss. Under the 3 feet of sea level rise scenario, 47.5% of the electrical generation facilities, and 10.7% of the transportation right-of-ways/nodes, would be at risk of inundation (see Table 10.1). These estimates do not include changes in water table elevations and do not account for the presence of hydrological connectivity between inland water bodies and coastal waters.

The areas at risk of inundation are not uniformly distributed throughout the study area. The southeast and southwest portions of the county are significantly more at risk of increased flooding due to a lower elevation. The analysis indicates that critical facilities such as schools, hospitals, libraries, and assisted living facilities are at relatively low risk with regard to the impacts of sea level rise. Other important public facilities such as sewage

Land use categories	Total area within the urban boundary (square miles)	Area of less than 1.5 feet elevation (square miles)	Percent of total area (%)	Area of less than 3 feet elevation (square miles)	Percent of total area (%)
Water	31.78	4.76	14.98	5.79	18.23
Conservation	4.93	1.14	23.17	1.74	35.27
Recreation	25.60	1.32	5.16	2.77	10.81
Low-density residential	131.66	3.51	2.67	10.64	8.08
Medium-density residential	40.97	1.10	2.68	2.86	6.99
High-density residential	56.67	0.79	1.39	4.54	8.01
Commercial/ office	30.37	0.26	0.85	1.45	4.78
Employment other	12.62	0.17	1.37	0.62	4.90
Transportation	19.47	0.64	3.29	1.75	9.00
Right-of-way/ nodes	**50.70**	**0.29**	**0.57**	**5.78**	**10.7**
Utilities	2.76	0.07	2.57	0.16	5.75
Electrical generation facility	**0.88**	**0.40**	**45.41**	**0.42**	**47.48**
Industrial	22.07	0.62	2.80	1.34	6.08
Total area at risk of inundation		**15.07**		**35.86**	
Total area above specific elevation		299.36		278.55	
Total area	430.52	314.43		314.42	

Table 10.1 Vulnerability of land use designations to 1.5 and 3 feet of sea level rise

treatment plants, water supply plants, and solid waste disposal sites are significantly more at risk, although not listed in the table. We can assume with 95% confidence that between 6 and 11 out of 21 sewage treatment sites would experience an increased flood hazard risk under the 1.5 feet SLR scenario. This figure may increase to 14 facilities (approximately 66.7%) if the sea level rises by almost 3 feet by the end of the century. Between 23.5% and 41% of the water supply plants (at the 95 % confidence level) lie at an elevation that indicates an increased risk of flooding under the 1.5 feet sea level rise scenario, while under the 3 feet scenario this proportion would increase to almost 80%. At least two out of five landfills may be at risk of flooding under the 1.5 feet scenario. We can assume with 95% confidence that a sea level rise of 3 feet would place an additional landfill at increased flood risk. A landfill inundation implies significant environmental and health hazards and requires significant resources for safeguarding, relocating, and mitigating such sites.

Mitigation and Adaptation Strategies Toward Resilience

Creating an overall climate change adaptation strategy for urban infrastructure poses considerable conceptual and operational challenges.

—Zimmerman and Farris (2010), p. 63

Climate hazards affect critical infrastructure and have adverse societal impacts and repercussions. Holistic systems thinking and collaborative approaches are fundamental aspects of mitigation and adaptation strategies to combat such adverse impacts. Resilience thinking necessitates collaboration and partnerships between public, private, and non-profit sectors. It is also important that public policy-makers formulate appropriate risk reduction measures that can benefit all sectors.

Cavelty (2005) notes that vulnerability assessment and other widely used risk analysis methodologies are not adequate for addressing issues of interdependencies and cascading effects of potential infrastructure failure from climate disasters. Chapter 13 covers this topic in more detail and presents more advanced tools and methods to address such limitations. Chapter 13 also includes a comprehensive discussion of other data and methodological issues. Part of the holistic systems thinking approach must account for the costs involved in implementing mitigation and adaptation projects. The cost of adaptation (e.g., retrofit and relocation) and the public-private partnerships required to move toward community resilience are reflected in climate action response plans for cities and communities. It is important, however, that the private sector and business community engage in similar efforts to assess the potential impact of climate change and climate hazards on their capital investments and modification plans (Zimmerman & Farris, 2010).

One main challenge in countries like the United States is that critical infrastructure (such as electricity production facilities) is predominantly owned by the private sector and user cooperatives (Zimmerman & Farris, 2010; Auerswald, Branscomb, La Porte & Michel-Kerjan, 2015; Austin, DiSera & Brooks, 2015). This creates problems for practitioners who perform vulnerability assessments that revolve around access to proprietary data. Furthermore, political-administrative boundaries (used by emergency managers and government agencies) and hazard boundaries (e.g., flood zones) do not coincide with utility operating boundaries. Electrical utilities are one example of this dilemma. Even when the data is eventually available, Austin, DiSera, and Brooks (2015) warn about the difficulty of aggregating large and disparate electricity utility data. For example, Hurricane Katrina impacted a large area that encompassed 33 electricity service providers.

There is also need to address the risk posed by natural hazard–triggered technological accidents when critical infrastructure such as power plants are unprotected from natural hazards (Krausmann & Cruz, 2013; Lipscy et al., 2013). Using the case of the Great East **Japan** Earthquake and tsunami, which damaged refineries, petrochemical facilities, and other types of chemical industry, Krausmann and Cruz (2013) have noted that at the most fundamental level, limiting of industrial development in natural hazard flood-prone areas must be carried out in addition to ongoing monitoring of building codes and safety regulations. Overall, mitigation and adaptation will have to account for such events given hazardous release and long-term environmental pollution and health impacts to impacted communities and first responders, as well as economic impacts. The importance of the sociopolitical context—laws, regulations, policies, and other economic, social, and national security considerations for the infrastructure environment—must be addressed as well (Cavelty, 2005, p. 266). Overall, types of impacts, their cascading effects, and their sociopolitical contexts, require continued examination and analysis of critical infrastructure interdependencies.

Note

1 Maps of horizontal inundation due to sea-level rise require consistency between the vertical accuracy of the elevation data and the increment of rise. The root mean square error is a measure of spatial data accuracy established by the Federal Geographic Data Committee to improve the testing and reporting requirements for topographic data (Climate Change Science Program [CCSP], 2009).

References

Auerswald, P.E., Branscomb, L.M., La Porte, T.M., Michel-Kerjan, E.O. 2006. *Seeds of Disaster, Roots of Response: How Private Action Can Reduce Public Vulnerability*, Cambridge, UK: Cambridge University Press.

Austin, R.F., DiSera, D.P., & Brooks, T.J. 2015. *GIS for Critical Infrastructure Protection*. Boca Raton, Florida: CRC Press.

Bach, C., Gupta, A.K., Nair, S.S., & Birkmann, J. 2013. Critical Infrastructures and Disaster Risk Reduction. *National Institute of Disaster Management and Deutsche Gesellschaft für internationale Zusammenarbeit GmbH (GIZ)*, New Delhi, 72p.

Bigger, J.E., Willingham, M.G., Krimgold, F., & Mili, L. 2009. Consequences of Critical Infrastructure Interdependencies: Lessons from the 2004 Hurricane

Season in Florida. *International Journal of Critical Infrastructures* 5(3), 199–219.

Cavelty, M.D. 2005. The Socio-Political Dimensions of Critical Information Infrastructure Protection (CIIP). *International Journal for Critical Infrastructure Protection* 2/3(1), 258–268.

Climate Change Science Program (CCSP). 2009. *Coastal Sensitivity to Sea-Level Rise: A Focus on the Mid-Atlantic Region*. Final Report: Synthesis and Assessment Product 4.1, Washington, DC. Retrieved from https://downloads.globalchange.gov/sap/sap4-1/sap4-1-final-report-all.pdf.

Cova, T.J., Theobald, D.M., Norman III, J.B., & Siebeneck, L.K. 2013. Mapping Wildfire Evacuation Vulnerability in the Western US: The Limits of Infrastructure. *GeoJournal* 78(2), 273–285.

Deshmukh, A., Ho Oh, E., & Hastak, M. 2011. Impact of Flood Damaged Critical Infrastructure on Communities and Industries. *Built Environment Project and Asset Management* 1(2), 156–175.

Flood, M. 2001. Laser Altimetry: From Science to Commercial LIDAR Mapping. *Photogrammetric Engineering & Remote Sensing* 67(11), 1209–1217.

Florida Division of Emergency Management (FDEM). 2017. Storm surge zones (last updated 8/14/2017). URL: http://geodata.floridadisaster.org/datasets/storm-surge-zones.

Florida Office of Insurance Regulation (FOIR). 2006. *Hurricane Data Summary*. Florida Insurance Study. Retrieved from www.floir.com/siteDocuments/HurricaneSummary20042005.pdf (accessed 07/2017).

Gornitz, V.M., White, T.W., & Daniels, R.C. 1992. *A Coastal Hazards Database for the U.S. East Coast*. Environmental Sciences Division Publication No. 3913. Prepared for the Global Change Research Program Environmental Sciences Division, Oak Ridge, Tennessee.

Heberger, M., Cooley, H., Herrera, P., Gleick, P.H., & Moore, E. 2009. *The Impacts of Sea-level Rise on the California Coast*. No. CEC-500–2009–024-F. Oakland, CA: Pacific Institute.

Herman, C. 2006. Katrina's Economic Impact: One Year Later. *ABC News*. Retrieved from http://abcnews.go.com/Business/HurricaneKatrina/story?id=2348619&page=1 (accessed 01/2017).

Jalayer, F., De Risi, R., De Paola, F., Giugni, M., Manfredi, G., Gasparini, P., Topa, M. E., Yonas, N., Yeshitela, K., Nebebe, A., Cavan, G., Lindley, S., Printz, A., & Renner, F. 2014. Probabilistic GIS-Based Method for Delineation of Urban Flooding Risk Hotspots. *Natural Hazards* 73(2), 975–1001.

Johnston, A., Slovinsky, P., & Yates, K.L. 2014. Assessing the Vulnerability of Coastal Infrastructure to Sea Level Rise Using Multi-Criteria Analysis in Scarborough, Maine (USA). *Ocean & Coastal Management* 95, 176–188.

Kourgialas, N.N., & Karatzas, G.P. 2016. A Flood Risk Decision Making Approach for Mediterranean Tree Crops Using GIS; Climate Change Effects and Flood-Tolerant Species. *Environmental Science & Policy* 63, 132–142.

Krausmann, E., & Cruz, A.M. 2013. Impact of the 11 March 2011, Great East Japan Earthquake and Tsunami on the Chemical Industry. *Natural Hazards* 67, 811–828.

Laucelli, D., Rajani, B., Kleiner, Y., & Giustolisi, O. 2014. Study on Relationships Between Climate-Related Covariates and Pipe Bursts Using Evolutionary-Based Modelling. *Journal of Hydroinformatics* 16(4), 743–757.

Lipscy, P.Y., Kushida, K.E., & Incerti, T. 2013. The Fukushima Disaster and Japan's Nuclear Plant Vulnerability in Comparative Perspective. *Environmental Science & Technology* 47(12), 6082–6088.

Little, R.G. 2003. Toward More Robust Infrastructure: Observations on Improving the Resilience and Reliability of Critical Systems. In: *System Sciences, 2003. Proceedings of the 36th Annual Hawaii International Conference* IEEE, 9.

McEvoy, D., Ahmed, I., & Mullett, J. 2010. The Impact of the 2009 Heat Wave on Melbourne's Critical Infrastructure. *Local Environment* 17(8), 783–796.

Mileti, D. 1999. *Disasters by Design: A Reassessment of Natural Hazards in the United States*. Washington, DC: The National Academies Press.

Mitsova, D., Li, Y., & Vos, J. 2011. Planning for Sea Level Rise in the Face of Uncertainty: Implications for Mitigating the Increased Risk of Flooding in Coastal Communities, Working Paper. Fort Lauderdale, FL: School of Urban and Regional Planning, Florida Atlantic University.

Molarius, R., Könönen, V., Leviäkangas, P., Rönty, J., Hietajärvi, A.M., & Oiva, K. 2014. The Extreme Weather Risk Indicators (EWRI) for the European Transport System. *Natural Hazards* 72(1), 189–210.

Oh, E.H., Deshmukh, A., & Hastak, M. (2010). Disaster Impact Analysis Based on Inter-Relationship of Critical Infrastructure and Associated Industries: A Winter Flood Disaster Event. *International Journal of Disaster Resilience in the Built Environment* 1(1), 25–49.

Oswald, M.R., & Treat, C. 2013. Identifying Sea Level Rise Vulnerability Using GIS: Development of a Transit Inundation Modeling Method. *International Journal of Geoinformatics* 9(1), 1–10.

Ouyang, M. 2014. Review on Modeling and Simulation of Interdependent Critical Infrastructure Systems. *Reliability Engineering & System Safety* 121, 43–60.

Parfomak, P.W. 2005. *Vulnerability of Concentrated Critical Infrastructure: Background and Policy Options*. Washington, D.C.: Congressional Research Service.

Paskal, C. 2010. The Vulnerability of Energy Infrastructure to Environmental Change. *China and Eurasia Forum Quarterly* 8(2), 149–163.

Petrova, E. 2011. Critical Infrastructure in Russia: Geographical Analysis of Accidents Triggered by Natural Hazards. *Environmental Engineering and Management* 10(1), 53–58.Streletskiy, D.A., Shiklomanov, N.I., & Nelson, F.E. 2012. Permafrost, Infrastructure, and Climate Change: A GIS-Based Landscape Approach to Geotechnical Modeling. *Arctic, Antarctic, and Alpine Research* 44(3), 368–380.

Tate, C.A., & Frazier, T.G. 2013. A GIS Methodology to Assess Exposure of Coastal Infrastructure to Storm Surge & Sea-level Rise: A Case Study of Sarasota County, Florida. *Journal of Geography & Natural Disasters* S1, 001.

Zhu, C., Shi, W., Li, O., Wang, G., Cheung, T.C.K., & Dai, E. 2005. Estimation of Average DEM Accuracy under Linear Interpolation Considering Random Error at the Nodes of TIN Model. *International Journal of Remote Sensing* 26, 5509–5523.

Zimmerman, R., & Farris, C. 2010. Infrastructure Impacts and Adaptation Challenges. *Annals of the New York Academy of Sciences* 1196, 63–85.

11 Assessing the Impact of Rising Temperatures and Urban Heat Islands on People and Places

Chapter Objectives

This chapter seeks to answer the following questions:

- What are urban heat islands and underlying factors?
- Why should we be concerned about urban heat islands?
- What are the range of geographic information system (GIS)-based studies used by scholars across the globe to assess the impacts of urban heat islands on people and places?
- What are some commonly encountered data collection, monitoring, and methodological issues when undertaking such analyses?
- What are some strategies to mitigate the impacts of urban heat islands?

Rising Temperatures and Urban Heat Islands: Contributing Factors

On a global scale, annual average temperatures have been rising since the mid-1970s with an increase of 0.2°C to 0.3°C. The National Oceanic and Atmospheric Administration (NOAA)'s Land & Ocean Temperature Percentiles maps (Figure 11.1) indicate record-high temperature anomalies over many parts of the world,

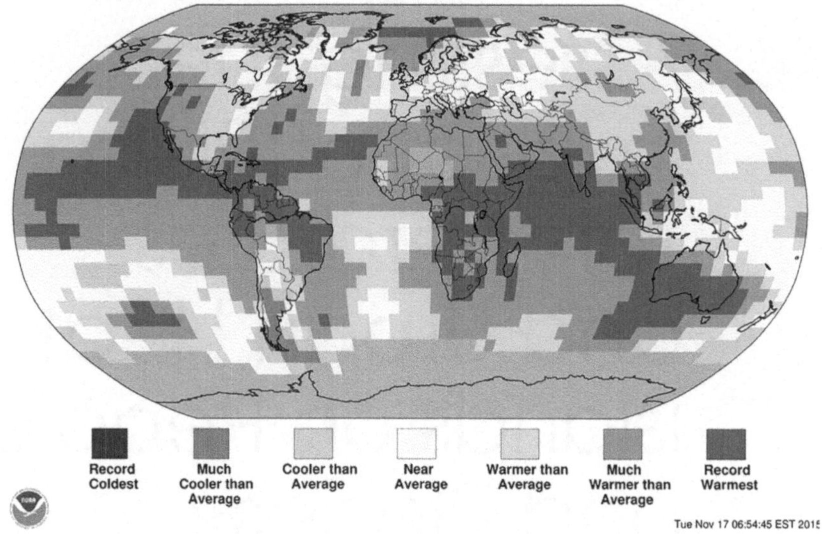

Figure 11.1 Land and ocean temperature percentiles (October 2015)
Source: NOAA-NCEI, www.ncdc.noaa.gov/sotc/global/201510

including Central America, Brazil and the northern part of Latin America, large portions of Africa and Australia, and east and Southeast Asia (National Oceanic and Atmospheric Administration–National Centers for Environmental Information [NOAA-NCEI], 2016). Record temperatures were also observed in the Scandinavian Peninsula, and much warmer-than-average temperatures were recorded in southern and western Australia.

The Intergovernmental Panel on Climate Change (IPCC) predicts, with a high level of certainty, that these trends will continue throughout the 21st century. The IPCC report on managing the risks of extreme events (IPCC, 2012, p. 11) states that "it is *virtually certain* that increases in the frequency and magnitude of warm daily temperature extremes and decreases in cold extremes will occur . . . on a global scale." Past studies have suggested that higher average global temperatures will result in increased evaporation and more intense precipitation events (Trenberth, 2011).

The phenomenon known as urban heat islands (UHIs) describes cities and urban areas with temperatures that are significantly higher than the surrounding rural areas (Taha, 1997; Bornstein et al., 2006; Gartland, 2008; Stone, Vargo, Liu, Hu & Russell, 2013; Kenward, Yawitz, Sanford & Wang, 2014). While cities are almost always hotter than the surrounding rural areas, urbanization and climate change could raise urban temperatures to levels that threaten human health, strain energy resources, and compromise economic productivity (Kenward, Yawitz, Sanford & Wang, 2014, p. 3). Zhao, Lee, Smith, and Oleson (2014) used mean-annual-midnight temperature differences between rural and urban areas across 65 of the largest American cities, and the UHI was found to positively correlate with the logarithm of population size, indicating that larger cities have larger UHI effects. Kenward, Yawitz, Sanford, and

Wang (2014) studied summer temperatures in 60 of the largest U.S. cities and found that 57 of them had measurable urban heat island effects over the 10-year period studied. They documented single-day urban temperatures in some metro areas to be 27°F higher than the surrounding rural areas (Kenward, Yawitz, Sanford & Wang, 2014). Urbanized areas, dense downtowns, and central business districts with high-rise buildings and other structures inhibit the outflow of the solar energy absorbed during day times, and impervious surfaces such as roads reduce processes of evaporative cooling, transpirational cooling, and evapotranspiration (Yuan & Bauer, 2007; United States Environmental Protection Agency [EPA], 2008; Declet-Barreto, Brazel, Martin, Chow & Harlan, 2013; Alhawiti & Mitsova, 2016).

Urban heat islands are generated from thermodynamic processes driven by albedo, vegetation, urban activity, building structure, and humidity (Taha, 1997; Buscail, Upegui & Viel, 2012; Zhao, Lee, Smith & Oleson, 2014). Albedo is a measure of the reflectivity of light off a surface. The higher the albedo the better because the more solar energy a material reflects as light, the cooler it remains. Dark, impermeable surfaces such as asphalt and concrete are less reflective and have lower albedos. The light energy that is not reflected back into space is absorbed and radiated from the surface back into the atmosphere in the form of thermal energy (heat). Vegetation has a higher albedo than urbanized areas, but vegetation and soil also enable evapotranspiration to moderate the local climate.

Albedo (also known as solar reflectance)

The ratio of reflected to incident sunlight, measured on a scale of 0 to 1 (or 0%–100%), where higher numbers indicate greater reflectance.

1: the highest reflectivity index (white surface)

0: the highest absorptive index (black surface)

Source: Lawrence Berkeley National Laboratory, https://heatisland.lbl.gov/glossary.

Jacobson and Hoeve (2012) studied the effects of albedo on urban heat islands at a global level. They specifically focused on the effect of raising the global albedo by making all the roofs in the world white in color. The researchers utilized Moderate Resolution Imaging Spectroradiometer (MODIS) satellite data and other data as inputs to 20-year meteorological models to account for feedback loops, such as cooling effects due to higher concentrations of aerosols due to deforestation over time, and additional heating resulting from fewer clouds driven by the reduction in humidity over impervious surfaces (Jacobson & Hoeve, 2012). Global land cover and urban surface area were determined from data on soil type, vegetation, albedo, and land use. The total contribution of global urban heat islands to overall global warming was estimated to be 2% to 4%, with white roofs having a fractional global cooling effect (Figure 11.2).

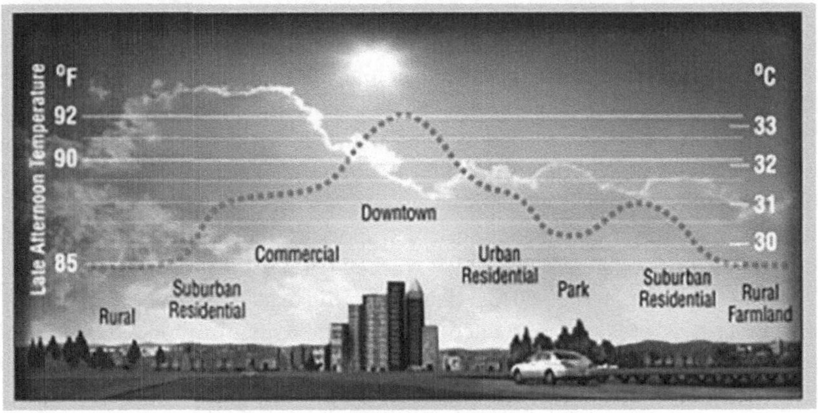

Figure 11.2 Diagram of an urban heat island

Source: Heat Island Group, Lawrence Berkeley National Laboratory, https://heatisland.lbl.gov/coolscience

Why Should We Care?

The urban heat island contributes about 2% to 4% to total global warming (Stone et al., 2014; Jacobson & Hoeve, 2012). As urban areas have expanded, UHIs have become a threat to human well-being as well as a public health issue (Shahmohamadia, Che-Ania, Etessamb, Mauludc, & Tawil, 2011; Sharma, 2016). Anthropogenic urban activities generate heat and are large contributors to the urban heat island (Taha, 1997; Zhao, Lee, Smith & Oleson, 2014). Some models predict anthropogenic heating to contribute 2°C to 3°C to the urban heat island heating effect (Taha, 1997). Building structure affects the urban heat island by impairing convection and reducing airflow. Zhao, Lee, Smith, and Oleson (2014) explored the role of convection as a contributor to the UHI and found that convection is the largest contributor to the UHI effect. Studies indicate that local and regional climates are influenced by changes in the urban landscape and surrounding areas (Kahn, 2009; Triantakonstantis & Mountrakis, 2012) and can potentially impact human health and the prosperity and well-being of communities (Patz, Campbell-Lendrum, Holloway & Foley, 2005; McMichael, Woodruff & Hales, 2006; Haines, Kovats, Campbell-Lendrum & Corvalan, 2006; Drewniak et al., 2013).

Government agencies have taken steps to compile datasets, geospatial tools, and models (such as meteorological models) to quantify, assess, and visualize urban heat island effects. For example, an Urban Heat Island Index (UHII) was developed to assist the California EPA (CalEPA) in evaluating the thermal impacts of urban areas in California (CalEPA, 2015). The study documents degree hours (DH) over several seasons; at the lower end of the scale for small urban areas, the UHII ranges from 2 to 20 DH/day (°C.hr/day), and at the higher end for larger areas, it reaches up to 125 DH/day or more (CalEPA, 2015, p. 1).

The study by Stone et al. (2014) attempted to project (to year 2050) the reduction in heat-related deaths as a result of urban heat island mitigation strategies in the metropolitan statistical areas (MSAs) of three American cities (Atlanta, Philadelphia, and Phoenix). Working at the census tract level, land cover change was modeled from 2010 to 2050 using data from the National Land Cover Database (NLCD). Seven land cover categories (forest, grass, barren land, water, wetlands, agriculture, impervious surfaces) were aggregated to 4-kilometer resolution and used as input for a meteorological model to compare a current path approach to alternatives with greening spaces (e.g., planting trees, shrubs) and albedo changes (increasing roof and road surface albedo). The resulting temperature changes were applied to an EPA model that predicts mortality based on temperature increase. Some projections estimated a mortality decrease of 40% to 99% with such albedo changes (Stone et al., 2014).

Urban heat islands also damage the environment in a number of ways—increased energy consumption and increased electricity demand for cooling buildings in cities, elevated emissions of air pollutants and greenhouse gases as a result of greater emissions of air pollutants, greater greenhouse gas emissions from power plants, and an elevation in the formation of ground-level ozone (EPA, 2016a; Rosenzweig, Solecki, & Slosberg, 2006). Urban heat islands can also result in changes to the local weather. Cities and metropolitan areas can create their own weather by making clouds, stirring up thunderstorms, and magnifying the area's smog problem (Rosenfeld, Akbari, Romm & Pomerantz, 1998; National Aeronautics and Space Administration [NASA], 2016). Stressed aquatic ecosystems and impaired water quality are additional adverse impacts of rapid temperature changes that can be caused by hot pavement and rooftop surfaces transferring their excess heat to stormwater, then into streams, rivers, ponds, and lakes (EPA, 2016a).

Assessing Vulnerability and Risk

Geospatial tools, techniques, and analysis allow for spatially explicit characterization of extreme heat vulnerability in urban environments (Wilhelmi & Hayden, 2010, p. 2). Remote sensing combined with census data, parcel data, and other administrative data is particularly useful for assessing and studying urban heat island impacts and pre-disposition to health and environmental vulnerabilities for neighborhoods, cities, and regions around the globe.

An increasing number of studies and scholarly research are focused on promoting mitigation and adaptation strategies to address adverse human health and comfort such as respiratory difficulties, heat cramps and exhaustion, non-fatal heat stroke, and heat-related mortality (Gosling, Lowe, McGregor, Pelling & Malamud, 2009; Huang, Barnett, Wang, Vaneckova, FitzGerald & Tong, 2011; Lin, Hsu, Van Zutphen, Saha, Luber & Hwang, 2012). Although it is the atmospheric conditions that generate a heat hazard, the heat island effect can be exacerbated during heat waves, causing a threat to human health especially for the elderly and disabled persons with low socioeconomic status and those lacking social networks (EPA, 2016a; Wilhelmi & Hayden, 2010). Research suggests that summer heat waves increase the mortality rate (Bustinza, Lebel, Gosselin, Bélanger & Chebana, 2013; Casati, Yagouti & Chaumont, 2013; Wang, Guo, FitzGerald, Aitken, Tippett, Chen & Tong, 2015). Heat waves can be amplified by the urban heat island, and the human toll can be compounded, particularly for these vulnerable populations (Hajat, Kovats & Lachowycz, 2007).

Linkage Between Convection and UHI

In a study by Zhao, Lee, Smith, and Oleson (2014), GIS was utilized to map the size of the UHI for day and night conditions, to illustrate the correlation with precipitation patterns in the **continental United States**, and to manipulate several datasets: MODIS (for surface temperature and vegetation), PRISM (Panchromatic Remote-sensing Instrument for Stereo Mapping)[1] (for precipitation), population from the 2010 US census, and the Community Earth System Model (CESM). The investigators generated energy models for the relative contribution of factors (radiation balance, evaporation, convection efficiency, heat storage, and anthropogenic heat) to the UHI based on climate zones. The models predict that cities in humid climates have dense vegetation in rural areas and that these areas produce aerodynamic friction, making convection more efficient in the rural areas than in the urban areas. The situation is reversed in dry climates, where the aerodynamic friction is lower in the sparsely vegetated rural areas relative to the friction produced by urban infrastructure. In humid climates, the rural area is 58% more efficient at removing heat from the surface through convection than the urban area, and in dry climates the urban area is 20% more efficent than the rural area (Zhao, Lee, Smith & Oleson, 2014). The result is a temperature increase of 3°C in humid cities versus a decline of 1.5°C as a result of differences in convective efficiency.

Societal Vulnerability and Heat-Related Health Risk Assessments

A study by Wilhelmi and Hayden (2010) generated an extreme heat vulnerability framework for urban areas to capture three interactive components and dynamic, spatially variable indicators of exposure, sensitivity, and adaptive capacity. The exposure variable captures the urban heat island effect of people in urban areas being exposed to higher temperatures, and it is measured from remote sensors on satellites of airplanes or meteorological stations (Wilhelmi & Hayden, 2010). The sensitivity indicator captures socioeconomic factors that can influence the ability to respond to heat and includes indicators such as age, obesity, medications, social isolation, poor/cannot afford air conditioning, and type of occupation (Wilhelmi & Hayden, 2010). The adaptive capacity indicator captures whether people know the risks and are able to counteract them such as by staying indoors with air conditioning (Wilhelmi & Hayden, 2010).

Tomlinson, Chapman, Thornes, and Baker (2011) utilized satellite data and socioeconomic data to better understand the potential for heat health risks in **Birmingham, UK**. Satellite data were used to remotely measure temperature using NASA's MODIS with a 1-kilometer resolution. Four social risk groups (elderly people, ill health, high population density, high-rise living) were created using Experian Mosaic 2009 data[2] and other supplemental data aggregated at the Lower Super Output Area (LSOA) level.[3] The elderly and ill groups were formed using Experian Mosaic, density was calculated, and high-rise buildings were generated using

Survey Mastermap, while sufficiently high-rise buildings were identified as over 10 households in one location. The study found a spatial pattern of exposed vulnerable groups and concluded that as the urban heat island effect increases, so does the number of "exposed and vulnerable" groups in the inner core of the city where the urban heat island effect is the largest (Tomlinson, Chapman, Thornes & Baker, 2011, p. 9).

Buscail, Upegui, and Viel (2012) used satellite data to measure temperature and build an index of health risk for the **city of Renne in France** using demographic and other census data. Landsat Enhanced Thematic Mapper (with a resolution of 60 meters) was used to obtain daytime temperatures. The image utilized in that study by Buscail, Upegui, and Viel (2012) coincided with a heat wave that affected Renne, France, on June 22, 2001. A control image was used with data captured on July 21, 2000. The census data included socioeconomic status, age, population density, and building obsolescence. When combined with temperature data, sparsely populated areas were not exposed to the heat hazard. Three index maps were created at the French IRIS level of aggregation (the equivalent of the census block group in the United States): (i) hazard index map under heat wave condition; (ii) vulnerability index map; and (iii) heat wave health risk map (Buscail, Upegui & Viel, 2012).

Impact of Land Use and Urban Development on Urban Heat Islands

An older study by Lo, Quattrochi, and Luvall (1997) utilized aircraft-based, remotely sensed thermal imagery at 5-meter resolution to measure the urban heat island effect over **Huntsville, Alabama (U.S.)** on September 7, 1994. The aircraft was a NASA Stennis Learjet equipped with the Advanced Thermal and Land Applications Sensor (ATLAS). Multiple observations were performed both during the day and night. GIS was used to assemble the data, including 10-meter-resolution aircraft data, 5-meter-resolution aircraft data, aerial color infrared photographs, land temperature measurement, meteorological station data, and radiosonde measurements. With irradiance used as a proxy for temperature, commercial-type land cover was the warmest of 10 land cover classifications studied and had the largest temperature difference from day to night. GIS was also utilized to map the irradiance zones. The output shows maximums in city centers and other small heat islands during the day and cooling of those areas at night. This research also showed the negative correlation between irradiance and the Normalized Difference Vegetation Index (NDVI) in urbanized areas, which is a proxy for vegetative cover. In other words, vegetative cover acts to reduce the temperature during the day and provide cooling at night.

Stone and Rodgers (2001) utilized NASA's high-resolution (10-meter) thermal imagery georeferenced to parcel-level data in **Atlanta, Georgia (U.S.)**, to determine the differences in the temperature of high-density versus low-density development. The radiant heat flux—the amount of energy transported from a surface—for each parcel was calculated. This heat flux was subtracted from the heat flux of a fully forested parcel to determine the excess heat flux produced by each parcel. Tree canopy cover for each parcel was determined from aerial imagery and operationalized as percent coverage of total parcel lot. Further, several control variables were geographically coded, including the year of construction, percent of impermeable and permeable surface area, and number of bedrooms. Finally, street density was calculated from U.S. Census Bureau TIGER files. Path analysis[4] was used to model the indirect effects of parcel design on thermal output properties of the parcel. The study found that more heat is contributed to the urban heat island per parcel in low-density single-family parcels than in denser development (Stone & Rodgers, 2001).

Tomlinson, Chapman, Thornes, and Baker (2012) used the MODIS sensor on a NASA's Aqua satellite to map the surface temperatures and to construct a profile for the urban heat island on **Birmingham, UK**, during the summer months from 2003 to 2009. The images were processed and georeferenced using an Environmental Systems Research Institute (ESRI) ArcMap plug-in named the Marine Geospace Ecology Tools (MGET). An urban heat island peak was detected of up to 5°C in the urban core, but parks were up to 7°C cooler than the surrounding urban space. The intensity of the urban heat island was most pronounced during heat waves and decreased in intensity as atmospheric instability increased. Land use classifications showed that dense urban areas had the highest UHI effect (Tomlinson, Chapman, Thornes & Baker, 2012).

Case Study: The Correlation Between Urban Heat Island and Urban Land Uses

Study Objective

As previously discussed, the formation and the extent of the urban heat islands are affected by urban land use, presence/absence of vegetation, and building type (Maloley, 2009; Zhang, Zhong, Feng & Wang, 2009; Wilhelmi & Hayden, 2010). Parking lots, high-rise buildings, asphalt roads, and industrial land use accumulate heat, decrease the moisture content of the air, and effectively block the cooling properties of the atmosphere (Zhang, Zhong, Feng & Wang, 2009). Previous studies have shown that urban environments dominated by the presence of vegetation and trees have lower temperatures than those dominated by concrete and impervious surfaces. Vegetation provides shade and maintains the cooling properties of the air by increasing its moisture content through the processes of evapotranspiration (Taha, 1997; Maloley, 2009; Zhang, Zhong, Feng & Wang, 2009). The objective of this case study is to analyze how patterns of land development and spatial distribution of land use affect the formation of urban heat islands in the city of Miami, Florida. See Figure 11.3 for a map of the study area.

Planning and Policy Applications

- Inform urban design and planning practices
- Inform strategies to mitigate temperature extremes

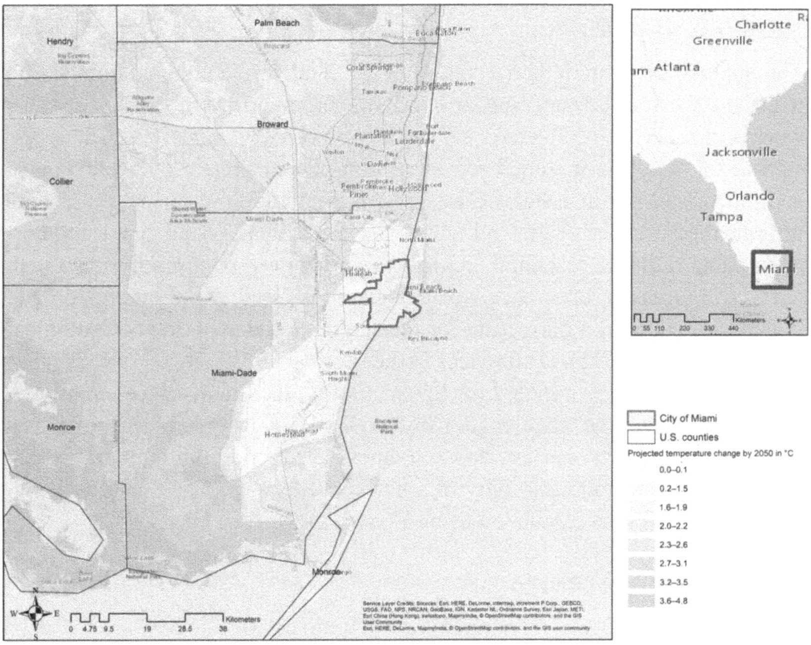

Figure 11.3 Projected temperature change in °C by 2050 for the study area: city of Miami in Miami-Dade County, Florida

Source: Created by authors

Data and Models

- Landsat 8 (Operational Land Imager [OLI] and Thermal Infrared Sensor [TIRS]) images (Row:015/Path:042) for **South Florida (U.S.)** were collected from the U.S. Geological Survey's Earth Resources Observation and Science (EROS). The images were acquired on March 23, 2014, April 23, 2014, October 17, 2014, and November 2, 2014. The data were projected in the Universal Transverse Mercator (UTM) coordinate system (Zone 17N) using the World Geodetic System (WGS) 1984 spheroid.

- Land use/land cover (LULC) data, obtained from the South Florida Water Management District (SFWMD), were reclassified into 10 urban land use classes that include: three residential use categories based on urban density, upland hardwood forests, transportation and utilities, services and commercial areas, industrial areas, parks and cemeteries, water bodies, and coastal vegetation (Alhawiti & Mitsova, 2016)

Summary of Analysis

The analysis steps include derivation of the land surface temperature (LST) using the thermal infrared sensor band 10. The majority of the calculations were performed using Raster Calculator in ArcGIS version 10.4.1. The first step in the analysis included retrieving top-of-atmosphere (TOA) radiance and reflectance (United States Geological Survey [USGS], 2015). TOA spectral radiance was derived from Landsat 8 TIRS and OLI band data using corresponding radiance rescaling factors specific to each band. In the next step, the TOA planetary reflectance was calculated using the OLI band 2–6 data, which also contains a correction for the sun angle (United States Geological Survey [USGS], 2015). Differences in the spectral emissivity of various types of land cover were accounted for by calculating the land surface emissivity, which is a function of wavelength (Snyder, Wan, Zhang & Feng, 1998). Composition, physical properties, and roughness are among the factors that can influence the spectral emissivity of a surface (Snyder, Wan, Zhang & Feng, 1998). The LST was calculated using Equation (1):

$$\text{TS}=\text{Trad1}+(\lambda\text{TRad}\,/\,\text{p})\text{In}\varepsilon \quad (1)$$

where *TS* is the LST and its unit is degrees Celsius (°C); *T*rad is the brightness temperature (BT) in Kelvin (K); λ is the center wavelength for band 10 (10.9 μm); p = h·, where h is the Planck constant (6.626 × 10–34 *J·s*), *c* is the velocity of light (2.998 × 108 *m/s*), and σ is the Boltzmann constant (1.38 × 10–23 *J/K*); ε is the surface emissivity (USGS, 2015).

The Normalized Difference Vegetation Index is the most commonly used satellite-based measure of vegetation coverage. It can be calculated in ArcGIS using the NDVI tool. In addition, Raster Calculator was used to calculate the Normalized Difference Built-up Index (NDBI), which is a measure of the intensity of imperviousness using satellite data (Bhatti & Tripathi, 2014).

Key Findings and Sample Output

The minimum, maximum, and mean land surface temperature values (in degrees Celsius) were compared among the 10 land use categories over the four time frames. Land use (LU) categories are shown in Figure 11.4. The results indicate that commercial land use is associated with the highest minimum, maximum, and mean LST during all four periods. The presence of large parking lots where concrete or asphalt surfaces are exposed to sunlight for many hours can explain the variation in the estimated LST. Land areas associated with transportation and utilities also exhibit significantly higher LST (with an average temperature of 38.7°C). Low-density residential areas were also found to have relatively high LST. The lowest LST has been estimated near coastal wetland vegetation, followed by upland hardwood forests and rivers and lakes.

Figures 11.5 and 11.6 display maps of the LST distribution across the city of Miami as well as the results for the NDVI and NDBI indices. Correlation

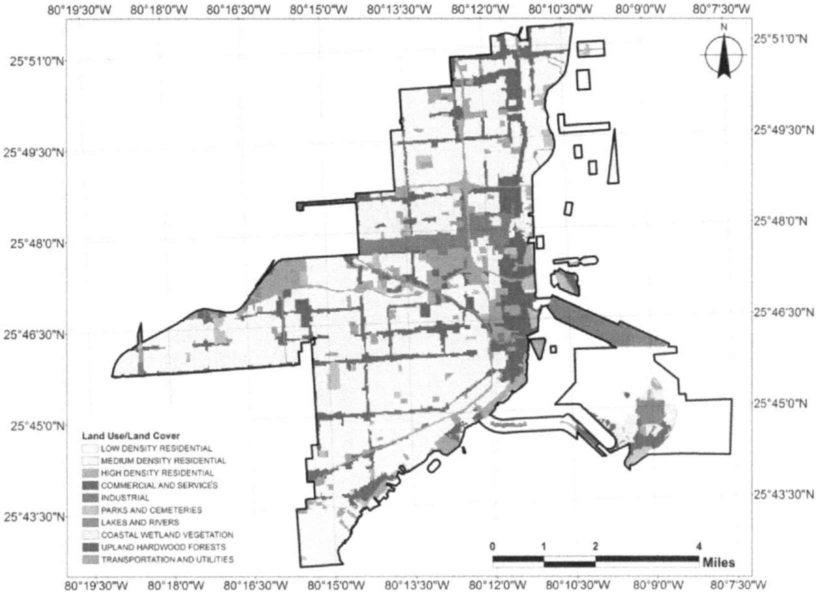

Figure 11.4 Land use map of the city of Miami, Florida

Source: Map prepared by Rayan Alhawiti

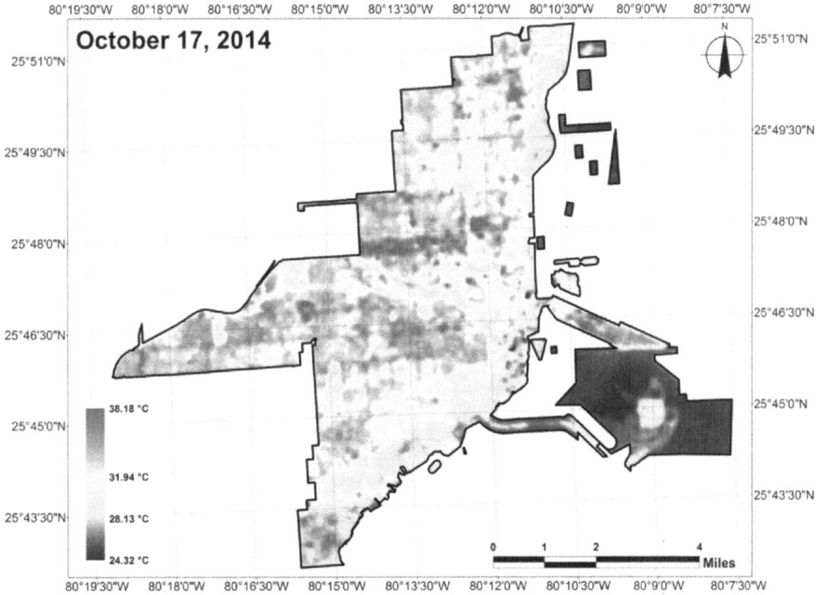

Figure 11.5 Land surface temperature for the city of Miami, Florida, derived from Landsat 8

Source: Maps prepared by Rayan Alhawiti

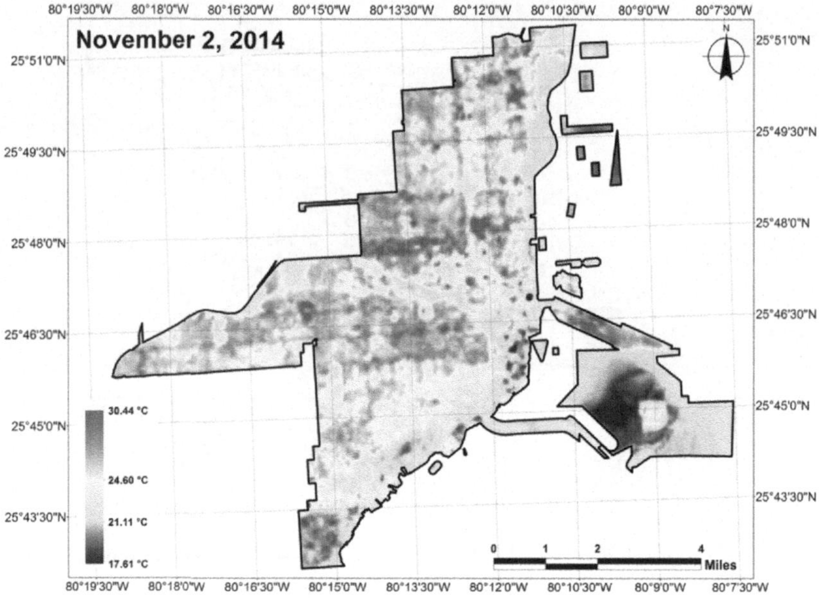

Figure 11.5 (Continued)

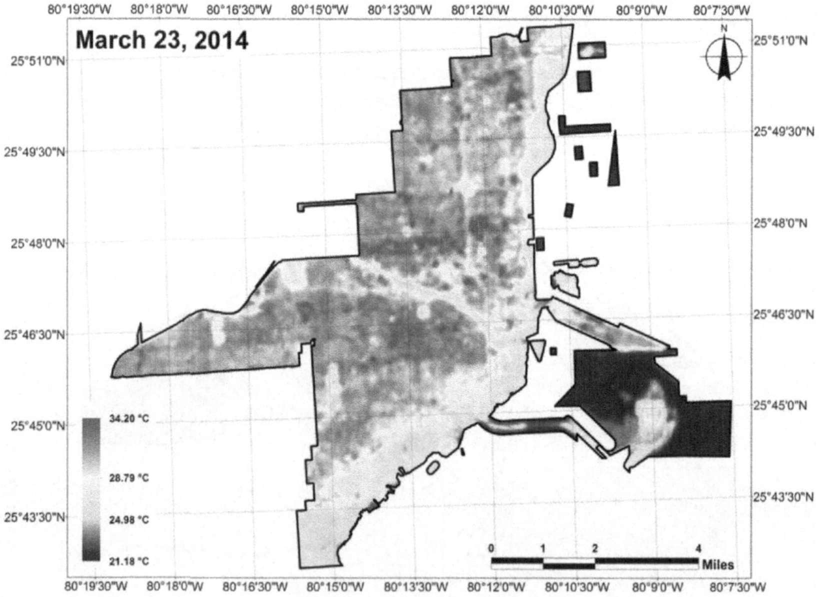

Figure 11.5 (Continued)

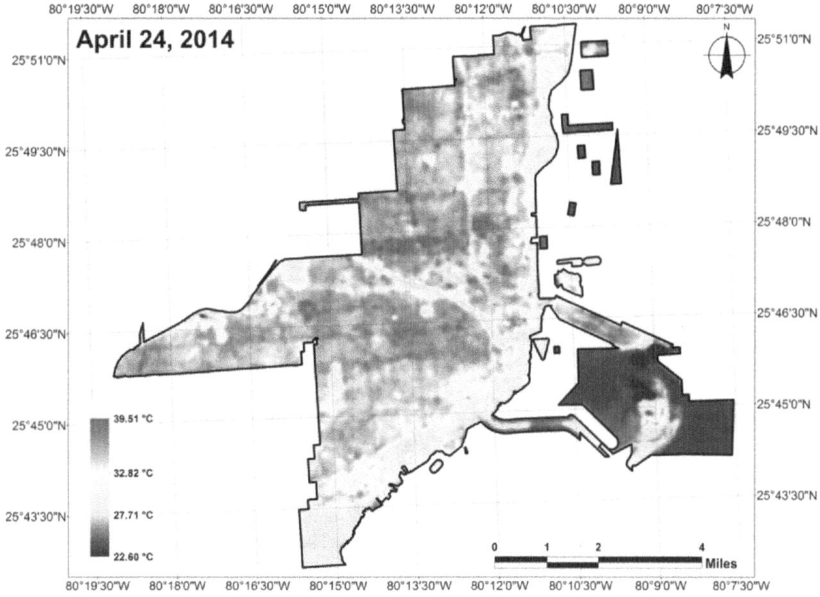

Figure 11.5 (Continued)

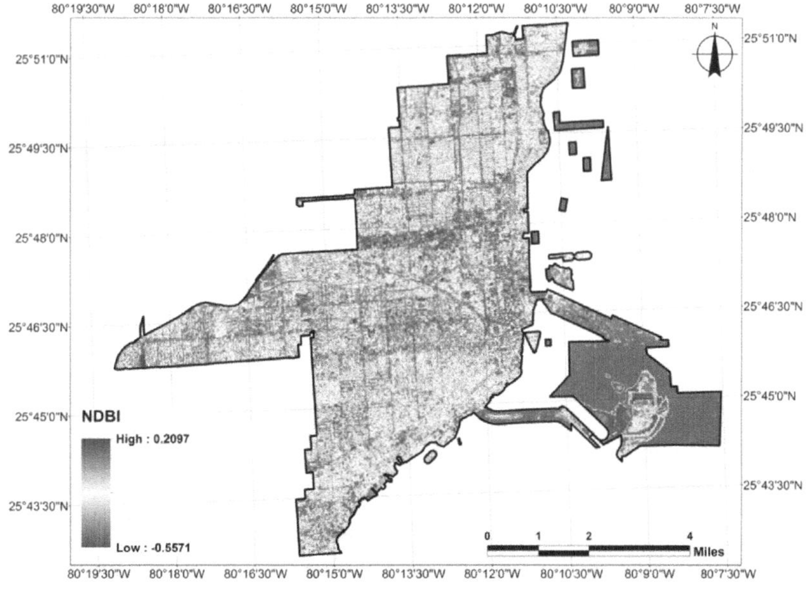

Figure 11.6 NDBI and NDVI indices for the city of Miami, Florida

Source: Maps prepared by Rayan Alhawiti

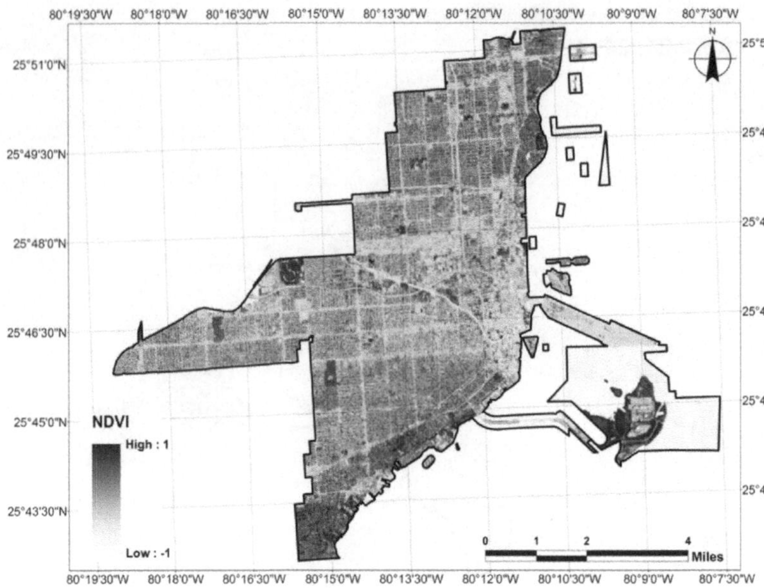

Figure 11.6 (Continued)

analysis using the Kruskal Wallis H test was conducted to understand the relationship between the thermal profile of various land use categories, vegetation coverage, and built-up profile (Alhawiti & Mitsova, 2016). The results indicate a positive correlation between LST and NDBI; that is, an increase in the values of the built-up index corresponds to an increase in the values of the land surface temperature. The correlation analysis also reveals a fairly strong negative relationship between NDBI and NDVI and LST and NDVI. This finding indicates that areas with more vegetation are statistically associated with lower land surface temperatures.

Monitoring and Measurement: Methodological Issues

Urban heat islands can be categorized into two distinct types: surface urban heat islands and atmospheric urban heat islands (EPA 2008). Surface urban heat islands are larger during the day, while atmospheric urban heat islands are larger at night (Yuan & Bauer, 2007). Surface temperatures are generally measured using remote sensing such as through airplanes or satellites. Given concerns about whether the Normalized Difference Vegetation Index alone can be used as the main indicator of urban climate, Yuan and Bauer (2007) studied surface urban heat island using measures of NDVI and impervious surface area. Landsat data were captured for the four seasons: February 27, 2001; May 21, 2002; July 16, 2002; and September 12, 2000. The Landsat images were reprojected to the UTM coordinate system and georeferenced to a base map of the **Minneapolis, Minnesota (U.S.)** area using about 60 reference points consisting of road network intersections. From a methdological perspective, the findings showed that a correlation between NDVI and irradiation (exposure to radiation) is problematic in the winter months, thereby concluding that impervious surface area seems to be a good complementary measure of surface temperature in urban heat island studies (Yuan & Bauer, 2007).

In addition to remote sensing, there are several other techniques and studies used to measure, monitor, and assess the effects of mitigation and adaptation strategies on UHI.

- **Transect study:** This type of study measures temperature changes across a transect, or sample area, often using hand-held devices or measurement equipment mounted on cars or aircraft for heat island monitoring processes (EPA, 2009).

- **Urban fabric analysis:** a method used to determine the proportions of vegetative, roofed, and paved surface cover relative to the total urban surface in the city (EPA, 2009). Urban fabric data are needed in order to estimate the impact of light-colored surfaces (roofs and pavements) and urban vegetation (trees, grass, shrubs) on the meteorology and air quality of a city and to design effective implementation programs (Akbari & Rose, 2001)

- **Solar reflectance index (SRI):** a metric for comparing the coolness of roof surfaces. The higher the SRI, the cooler the roof will be in the sun (Lawrence Berkeley National Laboratory, 2016). A clean black roof has an SRI of 0, while a clean white roof has an SRI of 100 (Urban & Roth, 2010; Lawrence Berkeley National Laboratory, 2016).

- **Photochemical modeling:** computer modeling of reactions in the atmosphere that produce ozone from nitrogen oxides and volatile organic compounds. Photochemical modeling can be used to evaluate the air quality impacts of heat island reduction strategies (EPA, 2009).

The atmospheric layer temperatures are typically measured using meteorological instrument stations. For these types of measurement and monitoring techniques, timing and conditions are important considerations. It is also important, such as in urban fabric analyses, that baseline conditions be recorded to ensure realistic simulations and estimates (EPA, 2009). The use of land-based monitoring equipment versus satellite data also needs to be assessed depending on the study area location and size. Given the sparse locations of land measurement facilities and stations, satellite data better capture urban heat islands, compared to land-based measurements. Depending on the size of the study area, sparse location of measuring stations means a lack of enough observations to precisely determine the location and magnitude of a temperature gradient. The EPA (2016b) provides important advice to researchers, including considering changes in instrumentation, sampling, data recording methods, and station microclimate, as well as documenting limitations of various surface and air temperature data capture techniques. For example, surface measurements taken by remote sensing have several limitations, such as not accurately capturing radiant emissions from vertical surfaces. As such, weather stations both within a city and in non-urban areas close to the city should be used to measure air temperature (EPA, 2016b).

An emerging field in climate research is the improvement of weather forecasts and urban climate modeling by developing high-resolution models at the interface of the urban land surface and atmospheric processes (Drewniak, Kotamarthi, Jacob, Chen, Catlett, Ching & Wu, 2013; Brown, Alexander & Rounsevell, 2017). Better representation of the properties of the built environment and understanding of urban climate zones (Stewart & Oke, 2012) are expected to enhance the simulation of larger-scale processes within the atmospheric boundary layer and improve the parametrization of smaller-scale physical processes related to cloud cover, hydrometeorology, aerosol formation, and air quality (Drewniak, Kotamarthi, Jacob, Chen, Catlett, Ching & Wu, 2013; Brown, Alexander & Rounsevell, 2017).

Mitigation and Adaptation Strategies for Addressing the Problem

Solutions to mitigating the urban heat island effect are critical especially given that the majority of the world's population will live in cities by 2050 (Sharma, 2016). Mitigation is associated with changing the conditions that create heat, such as changing the albedo of buildings and ground cover or increasing the vegetative cover. As noted earlier, higher albedo materials reflect more sunlight than lower albedo materials. The sunlight that is not reflected is absorbed, causing nighttime temperature to rise. Vegetative cover reduces heat through evapotranspiration and shade. There are increasing efforts to highlight the value of incorporating heat-mitigation strategies into climate action plans (Stone, Vargo & Habeeb, 2012). Common mitigation strategies include: increasing tree and vegetative cover, installing green roofs (also called "rooftop gardens" or "eco-roofs"), installing cool—mainly reflective—roofs, and using cool pavements (Gartland, 2008; Sharma, 2016; EPA, 2016a).

Solecki et al. (2005) also reported on the mitigating effects of vegetation and reflective roofs. The study by Solecki et al. (2005) utilized GIS to model heat-mitigating effects of vegetation and reflective rooftops in the **cities of Newark and Camden, New Jersey**. A total of six sites (three in each case study city) were used to evaluate the benefits of urban heat-mitigation strategies. CITYgreen[5] was used to input climatological variables and building energy usage variables, as well as type of surface, presence of A/C in homes, and buildings and trees. To calculate the impact of roof color, CITYgreen used the location and albedo of each building in the study area. Aerial photographs and site visits were conducted to identify site characteristics such as building heights and characteristics of vegetation. The study found cooling effects of vegetation, but the software was not able to directly translate roof albedo to outdoor temperature (Solecki et al., 2005). Stone, Vargo, and Habeeb (2012) also found tree planting and other vegetative strategies to be most effective compared to albedo enhancement strategies. Ongoing research is necessary at multiple scales, such as a micro-scale study by Declet-Barreto, Brazel, Martin, Chow, and Harlan (2013) that assessed energy fluxes among vegetated, non-vegetated, and impervious surfaces in inner-city **Phoenix, Arizona (U.S.)**. Their study showed that creating localized park cool islands (i.e., irregular patterns of cooler areas nested with generally warmer areas) can mitigate local effects of UHIs (Declet-Barreto, Brazel, Martin, Chow & Harlan, 2013).

Bierbaum et al. (2013) noted the need for transformational changes to adapt to changes in climate. The reality, however, is that adaptation changes across scales and across the various private, public, and non-profit sectors are incremental. Some of these transformational changes require policy changes that promote different building and construction standards. For example, Lawrence Berkeley National Laboratory (2016) notes that many building codes and utility rebates now specify minimum values of SRI. Additionally, the United States Green Building Council (USGBC)'s Leadership in Energy and Environmental Design (LEED) program currently uses minimum SRI values of 78 and 29 for low and steep-sloped cool roofs respectively (Urban & Roth, 2010, p. 5). These initiatives are not new. In the late 1990s, Rosenfeld, Akbari, Romm, and Pomerantz (1998) commented on codes that specified performance thresholds that new buildings should meet. In referring to Los Angeles, Rosenfeld, Akbari, Romm, and Pomerantz (1998) discussed incorporation of cool roofs and cool surfaces into revised building standards and tradable smog-offset credit initiatives. The good news is that cities are paying attention to the adverse effect of UHIs and are taking steps toward mitigation and adaptation of our physical urban environments.

Notes

1 The PRISM Climate Group, based at Oregon State University, compiles climate observations from a range of sources and develops spatial climate datasets—www.prism.oregonstate.edu.

2 Experian Mosaic is a for-profit data source for household-based consumer lifestyle segmentation data. See their website for more details.

3 LSOA (Lower Super Output Area) are census aggregation units used in England and Wales with populations averaging 1,500 people. There are 641 LSOA units in Birmingham.

4 Path analysis allows for the study of direct effects of independent variables on a dependent variable, but also the indirect effects through intervening variables.

5 At the time that the analysis was completed, CITYgreen was an add-on program to ArcGIS for the purpose of analyzing the economic and ecological benefits of green space and tree cover. This program seems to have been replaced by InVEST (Integrated Valuation of Ecosystem Services and Trade-offs) (www.americanforests.org/our-programs/urbanforests/urban-forests-tools-resources/urban-forest-assessments-resource-guide/urban-forest-assessment-tools/top-down-urban-forest-assessment-tools/#API). Both were produced by American Forests

(americanforests.org). These programs, based on user input of land cover, topography, and land use, allow for the modeling of alternative scenarios in the study of human use of ecosystems.

References

Akbari, H., & Rose, L.S. 2001. *Characterizing the Fabric of the Urban Environment: A Case Study of Salt Lake City, Utah*. Heat Island Group Environmental Energy Technologies Division Lawrence Berkeley National Laboratory University of California Berkeley, California.

Alhawiti, R.H., & Mitsova, D. 2016. Using Landsat-8 Data to Explore the Correlation Between Urban Heat Island and Urban Land Uses. *International Journal of Research in Engineering and Technology* 5(3), 457–466.

Bhatti, S., & Tripathi, N. 2014. Built-Up Area Extraction Using Landsat 8 OLI Imagery. *GIScience & Remote Sensing* 51(4), 445–467.

Bierbaum, R., Smith, J.B., Lee, A., Blair, M., Carter, L., Chapin III, F.S., Fleming, P., Ruffo, S., Stults, M., McNeeley, S., Wasley, E., & Verduzco, L. 2013. A Comprehensive Review of Climate Adaptation in the United States: More Than Before, But Less Than Needed. *Mitigation and Adaptation Strategies for Global Change* 18(3), 361–406.

Bornstein, R., Balmori, R.T.F., Taha, H., Byun, D., Cheng, B., Nielsen-Gammon, J., & Smith, P. 2006. Modeling the Effects of Land-Use/Land-Cover Modifications on the Urban Heat Island Phenomena in Houston, Texas. *Final report to David Hitchcock Houston Advanced Research Center*. Retrieved from www.researchgate.net/publication/266214566_MODELING_THE_EFFECTS_OF_LAND-USELAND-COVER_MODIFICATIONS_ON_THE_URBAN_HEAT_ISLAND_PHENOMENA_IN_HOUSTON_TEXAS_Final_Report_to.

Brown, C., Alexander, P., Holzhauer, S., & Rounsevell, M.D.A. 2017. Behavioral Models of Climate Change Adaptation and Mitigation in Land-based Sectors. *Wiley Interdisciplinary Reviews: Climate Change* 8(2), e448.

Buscail, C., Upegui, E., & Viel, J.F. 2012. Mapping Heatwave Health Risk at the Community Level for Public Health Action. *International Journal of Health Geographics* 11(38), 1–9.

Bustinza, R., Lebel, G., Gosselin, P., Bélanger, D., & Chebana, F. 2013. Health Impacts of the July 2010 Heat Wave in Quebec, Canada. *BMC Public Health* 13(1), 56.

California Environmental Protection Agency (CalEPA). 2015. Creating and Mapping an Urban Heat Island Index for California. Report prepared by Altostratus Inc. Agreement No. 13–001. Retrieved from www.calepa.ca.gov/files/2016/10/UrbanHeat-Report-Report.pdf

Casati, B., Yagouti, A., & Chaumont, D. 2013. Regional Climate Projections of Extreme Heat Events in Nine Pilot Canadian Communities for Public Health Planning. *Journal of Applied Meteorology and Climatology* 52(12), 2669–2698.

Declet-Barreto, J., Brazel, A.J., Martin, C.A., Chow, W., & Harlan, S.L. 2013. Creating the Park Cool Island in an Inner-City Neighborhood: Heat Mitigation Strategy for Phoenix, AZ. *Urban Ecosystems* 16(3), 617–635.

Drewniak, B., Kotamarthi, R., Jacob, R., Chen, F., Catlett, C., Ching, J., & Wu, W. 2013. Urban: Landscape and Climate Change, Workshop Summary, August 28–29, 2013. Lemont, IL: Argonne National Laboratory.

Environmental Protection Agency (EPA). 2009.

Environmental Protection Agency (EPA). 2016a.

Environmental Protection Agency (EPA). 2016b.

Gartland, L.M. 2008. *Heat Islands: Understanding and Mitigating Heat in Urban Areas*. New York, NY: Earthscan.

Gosling, S.N., Lowe, J.A., McGregor, G.R., Pelling, M., & Malamud, B.D. 2009. Associations Between Elevated Atmospheric Temperature and Human Mortality: A Critical Review of the Literature. *Climatic Change* 92(3–4), 299–341.

Haines, A., Kovats, R.S., Campbell-Lendrum, D., & Corvalan, C. 2006. Climate Change and Human Health: Impacts, Vulnerability and Public Health. *Public Health* 120, 585–596.

Hajat, S., Kovats, R.S., & Lachowycz, K. 2007. Heat-Related and Cold-Related Deaths in England and Wales: Who Is at Risk? *Occupational and Environmental Medicine* 64(2), 93–100.

Hoekstra, J.M., Molnar, J.L., Jennings, M., Revenga, C., Spalding, M.D., Boucher, T.M., Robertson, J.C., Heibel, T.J., & Ellison, K. 2010. In: Molnar, J.L. (Ed.) *The Atlas of Global Conservation: Changes, Challenges, and Opportunities to Make a Difference*. Berkeley, CA: University of California Press.

Huang, C., Barnett, A.G., Wang, X., Vaneckova, P., FitzGerald, G., & Tong, S. 2011. Projecting Future Heat-Related Mortality Under Climate Change Scenarios: A Systematic Review. *Environmental Health Perspectives* 119(12), 1681–1690.

Intergovernmental Panel on Climate Change (IPCC). 2012. Summary for Policymakers. In: Field, C.B., Barros, V., Stocker, T.F., Qin, D., Dokken, D.J., Ebi, K.L., Mastrandrea, M.D., Mach, K.J., Plattner, G.-K., Allen, S.K., Tignor, M., & Midgley, P.M. (Eds.) *Managing the Risks of Extreme Events and Disasters to Advance Climate Change Adaptation*. A Special Report of Working Groups I and II of the Intergovernmental Panel on Climate Change, 1–19. Cambridge, UK and New York, NY: Cambridge University Press.

Jacobson, M.Z., & Ten Hoeve, J.E. 2012. Effects of Urban Surfaces and White Roofs on Global and Regional Climate. *Climate* 25(3), 1028–1044.

Kahn, M.E. 2009. Urban Growth and Climate Change. *Annual Review of Resources Economics* 1, 16.1–16.7.

Kenward, A., Yawitz, D., Sanford, T., & Wang, R. 2014. *Summer in the City: Hot and Getting Hotter*. Princeton, NJ: Climate Central. Retrieved from http://assets.climatecentral.org/pdfs/UrbanHeatIsland.pdf.

Lawrence Berkeley National Laboratory. 2016. *Glossary*. Heat Island Group Environmental Energy Technologies Division Lawrence Berkeley National Laboratory University of California Berkeley, California. Retrieved from https://heatisland.lbl.gov/glossary#letter_s.

Lin, S., Hsu, W.H., Van Zutphen, A.R., Saha, S., Luber, G., & Hwang, S.A. 2012. Excessive Heat and Respiratory Hospitalizations in New York State: Estimating Current and Future Public Health Burden related to Climate Change. *Environmental Health Perspectives* 120(1), 1571–1577.

Lo, C.P., Quattrochi, D.A., & Luvall, J.C. 1997. Application of High-Resolution Thermal Infrared Remote Sensing and GIS to Assess the Urban Heat Island Effect. *International Journal of Remote Sensing* 18(2), 287–304.

Maloley, M. 2009. *Thermal Remote Sensing of Urban Heat Island Effects: Greater Toronto Area; Report; Enhancing Resilience to Climate Change Program.* Ottawa, ON, Canada: Natural Resources Canada.

Maurer, E.P., Adam, J.C., and Wood, A.W. 2009. Climate model based consensus on the hydrologic impacts of climate change to the Rio Lempa basin of Central America. *Hydrology and Earth System Sciences* 13, 183–194.

McMichael, A., Woodruff, R.E., & Hales, S. 2006. Climate Change and Human Health: Present and Future Risks. *The Lancet* 367(9513), 859–869.

National Aeronautics and Space Administration (NASA). 2016. Atlanta's Urban Heat Island. *Scientific Visualization Studio*. Retrieved from http://svs.gsfc.nasa.gov/stories/Landsat/atlanta_heat_background.html.

National Center for Atmospheric Research-Community Climate System Model (NCAR-CCSM4). 2016. *World Temperature Change 2050 Scenario 6.0*. URL: http://www.arcgis.com/home/item.html?id=5f989ab94fb34b91ab1a44a73a9d5e72.

National Oceanic and Atmospheric Administration–National Centers for Environmental Information (NOAA-NCEI). 2016. State of the Climate: Global Climate Report-October 2015. Retrieved from www.ncdc.noaa.gov/sotc/global/201510.

Patz, J.A., Campbell-Lendrum, D., Holloway, T., & Foley, J.A. 2005. Impact of Regional Climate Change on Human Health. *Nature* 438, 310–317.

Rosenfeld, A.H., Akbari, H., Romm, J.J., & Pomerantz, M. 1998. Cool Communities: Strategies for Heat Island Mitigation and Smog Reduction. *Energy and Buildings* 28(1), 51–62.

Rosenzweig, C., Solecki, W.D., & Slosberg, R.B. 2006. *Mitigating New York City's Heat Island with Urban Forestry, Living Roofs and Light Surfaces*. Final Report prepared for New York State Energy Research and Development Authority.

Shahmohamadia, P., Che-Ania, A.I., Etessamb, I., Mauludc, K.N.A., & Tawil, N.M. 2011. The Need to Mitigate Urban Heat Island Effects on Human Health. *Procedia Engineering* 20, 61–70.

Sharma, A. 2016. Green and Cool Roofs Provide Relief for Hot Cities, but Should Be Sited Carefully. *The Conversation* (June 30, 2016). Retrieved from https://theconversation.com/green-and-cool-roofs-provide-relief-for-hot-cities-but-should-be-sited-carefully-60766.

Snyder, W., Wan, Z., Zhang, Y., & Feng, Y.-Z. 1998. Classification-Based Emissivity for the EOS/MODIS Land Surface Temperature Algorithm. *International Journal of Remote Sensing* 19(14), 2753–2774.

Solecki, W.D., Rosenzweig, C., Parshall, L., Pope, G., Clark, M., Cox, J., & Wiencke, M. 2005. Mitigation of the Heat Island Effect in Urban New Jersey. *Global Environmental Change Part B: Environmental Hazards* 6(1), 39–49.

Stewart, I.D., & Oke, T.R. 2012. Local Climate Zones for Urban Temperature Studies. *Bulletin of American Meteorological Society* 93, 1879–1900.

Stone, B., & Rodgers, M.O. 2001. Urban Form and Thermal Efficiency: How the Design of Cities Influences the Urban Heat Island Effect. *Journal of American Planning Association* 67(2), 186–198.

Stone, B., Vargo, J., & Habeeb, D. 2012. Managing Climate Change in Cities: Will Climate Action Plans Work? *Landscape and Urban Planning* 107(3), 263–271.

Stone, B., Vargo, J., Liu, P., Habeeb, D., DeLucia, A., Trail, M., Hu, Y., & Russell, A. 2014. Avoided Heat-Related Mortality Through Climate Adaptation Strategies in Three US Cities. *PLoS ONE* 9(6), e100852.

Stone, B., Vargo, J., Liu, P., Hu, Y., & Russell, A. 2013. Climate Change Adaptation Through Urban Heat Management in Atlanta, Georgia. *Environmental Science & Technology* 47(14), 7780–7786.

Taha, H. 1997. Urban Climates and Heat Islands: Albedo, Evapotranspiration, and Anthropogenic Heat. *Energy and Buildings* 25(2), 99–103.

Tomlinson, C.J., Chapman, L., Thornes, J.E., & Baker, C.J. 2011. Including the Urban Heat Island in Spatial Heat Health Risk Assessment Strategies: A Case Study for Birmingham, UK. *International Journal Health Geographics* 10(1), 42.

Tomlinson, C.J., Chapman, L., Thornes, J.E., & Baker, C.J. 2012. Derivation of Birmingham's Summer Surface Urban Heat Island from MODIS Satellite Images. *International Journal of Climatology* 32(2), 214–224.

Trenberth, K.E. 2011. Changes in Precipitation with Climate Change. *Climate Research* 47, 123–148.

Triantakonstantis, D., & Mountrakis, G. 2012. Urban Growth Prediction: A Review of Computational Models and Human Perceptions. *Journal of Geographic Information System* 4, 555–587.

United States Environmental Protection Agency (EPA). 2008. Urban Heat Islands Basic. In: *Reducing Urban Heat Islands: Compendium of Strategies*. Washington, DC: U.S. Environmental Protection Agency. Retrieved from www.epa.gov/heat-islands/reducing-urban-heat-islands-compendium-strategies.

United States Environmental Protection Agency (EPA). 2009. Vocabulary Catalog: Heat Island Effect Glossary. *Office of Air and Radiation*. Retrieved from http://ofmpub.epa.gov/sor_internet/registry/termreg/searchandretrieve/glossariesandkeywordlists/search.do?details=&vocabName=Heat%20Island%20Effect%20Glossary.

United States Environmental Protection Agency (EPA). 2016a. *Heat Island Effect*. Retrieved from www.epa.gov/heat-islands.

United States Environmental Protection Agency (EPA). 2016b. *Heat Islands: Measuring Heat Islands*. Retrieved from www.epa.gov/heat-islands/measuring-heat-islands.

United States Geological Survey (USGS). 2015. *Landsat 8 (L8) Data Users Handbook, LSDS-1574*. Version 1.0. Retrieved from http://landsat.usgs.gov/Landsat8_Using_Product.php.

Urban, B., & Roth, K. 2010. *Guidelines for Selecting Cool Roofs*. U.S. Department of Energy. Retrieved from https://heatisland.lbl.gov/sites/all/files/coolroof-guide_0.pdf.

Wang, X., Guo, Y., FitzGerald, G., Aitken, P., Tippett, V., Chen, D., & Tong, S. 2015. The Impacts of Heatwaves on Mortality Differ with Different Study Periods: A Multi-City Time Series Investigation. *PloS One* 10(7), e0134233.

Wilhelmi, O.V., & Hayden, M.H. 2010. Connecting People and Place: A new Framework for Reducing Urban Vulnerability to Extreme Heat. *Environmental Research Letters* 5(1), 014021.

Yuan, F., & Bauer, M.E. 2007. Comparison of Impervious Surface Area and Normalized Difference Vegetation Index as Indicators of Surface Urban Heat Island Effects in Landsat Imagery. *Remote Sensing of Environment* 106(3), 375–386.

Zhang, X., Zhong, T., Feng, X., & Wang, K. 2009. Estimation of the Relationship Between Vegetation Patches and Urban Land Surface Temperature with Remote Sensing. *International Journal of Remote Sensing* 30(8), 2105–2118.

Zhao, L., Lee, X., Smith, R.B., & Oleson, K. 2014. Strong Contributions of Local Background Climate to Urban Heat Islands. *Nature* 511(7508), 216–219.

12 Climate Hazards and Impact on Public Health

Chapter Objectives

This chapter focuses on answering the following questions:

- What are common health impacts caused or exacerbated by climate change and climate hazards?
- Why should we be concerned as a society?
- What is the range of geospatial tools and techniques used for vulnerability and risk assessments?
- What are some commonly encountered data issues when undertaking such analyses?
- What are some strategies to mitigate health impacts from climate hazards?

Health Impacts of Climate Change and Weather-Related Disasters

Increases in global temperatures and predicted increases in extreme rainfall and tornadic and hurricane activity, as well as prolonged drought and forest fires, can have adverse impacts on human health. This section provides an overview of several categories of health impacts from weather changes: (i) heat stress from exposure to extreme heat events; (ii) respiratory illness associated with increased wildfire exposure but also air pollutants associated with increased ambient air temperature; (iii) disaster-related injuries; (iv) food- and waterborne diseases that have a linked climate variability component; (v) vector-borne and zoonotic illnesses; and (vi) issues around disaster exposure and mental health implications. These health impacts and outcomes are largely driven by geographical context (Cromley & McLafferty, 2012), including natural hazard exposure.

Heat Stress

Climate change poses a direct threat to human health and potential increases in heat-related mortality and morbidity in the future (Hondula, Balling, Vanos & Georgescu, 2015). Rising temperatures are projected to produce an increased incidence of heat waves throughout the planet (Forsberg et al., 2012). The Intergovernmental Panel on Climate Change (IPCC) reports that "it is *virtually certain* that there will be more frequent hot, and fewer cold temperature, extremes over most land areas on daily and seasonal timescales, as global mean surface temperature increases" (IPCC, 2014, p. 10). Heat waves occur when consecutive nights of elevated nighttime temperatures are preceded by stagnant air masses (Luber & McGeehin, 2008). Exposure to prolonged heat waves can cause death from heat stroke, as well as serious stress on the body, such as renal failure due to dehydration and other long-lasting health impacts (Rosen, 2016).

Researchers found that emergency medical services faced greater demands during heat waves in **King County, Washington** (Calkins, Isaksen, Stubbs, Yost & Fenske, 2016). Another study found a 27% increase in early-term pregnancy deliveries during periods of excessive heat from 1981–2010 in **Montreal, Canada** (Auger, Naimi, Smargiassi, Lo & Kosatsky, 2014). Excessive heat stress can also lead to mortality. During the June 2015 heat wave in **Karachi, Pakistan**, residents were 17 times more likely to die of heat-related causes compared to the referent period the previous year (Ghumman & Horney, 2016).

A study by Wang et al. (2015) assessed the relationship between heat waves and mortality in three Australian cities: **Brisbane**, **Melbourne**, and **Sydney**. The study found increases in cardiovascular mortality during all heat wave periods (Wang, Armstrong, Zuza, Gasparrini, Linares & Diaz, 2015). In a similar multi-city study of the 2003 heat wave in Spain, Tobias et al. (2012) found a 25% increase in relative risk of heat-related deaths. Another study projected heat-related mortality would rise by 70% from 1990 to 2050 in the **New York City** region (Knowlton et al., 2007).

Respiratory Illness

Rising temperatures can impact respiratory health through the increase in climate-sensitive air pollutants, wildfires, extreme weather systems, and aeroallergens. Higher temperatures are likely to increase fungal growth and fungal spore transmission, which in turn intensify asthma and other respiratory illnesses (Rosen, 2016). Furthermore, higher atmospheric CO_2 concentrations may induce growth of plants (e.g., ragweed), and extremes in temperature are associated with increased risk of asthma hospitalizations (Soneja, Jiang, Fisher, Upperman, Mitchell & Sapkota,

2016). With increased wildfire activity under a changing climate, studies show particulate matter released during wildfires is associated with increased respiratory infections and possible links to all-cause mortality (Delfino et al., 2009; Wegesser, Pinkerton & Last, 2009; Henderson, Brauer, MacNab & Kennedy, 2011; Takaro, Knowlton & Balmes, 2013; Reid, Brauer, Johnston, Jerrett, Balmes & Elliott, 2016). Higher temperatures are also associated with longer ozone-exceedance days. Ozone is an air pollutant linked to elevated asthma and chronic obstructive pulmonary disease hospitalization (Rosen, 2016). Several climate models predict increases in ozone concentration. As Kinney (2008) notes, more work with integrated modeling around air pollutants and climate change simulation is needed, specifically around particulate matter composition and aeroallergens (Kinney, 2008).

Injury

Hurricane, flooding, and drought extremes can result in direct and indirect forms of injury. Storm surge can lead to immediate mortality through exposure to floodwaters (Lane, Charles-Guzman, Wheeler, Abid, Graber & Matte, 2013). Migration and displacement may also indirectly lead to increased risk of injury and conflict. One study conducted 12 months after Hurricane Katrina found that compared to non-displaced individuals, displaced older adults in new and inferior housing had increased risk of hip fractures, sprains, and lacerations (Uscher-Pines, Vernick, Curriero, Lieberman & Burke, 2009). Another study found increased risk of intimate partner violence and use of violence to resolve conflicts among individuals exposed to Hurricane Katrina (Harville, Taylor, Tesfai, Xu & Buekens, 2011). Displacement and conflict also lead to malnutrition (Mason et al., 2012). Patz and Hatch (2014) estimate an increase in the population at risk of hunger from 34% to 64%–72% by 2050 due to drought and crop failure. These examples can be considered indirect health effects, yet there are knowledge gaps in understanding how climate change, migration, and conflicts interrelate (Jacobs & Harville, 2015).

Food- and Waterborne Diseases

Heavy rainfall and flooding facilitate the movement of pathogens (such as *E. coli*, cryptosporidium, and *Campylobacter jejuni*) from sewage and other sources into the water system (Patz & Hatch, 2014). As such, increased rainfall promotes the transmission of water- and food-borne diseases. Increased incidence of salmonellosis has been connected with increased temperatures (Patz & Hatch, 2014), and other diarrheal diseases such as cholera are expected to become more prevalent. Warmer ocean water is associated with the spread of bacteria in raw oysters, causing diarrheal diseases (Weber, 2010). Small island states are particularly vulnerable to the heat effect of climate change (Ebi, Lewis & Corvalan, 2006). In tropical regions, such as the **Federated States of Micronesia** in the Pacific Ocean, temperature increases have been associated with increased incidence of diarrheal disease (McIver et al., 2015).

A **Taiwanese** study found that climatic variability in ocean and surface temperature, rainfall patterns, and other variables was associated with increased incidences of food poisoning (Hsiao, Jan & Chi, 2016). The researchers obtained monthly outbreak data of *Vibrio parahaemolyticus*[1] for the period 2000 to 2011 from the Taiwanese Department of Health. Climatic data for that study was obtained from the Ministry of Science and Technology and the Central Weather Bureau and included temperature, relative humidity, daily rainfall, ocean temperature, and ocean salinity. The authors found that ambient temperature, ocean salinity, and ocean temperature were positively associated with *Vibrio parahaemolyticus* outbreaks and that these climate variables, compared to rainfall patterns, may have a higher impact on microbial concentrations than rainfall patterns (Hsiao, Jan & Chi, 2016).

Vector-borne and Zoonotic Diseases

The spread of vector-borne and zoonotic diseases is indirectly related to changing climatic conditions. Insects transmit many pathogens, such as malaria and chikungunya, that affect human health. Malaria spread by mosquitos and schistosomiasis spread by aquatic snails have been linked to flooding and increasing temperatures (IPCC, 2008). Tick-borne diseases, such as Lyme disease, are forecasted to increase in Canada (where they are uncommon) as temperatures continue to increase (Greer, Ng & Fisman, 2008). Animals that carry disease transmissible to humans also pose a danger to health. The climate-sensitive drivers of disease transmission include the expansion of suitable habitats for the various disease carriers, shortened incubation periods, host-seeking behaviors, and increased transmission rates (Gage, Burkot, Eisen & Hayes, 2008; Hii et al., 2009; Patz & Hatch, 2014; Weber, 2010).

Hii et al. (2009) found that increases in heat and precipitation generated an increase in the transmission of dengue fever with a lag time of up to 20 weeks. Nakazawa et al. (2007) modeled the geographic shift, although subtle, of tularemia and plague into the northern limits of the United States due to climate change (Nakazawa, Williams, Peterson, Mead, Staples & Gage, 2007). In **Uganda**, researchers found rainfall patterns predictive of human cases of plague (Moore, Monaghan, Griffith, Apangu, Mead & Eisen, 2012).

The Zika outbreak in 2016 was a global public health emergency. The Zika virus is transmitted by the *Aedes aegypti* mosquito, which has the ability to transmit yellow fever and dengue fever. The mosquito is expected to spread during periods of above average rainfall, such as during an El Niño event or projected global warming (World Health Organization [WHO], 2016). In South America, El Niño is expected to cause increases in precipitation, which in turn lead to flooding and an increase in vector-borne diseases (WHO, 2016). The 2015–2016 El Niño is comparable to the 1998–1999 event, during which outbreaks of malaria increased by 440% in **Ecuador** (WHO, 2016). It is important to note, however, that the spread of infectious diseases is also subject to non-climatic factors as well—such as the built environment, migration, and the lack of public health infrastructure.

Mental Health Sequelae

The risk of psychological trauma is expected to increase as storms become more extreme under a changing climate. Post-traumatic stress disorder (PTSD) is the most common mental health illness associated with exposure to natural disasters (Caffo & Belaise, 2003). After controlling for demographic factors, one study found that **New York City** residents exposed to Hurricane Sandy had a greater likelihood of depression and PTSD (Schwartz et al., 2015). These stressors may have longer lasting impacts compared to other health stressors associated with natural disasters. A study in **China** found that 15% of the flood victims surveyed were initially diagnosed with PTSD and were still diagnosed with PTSD more than a decade after the event (Dai et al., 2016). Other mental health problems associated with traumatic events include substance abuse and suicidal ideations (Krug et al., 1998; Anastario, Larrance & Lawry, 2008; Lowe, Sampson, Gruebner & Galea, 2015). Similar to the spread of infectious diseases, mental health issues are indirectly related to climate change. Additional drivers of mental health include biology, social support systems, and adequate access to mental healthcare, as well as a host of other community factors (Acierno et al., 2007; Lowe, Sampson, Gruebner & Galea, 2015; Gruebner et al., 2016; Platt, Lowe, Galea, Norris & Koenen, 2016).

Low-income countries will likely suffer more than high-income countries given that they do not have the health system infrastructure to be able to respond to the health needs of citizens. Rural areas may not have the financial resources to meet the growing economic costs of more extreme weather conditions. As cities continue to develop in sprawling unstainable patterns, increasing temperatures and extreme precipitation events will begin to tax the ability of public health systems to adapt to a changing climate.

Assessing Vulnerability and Risk

> *At its most basic level, a geographic foundation for public health looks at the question "Where?" Where do people live? Where are the agents of disease? Where can we intervene to eliminate risks or to improve health services delivery? Both people and agents that cause disease in humans are dispersed, often unevenly, across communities and regions. The processes that bring people into contact with disease agents and that impact their access to social and material resources are also geographically variable.*
>
> —Cromley and McLafferty (2012), pp. 1–2

Geospatial technologies are commonplace in the public health field, widely used to inform public health intervention activities. GIS (geographic information system) facilitates exploration and analysis of large databases of health events and risk factors, crisis mapping, and modeling of temporal and spatial patterns of hazards and risks (Hay, Randolph & Rogers, 2000; Tanser & Le Sueur, 2002; Kulldorff, Tango & Park, 2003; Jarup, 2004; Waller & Gotway, 2004; Waring, Zakos-Feliberti, Wood, Stone, Padgett & Arafat, 2005; Mills & Curtis, 2008; Elliott & Pais, 2010; Cromley & McLafferty, 2012). Other applications range from the use of GIS to support post-disaster rapid epidemiological assessments to spatial inferential statistics such as cluster analyses and spatially weighted regression modeling. This section provides examples of various tools and geospatial applications for a range of analysis units, from country to neighborhood.

Nations and Large Regions

El Morjani Zel, Ebener, Boos, Abdel Ghaffar, and Musani (2007) discussed the development of the Atlas of Disaster Risk for the World Health Organization. The tool mapped the potential distribution of flood, landslide, wind speed, heat, and seismic hazards by country within the eastern Mediterranean region. Population size, climatic parameters such as ambient temperature, and measures of hazard vulnerability were used to create a risk index methodology with the purpose of assisting emergency preparedness and disaster response practitioners (El Morjani Zel, Ebener, Boos, Abdel Ghaffar & Musani, 2007). The Community Assessment for Public Health Emergency Response (CASPER) designed by the Centers for Disease Control and Prevention is a cluster sampling methodology to assess community needs after a disaster. The toolkit, which has been deployed in major U.S. disasters, has a mapping component to aid in sample household survey selection through U.S. Census data (Bayleyegn et al., 2015). In addition to assessing community needs from natural and man-made disasters, CASPERs have been deployed in non-emergency settings to support public health action and to inform household-level disaster preparedness and response (Bayleyegn et al., 2015).

Anno et al. (2015) used remote sensing and GIS analyses to understand the ecological and sociodemographic factors of dengue fever spread over space and time in northern **Sri Lanka**. The spatial unit of analysis was at the health administrative division, with socioeconomic and demographic data provided by the Sri Lanka Ministry of Health. The administrative divisions provided data on confirmed dengue fever cases. Satellite data were obtained through the Earth Observation Research Center for the ecological predictors of rainfall, humidity, temperature, and urbanization patterns. The authors tested the temporal and spatial association between the predictors and outbreaks (Anno et al., 2015). Spatiotemporal clustering was then identified using the Kulldorff cluster detection method. The authors found statistically significant hotspots of dengue fever cases in areas of increased temperature, rainfall, and urbanization as well as multiple space-time clustering of dengue fever outbreaks.

In 2013, a large-scale dengue fever outbreak occurred in northeast **Pakistan**. Wesolowski et al. (2015) combined dengue fever spread with patterns of population mobility in Pakistan. The authors used mobile phone data linked to climate data to predict the spatial and temporal trends of dengue fever outbreaks in a retrospective epidemiological study. Subscriber call data records from the largest mobile network operator in the region were aggregated to small administrative boundaries. Data for dengue fever cases were provided by the provincial health department. The authors compared estimates of dengue fever outbreaks using an ordinary differential equation model and compared those estimates to gravity models of imported dengue fever cases through mobile phone movements. Wesolowski et al. (2015) found that mobile-based estimates did accurately predict the large-scale dengue fever outbreak in Pakistan. As a result, the authors were able to develop dynamic risk maps based on climate suitability and the influx of travelers from endemic areas.

Yao, Brauer, and Henderson (2013) utilized GIS methods to evaluate a wildfire smoke forecasting system in **British Columbia, Canada**. The researchers focused on the time period between July 24 and August 29, 2010, when 330,000 hectares of forest burned. Using local health areas (LHAs) as the unit of analysis, the data compiled included daily BlueSky 48 hour forecasts for daily average particulate matter, actual PM2.5 concentrations from 36 monitoring stations,

Hazard Mapping System (HMS) smoke plume forecasts, daily counts of asthma drug salbutamol sulfate dispersions in each LHA, and outpatient physician visits for asthma complaints. The data were georeferenced and overlaid on a map, and statistical analysis was performed. BlueSky was found to provide a better forecast than other forecast systems and performed best during intense fire periods. BlueSky also provided estimates consistent with observed health outcome data (Yao, Brauer & Henderson, 2013).

Cities

Health impacts of climate change can affect the resilience of cities. Research by Owrangi, Lannigan, and Simonovic (2015) presents a methodology to identify and assess climate-related health impacts in large coastal cities. The human health impact indicator was composed for **Vancouver, Canada**, using a combination of three years of data (2001, 2006, 2011) on physical attributes (i.e., land elevation), disease burden, and social vulnerability. The index was generated from a weighted average of physical and non-physical impacts, and the components of the non-physical sub-index were equally weighted. GIS was used to normalize and integrate, then display the data, and to generate the final human health impact map.

A study by Lane et al. (2013) mapped health vulnerability indicators in **New York City** neighborhoods based on lessons learned from Hurricane Sandy, Hurricane Katrina, and other storms that severely impacted urban coastal areas in the United States. Numerous pathways for storm exposure health effects were identified. A path diagram was assembled and included exposure to secondary hazards such as contaminated drinking water, contact with contaminated floodwaters, mold and moisture in housing, and mental health effects from traumatic or stressful experiences during and after the storms (Lane, Charles-Guzman, Wheeler, Abid, Graber & Matte., 2013). GIS was utilized to map vulnerability indicators that included poverty level, housing condition, social issues, health issues, age, disability, health status, insurance, race, and language (Lane, Charles-Guzman, Wheeler, Abid, Graber & Matte, 2013).

Gruebner et al. (2016) used longitudinal data to examine whether geographical variations of mental health outcomes existed over time for populations in **Galveston, Texas**, affected by Hurricane Ike. The authors investigated spatial clustering, predictors of mental health trajectories, and the endurance, or persistence, of those outcomes. Gruebner et al. (2016) used address-level data for their study to examine post-traumatic stress symptoms and depression from 1 month prior to Hurricane Ike and up to 8 months after the storm. They adjusted for education level, age, gender, and ethnicity. Measures of social support, pre-disaster trauma, and hurricane-related trauma and stressors were also included, and unadjusted and predictor-adjusted spatial scan statistics[2] were used to identify spatial clustering of the mental health trajectories. The authors found statistically significant clustering of chronic post-traumatic stress symptoms, particularly among the elderly, Hispanics, and those who experienced sentimental or financial loss. Delayed onset of post-traumatic stress symptoms and delayed onset depression were highest among non-Hispanic blacks (Gruebner et al., 2016).,

Counties and Neighborhoods

A study by Harlan, Declet-Barreto, Stefanov, and Petitti (2012) focused on the temperature-mortality relationship in urbanized areas in **Maricopa County, Arizona**. The factors used for population vulnerability included: the very young and very old, people with illness, the disabled, pregnant women, those who did not have access to air conditioning, and the poor. The authors utilized remotely sensed data (temperature and Normalized Difference Vegetation Index [NDVI]) from Landsat with 30-meter resolution, along with tax parcel data, death records from Maricopa County Department of Public Health, and 2000 U.S. Census data aggregated at the census block group level. The data were used to operationalize neighborhood vulnerability into several indicators: ethnic minority, Latino immigrant, below the poverty line, less than high school education, no central air conditioning, over 65 years of age, living alone, and unvegetated area. These variables produced factors of socioeconomic vulnerability. The elderly and those in isolation were the factors identified as predictive of heat-related deaths. More deaths were found in inner-city areas with vulnerable populations than in higher income, suburban areas (Harlan, Declet-Barreto, Stefanov & Petitti, 2012). GIS was used to spatially aggregate tax parcel data to census block group level in order to: (i) determine the percentage of homes in each block group that did not have air conditioning; (ii) aggregate NDVI and temperature data to the block group level; (iii) geocode death locations from the electronic death certificate database; and (iv) display the vulnerability data.

An older study by Harlan, Brazel, Prashad, Stefanov, and Larsen (2006) examined the microclimates of eight neighborhoods in **Phoenix, Arizona**, in order to understand the social justice aspect of coping with heat risk. The neighborhood attributes included: distance from city center, population density, amount of open space, and vegetation density. Socioeconomic data were obtained from the 2000 U.S. Census at the block group level; open space was obtained from land use characteristics data; vegetation density was obtained from Soil-Adjusted Vegetation Index (SAVI) with a resolution of 30 square meters; and qualitative social ties were measured using interview data. The authors developed the Human Thermal Comfort Index (HTCI) by neighborhood. The findings indicate that not only are neighborhoods with lower socioeconomic populations warmer than other neighborhoods, but the residents are the least equipped to deal with heat stress due to financial factors and very small social networks (Harlan, Brazel, Prashad, Stefanov & Larsen, 2006).

Case Study: Heat-Related Illnesses in California: Tracking and Distribution of Emergency Visits and Hospitalizations

Study Objective

As noted earlier in the chapter, climate change is expected to adversely affect or exacerbate pre-existing health problems such as asthma and cardiovascular diseases. Previous studies have shown that during days with excessive heat, emergency department visits for heat-related illnesses and chronic diseases increase disproportionately (Guirguis, Gershunov, Tardy & Basu, 2014), and, therefore, emergency medical services face greater demands during heat waves (Calkins, Isaksen, Stubbs, Yost & Fenske, 2016; Auger, Naimi, Smargiassi, Lo & Kosatsky, 2014). In this case study, we seek to understand the relationship between past records of heat wave days, the proportion of the socially vulnerable populations, the number of workers employed in the agricultural sector, and the number of emergency department visits and hospitalizations due to heat illness.

We focus on California given a recent trend of heat waves and record-high temperatures—a trend that is expected to become the norm in the West (McPhate, 2017). The state also has a history of prolonged drought periods, with severe drought conditions experienced most recently between 2011 and 2014. The 2006 heat wave is especially noteworthy for the magnitude and duration of high temperatures (Gershunov, Cayan & Iacobellis, 2009). The case study seeks to highlight the value of using GIS to assess health impacts of extreme heat events using time series of extreme heat days (i.e., days more than 95°C), socioeconomic variables, and hospitalizations for heat-related illnesses and deaths. This combination of data can provide important insights into physical and socioeconomic vulnerability.

Planning and Policy Applications

- Highlight the value and importance of environmental health tracking systems in climate adaptation planning.
- Assess populations vulnerable to heat-related illnesses in the context of heat waves.
- Identify patterns (space and time) of exposure to heat-related health impacts to facilitate prioritization of mitigation and adaptation initiatives.

Data and Models

- Number of heat wave days from May to September (1981–2010) for all counties in the U.S. The text file was downloaded from the Centers for Disease Control and Prevention (CDC) WONDER online database, released 2015 (http://wonder.cdc.gov/NCA-heatwavedays-historic.html).
- The California Environmental Health Tracking Program (www.cehtp.org/page/hri/query) website provides a key source of data on (i) crude rate[3] of emergency department visits due to heat (years 2005–2015 data available) and (ii) age-adjusted rate[4] of hospitalizations due to

heat (years 2000–2015 data available). For the purposes of this analysis, the crude rate of emergency department visits is most suitable as it accounts for the overall exposure. The age-adjusted rate provides the risk exposure of any given age group and in this case was used to map the rate of hospital admissions. See the following website for a full description of these two classifications: www.cehtp.org/faq/climate_change/heatrelated_illness_and_death_data_methods_and_limitations. A spreadsheet was compiled at the statewide level and ***for each county*** in California.

- Socioeconomic data and shapefiles for California counties downloaded from the American FactFinder of the U.S. Census Bureau: https://factfinder.census.gov/faces/nav/jsf/pages/index.xhtml.
- Agricultural employment in the state of California data downloaded from U.S. Census Bureau (*2010 Geography Area Series: County Business Patterns by Employment Size Class, Industry Statistics Portal, Economic Data for Programs published on American FactFinder*, www.census.gov/econ/isp).

Summary of Analysis

A simplified index of social vulnerability was constructed at the county level using the percent of people with disabilities, percent of the population below the federal poverty level, percent of the population less than 5 years old and 65 and older, and proportion of people without health insurance. The linear transformation approach, also known as min-max normalization (Equation 1), is used to standardize the raw data:

$$R_i = \frac{V_i - V_{min}}{V_{max} - V_{min}} \qquad \text{(Eq. 1)}$$

Where R_i is the standardized variable of interest, V_i is the value of the variable we seek to normalize, V_{min} is the minimum value of the variable of interest for all counties in California, and V_{max} is the maximum value in the dataset. The ordinary least squares (OLS) approach is employed to determine which of the variables of interest is a better predictor of the rate of emergency department visits and hospital admissions related to heat illness.

Key Findings and Sample Output

The results indicate that the rate of emergency department visits related to heat illness is not uniformly distributed throughout the State of California (Figure 12.1). Imperial and Tehama counties have the highest rates of emergency

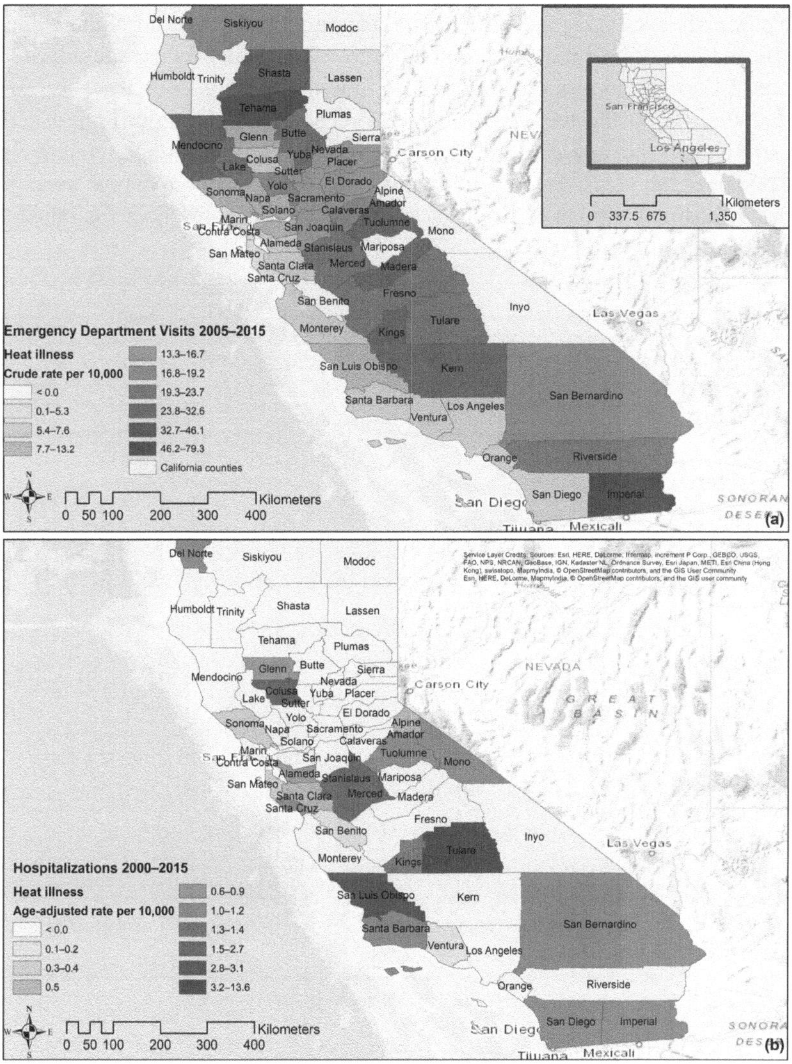

Figure 12.1 Heat-related illness in California: (a) annual average rate of emergency department visits for the period 2005–2015 (crude rate per 10,000); and (b) annual average rate of hospitalizations for the period 2000–2015 (age-adjusted rate per 10,000)

Source: California Environmental Health Tracking Program

department visits (a crude rate of 79.4 and 46.2 per 10,000, respectively) (Figure 12.2) while Tulare and San Luis Obispo counties exhibit the highest rates of heat illness–related hospitalizations (Figure 12.3). During the infamous 2006 heat wave, Imperial County had the highest rate of emergency department visits (Figure 12.2) while Tulare County had the highest age-adjusted rate of hospital admissions (Figure 12.3). The average age-adjusted rate of hospital admissions due to heat illness in Tulare County is 13.6 per 10,000 (23.3 during the 2006 heat wave). The graph shown in Figure 12.3 also suggests that age-adjusted rates of hospital admissions remain relatively stable over the years. Tulare County maintains an anomalously high average rate of hospitalizations due to heat exposure, much higher than the rates of other California counties.

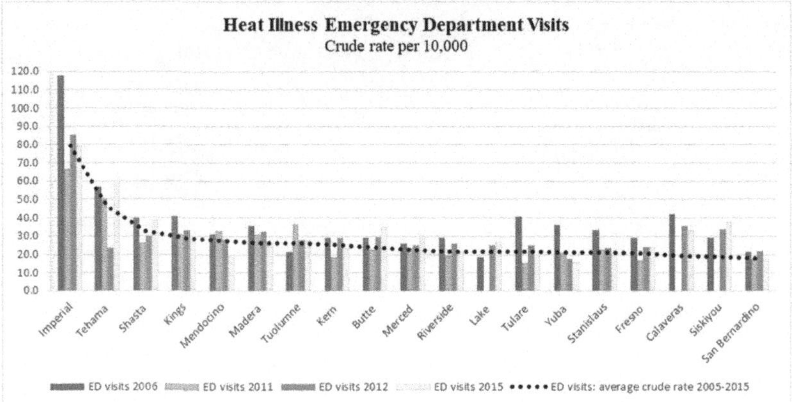

Figure 12.2 Emergency department visits for heat-related illnesses for the top tier counties

Source: California Environmental Health Tracking Program

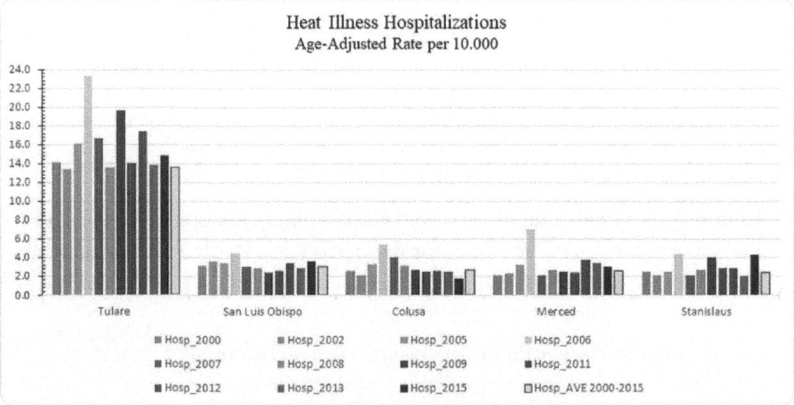

Figure 12.3 Hospital admissions due to heat-related illnesses for selected years for the top five counties

Source: California Environmental Health Tracking Program

To gain an insight into the contributing factors for these alarming trends, the relationship between the rates of ED (emergency department) visits for heat-related illnesses and three socioeconomic and physical variables was explored including the number of heat wave days, the level of social vulnerability, and the number of agricultural workers in each county. Figure 12.4a

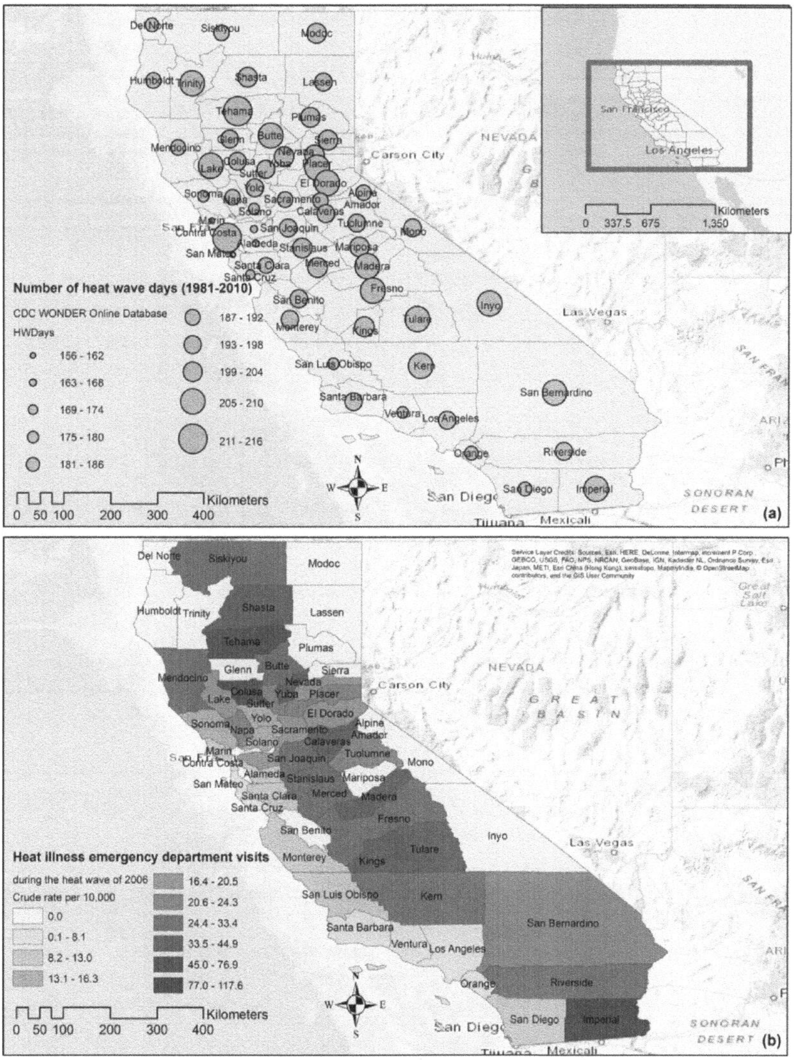

Figure 12.4 Heat waves in California: (a) the number of peak heat wave days over a 30-year period; and (b) emergency department visits for heat-related illnesses during the 2006 heat wave (crude rate per 10,000)

Source: CDC Wonder; California Environmental Health Tracking Program

indicates that approximately 30% of the California counties have experienced a total of 180 or more heat wave days over the past 30 years. A comparison between Figures 12.4a and 12.5b reveals that counties with a higher number of heat wave days also have a higher number of agricultural workers. These counties also have higher levels of social vulnerability (Figure 12.5a).

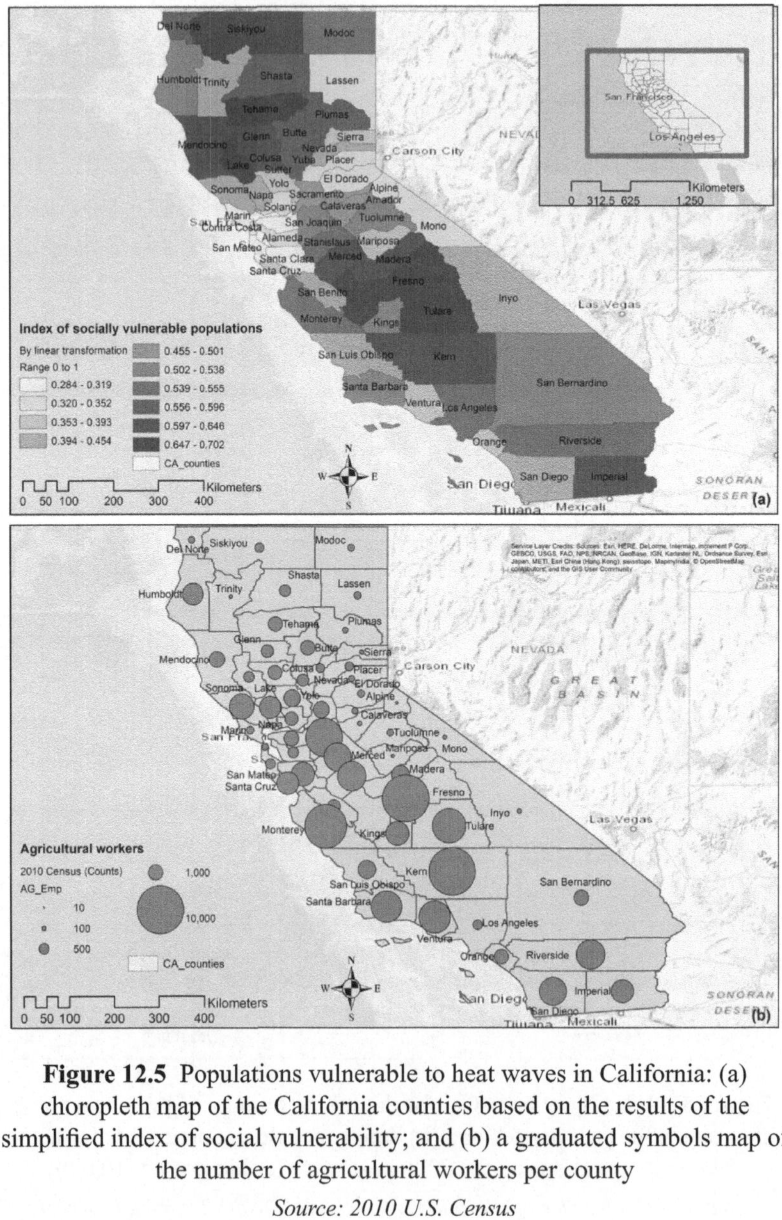

Figure 12.5 Populations vulnerable to heat waves in California: (a) choropleth map of the California counties based on the results of the simplified index of social vulnerability; and (b) a graduated symbols map of the number of agricultural workers per county

Source: 2010 U.S. Census

The rate of emergency department visits per county was regressed on social vulnerability, heat wave days, and the number of agricultural as predictor variables. The t-test for the β coefficients reveals that both the heat wave days and the number of agricultural workers per county are not statistically significant. A model with only the simplified index of social vulnerability as a predictor variable indicates a good fit to the data ($F = 17.48$, p value = 0.000). The t-statistic for the β coefficient is statistically significant at virtually any confidence level (Table 12.1).

ANOVA						
	df	SS	MS	F	Significance F	
Regression	1.00	5941.54	5941.54	17.48	0.000	
Residual	55.00	18693.52	339.88			
Total	56.00	24635.06				
Dependent variable: **ED visits during the 2006 heat wave**						
	Coefficients	**Standard Error**	**t Stat**	**P value**	**Lower 95%**	**Upper 95%**
Intercept	−22.05	10.35	−2.13	0.04	−42.78	−1.31
Index of socially vulnerable populations	86.82	20.76	4.18	0.00	45.20	128.43

Table 12.1 Results of the regression analysis linking social vulnerability to the number of ED visits during the 2006 heat wave.

Data Issues: Availability and Access

The range of applications presented in previous sections highlights the use of many different models and geographic scales. Furthermore, linking public health information to the geospatial data raises issues of spatial mismatch and privacy concerns as well as data availability, particularly for longitudinal studies. In analyzing long-term diabetes control after the earthquake, tsunami, and nuclear disaster in **Japan**, researchers found deteriorating health effects in particular sociodemographic groups. However, the cohort study was limited by the lack of access to hospital data (Leppold et al., 2016).

Satellite data can be used to track disease outbreaks and toxic harmful algae blooms. Data reliability concerns are raised when attempting to blend satellite imagery with ground source data (Seltenrich, 2014). While the use of satellite-based remote sensing data provides the long-term time frame needed in climate forecasting, there are inherent limitations, such as the inability to provide absolute values in air pollutant concentrations and reflect particulate matter in certain highly reflective environments (Seltenrich, 2016).

Web-based geospatial technologies are also growing in the field of disaster response. Researchers used these technologies to assist in disaster relief after the 2008 **Sichuan (China)** and 2010 **Haiti** earthquakes. Aerial and satellite imagery was provided by private vendors free of charge to be deployed in differing relief agencies' web mapping platforms. Researchers saw an explosion in the use of crowdsource mapping as citizens contributed data to open-source platforms such as MapGive, which is the U.S. State Department's program for supporting OpenStreetMap.[5] However, the use of crowdsource mapping has issues yet to be resolved. These include data security and the inability to feed updated information about the situation on the ground directly to on-site responders (Kawasaki, Berman & Guan, 2013, p. 217). Overall, conflicting missions between private, government, and non-government entities may delay access to critical geospatial information during disasters.

Mitigation and Public Health Adaptation Strategies

Efforts toward minimizing the impact of climate change on environmental health and public health require a multidisciplinary approach. First is the role of planners and engineers in shaping land use, land development, and transportation infrastructure policy. As noted by Rosen (2016, p. 3), "land use planning may be able to alter environmental health outcomes as risks change." Specifically, we should adopt strategies that minimize the effects of urban heat islands (see discussion in Chapter 11), encourage transportation options aimed at improving air quality, promote development patterns and building designs that reduce stormwater runoff and related water pollution problems, and optimize energy use. Geospatial technologies combined with modeling and forecasting techniques can help inform planners in such mitigation planning efforts. Rosen (2016) warns, however:

> While action is imperative and can mitigate negative outcomes, planning should approach interventions with caution, to avoid creating new health risks. Acting to address one climate change hazard can contribute to negative health outcomes from other hazards.
>
> (Rosen, 2016, p. 15)

Mitigation strategies in the form of reducing greenhouse gas emissions associated with climate change are often outside the purview of the core mission of public health agencies. However, public health professionals can play a role in advocating for mitigation strategies by leveraging exogenous health benefits associated with specific strategies. These include cleaner air and fewer traffic-related injuries and fatalities with the reduction of vehicle miles traveled, reducing obesity and mental health issues with increased walkable neighborhoods and green spaces, and better access to healthier nutritious foods associated with more sustainable farming practices. Inclusion of public health expertise through health impact assessments of proposed mitigation strategies may expand the greater possibility of policy acceptance and implementation.

Mitigation and adaptation strategies need to be jurisdictionally specific. For instance, **Spain's** national public health adaptation strategy places heavy emphasis on spatial assessments of vector-borne disease spread, the creation of early warning systems, and increased surveillance (Panic & Ford, 2013). Paterson et al. (2012) reported on issues related to climate adaptation within the public health system of **Ontario, Canada**. Using semi-structured interviews with public health officials, the authors identified perceptions, barriers, and opportunities for integrating adaptation efforts into ongoing public health system functions. In the United States, the **city of Philadelphia's** program around extreme heat events revolves around innovative partnerships with non-profits and the weather services to provide targeted communications to at-risk communities in order to reduce vulnerability (White-Newsome et al., 2014). Also important is the need for public health plans for heat to be locally tailored and not assume that pre-identified vulnerability indicators are universally applicable (Hondula, Davis, Saha, Wegner & Veazey, 2015).

Overall, there is need for targeted policy interventions, increased preparedness, and better governance of public health systems. Standardized metrics toward evaluating adaptation interventions, and increasing the role of vulnerability assessments for targeted interventions, were all cited by Paterson et al. (2012) as needed inputs in creating a public health system adaptive to climate change. Whether climate change will augment existing threats to the system or present new distinct threats is still unclear (Hess, McDowell & Luber, 2012). What is clear is that public health systems must play a major role in reducing the burdens of climate-sensitive health outcomes through routine agency functions.

Notes

1 *Vibrio parahaemolyticus* (*V. parahaemolyticus*) are bacteria that occur naturally in warm coastal areas and typically cause diarrhea. See www.foodsafety.gov/poisoning/causes/bacteriaviruses/vibrio_infections.

2 The spatial scan statistic is one of the main epidemiological tools to test for the presence of disease clusters in a geographical region (Zhang, Assuncao & Kulldorff, 2010)

3 Crude rates (i.e., unadjusted) are calculated by taking the total number of events for a given time period (e.g., 2000–2012) and geography (e.g., county) and dividing by the total underlying population for the same time period and geography. The rates are then multiplied by 100,000 and expressed as X events per 100,000 California residents (source of definition: California Environmental Public Tracking Network, 2014).

4 Age-adjusted rates take into account the age distribution of a population and are calculated to allow for direct comparisons between two or more populations at one point in time or between a single population at two or more points in time. Crude rates measure the true risk for a population, while age-adjusted rates are useful as a relative index of risk (source of definition: California Environmental Public Tracking Network, 2014).

5 Please see the following website for more details: https://mapgive.state.gov.

References

Acierno, R., Ruggiero, K.J., Galea, S., Resnick, H.S., Koenen, K., Roitzsch, J., de Arellano, M., Boyle, J., & Kilpatrick, D.G. 2007. Psychological Sequelae Resulting from the 2004 Florida Hurricanes: Implications for Postdisaster Intervention. *American Journal of Public Health* 97(Suppl 1), S103–108.

Anastario, M.P., Larrance, R., & Lawry, L. 2008. Using Mental Health Indicators to Identify Postdisaster Gender-Based Violence Among Women Displaced by Hurricane Katrina. *Journal of Women's Health* 17(9), 1437–1444.

Anno, S., Imaoka, K., Tadono, T., Igarashi, T., Sivaganesh, S., Kannathasan, S., Kumaran, V., & Surendran, S.N. 2015. Space-time Clustering Characteristics of Dengue based on Ecological, Socio-economic and Demographic Factors in Northern Sri Lanka. *Geospatial Health* 10(2), 376.

Auger, N., Naimi, A.I., Smargiassi, A., Lo, E., & Kosatsky, T. 2014. Extreme Heat and Risk of Early Delivery Among Preterm and Term Pregnancies. *Epidemiology* 25(3), 344–350.

Bayleyegn, T.M., Schnall, A.H., Ballou, S.G., Zane, D.F., Burrer, S.L., Noe, R.S., & Wolkin, A.F. 2015. Use of Community Assessments for Public Health Emergency Response (CASPERs) to Rapidly Assess Public Health Issues—United States, 2003–2012. *Prehospital Disaster Medicine* 30(4), 374–381.

Caffo, E., & Belaise, C. 2003. Psychological Aspects of Traumatic Injury in Children and Adolescents. *Child and Adolescent Psychiatric Clinics of North America* 12(3), 493–535.

California Environmental Public Tracking Network. 2014. *Heat-Related Inpatient Hospitalizations and Emergency Room Visits Among California Residents*, May–September, 2000–2012. Retrieved from www.cehtp.org/faq/climate_change/heat_related_inpatient_hospitalizations_and_emergency_room_visits_metadata (accessed 09/2017).

Calkins, M.M., Isaksen, T.B., Stubbs, B.A., Yost, M.G., & Fenske, R.A. 2016. Impacts of Extreme Heat on Emergency Medical Service Calls in King County, Washington, 2007–2012: Relative Risk and Time Series Analyses of Basic and Advanced Life Support. *Environmental Health* 15(1), 13.

Cromley, E.K., & McLafferty, S.L. 2012. *GIS and Public Health* (2nd edition). New York: The Guilford Press.

Dai, W., Chen, L., Tan, H., Wang, J., Lai, Z., Kaminga, A.C., Li, Y., & Liu, A. 2016. Association Between Social Support and Recovery from Post-Traumatic Stress Disorder After Flood: A 13–14 Year Follow-up Study in Hunan, China. *BMC Public Health* 16, 194.

Delfino, R.J., Brummel, S., Wu, J., Stern, H., Ostro, B., Lipsett, M., Winer, A., Street, D.H., Zhang, L., Tjoa T., & Gillen, D.L. 2009. The Relationship of Respiratory and Cardiovascular Hospital Admissions to the Southern California Wildfires of 2003. *Occupational and Environmental Medicine* 66(3), 189–197.

Ebi, K.L., Lewis, N.D., & Corvalan, C. 2006. Climate Variability and Change and Their Potential Health Effects in Small Island States: Information for Adaptation Planning in the Health Sector. *Environmental Health Perspectives* 1957–1963.

El Morjani Zel, A., Ebener, S., Boos, J., Abdel Ghaffar, E., & Musani, A. 2007. Modelling the Spatial Distribution of Five Natural Hazards in the Context of the WHO/EMRO Atlas of Disaster Risk as a Step Towards the Reduction of the Health Impact Related to Disasters. *International Journal of Health Geography* 6, 8.

Elliott, J.R., & Pais, J. 2010. When Nature Pushes Back: Environmental Impact and the Spatial Redistribution of Socially Vulnerable Populations. *Social Science Quarterly* 91(5), 1187–1202.

Forsberg, B., Braback, L., Keune, H., Kobernus, M., Krayer von Krauss, M., Yang, A., & Bartonova, A. 2012. An Expert Assessment on Climate Change and Health—with a European Focus on Lungs and Allergies. *Environment Health* 11(Suppl 1), S4.

Gage, K.L., Burkot, T.R., Eisen, R.J., & Hayes, E.B. 2008. Climate and Vector-Borne Diseases. *American Journal of Preventive Medicine* 35(5), 436–450.

Gershunov, A., Cayan, D.R., & Iacobellis, S.F. 2009. The Great 2006 Heat Wave Over California and Nevada: Signal of an Increasing Trend. *Journal of Climate* 22, 6081–6203.

Ghumman, U., & Horney J. 2016. Characterizing the Impact of Extreme Heat on Mortality, Karachi, Pakistan, June 2015. *Prehospital Disaster Medicine* 31(3), 263–266.

Greer, A., Ng, V., & Fisman, D. 2008. Climate Change and Infectious Diseases in North America: The Road Ahead. *Canadian Medical Association Journal* 178(6), 715–722.

Gruebner, O., Lowe, S.R., Tracy, M., Joshi, S., Cerda, M., Norris, F.H., Subramanian, S.V., & Galea, S. 2016. Mapping Concentrations of Posttraumatic Stress and Depression Trajectories Following Hurricane Ike. *Scientific Reports* 6, 32242.

Guirguis, K., Gershunov, A., Tardy, A., & Basu, R. 2014. The Impact of Recent Heat Waves on Human Health in California. *Journal of Applied Meteorology and Climatology* 53, 3–19.

Harlan, S.L., Brazel, A.J., Prashad, L., Stefanov, W.L., & Larsen, L. 2006. Neighborhood Microclimates and Vulnerability to Heat Stress. *Social Science and Medicine* 63(11), 2847–2863.

Harlan, S.L., Declet-Barreto, J.H., Stefanov, W.L., & Petitti, D.B. 2012. Neighborhood Effects on Heat Deaths: Social and Environmental Predictors of Vulnerability in Maricopa County, Arizona. *Environmental Health Perspectives* 121(2), 197–204.

Harville, E.W., Taylor, C.A., Tesfai, H., Xu, X., & Buekens, P. 2011. Experience of Hurricane Katrina and Reported Intimate Partner Violence. *Journal of Interpersonal Violence* 26(4), 833–845.

Hay, S.I., Randolph, S.E., & Rogers, D.J. 2000. *Remote Sensing and Geographical Information Systems in Epidemiology*. London, UK: Academic Press.

Henderson, S.B., Brauer, M., MacNab, Y.C., & Kennedy, S.M. 2011. Three Measures of Forest Fire Smoke Exposure and Their Associations with Respiratory and Cardiovascular Health Outcomes in a Population-Based Cohort. *Environmental Health Perspectives* 119(9), 1266–1271.

Hess, J., McDowell, J.Z., & Luber, G. 2012. Integrating Climate Change Adaptation into Public Health Practice: Using Adaptive Management to Increase Adaptive Capacity and Build Resilience. *Environ Health Perspectives* 120, 171–179.

Hii, Y.L., Rocklöv, J., Ng, N., Tang, C.S., Pang, F.Y., & Sauerborn, R. 2009. Climate Variability and Increase in Intensity and Magnitude of Dengue Incidence in Singapore. *Global Health Action* 2, 10.3402/gha.v2i0.2036. http://doi.org/10.3402/gha.v2i0.2036.

Hondula, D.M., Balling Jr., R.C., Vanos, J.K., & Georgescu, M. 2015. Rising Temperatures, Human Health, and the Role of Adaptation. *Current Climate Change Reports* 1(3), 144–154.

Hondula, D.M., Davis, R.E., Saha, M.V., Wegner, C.R., & Veazey, L.M. 2015. Geographic Dimensions of Heat-Related Mortality in Seven US Cities. *Environmental Research* 128(April), 439–452.

Hsiao, H., Jan, M., & Chi, H. 2016. Impacts of Climatic Variability on Vibrio *Parahaemolyticus* Outbreaks in Taiwan. *International Journal of Environment Research and Public Health* 13(2), 188.

Intergovernmental Panel on Climate Change (IPCC). 2008. Climate Change and Water. *Technical Paper of the Intergovernmental Panel on Climate Change*. Bates, B.C., Kundzewicz, Z.W., Wu, S., & Palutikof, J.P. (Eds.). IPCC Secretariat, Geneva, 210.

Intergovernmental Panel on Climate Change (IPCC). 2014. Climate Change 2014: Synthesis Report. *Contribution of Working Groups I, II and III to the*

Fifth Assessment Report of the Intergovernmental Panel on Climate Change. Pachauri, R.K., & Meyer, L.A. (Eds.). IPCC, Geneva, Switzerland, 151 pp.

Jacobs, M.B., & Harville, E.W. 2015. Long-Term Mental Health Among Low-Income, Minority Women Following Exposure to Multiple Natural Disasters in Early and Late Adolescence Compared to Adulthood. *Child Youth Care Forum* 44(4), 511–525.

Jarup, L. 2004. Health and Environment Information Systems for Exposure and Disease Mapping, and Risk Assessment. *Environmental Health Perspectives* 112(9), 995–997.

Kawasaki, A., Berman, M.L., & Guan, W. 2013. The Growing Role of Web-Based Geospatial Technology in Disaster Response and Support. *Disasters* 37, 201–221.

Kinney, P.L. (2008). Climate Change, Air Quality, and Human Health. *American Journal of Preventative Medicine* 35(5), 459–467.

Knowlton, K., Lynn, B., Goldberg, R.A., Rosenzweig, C., Hogrefe, C., Rosenthal, J.K., & Kinney, P.L. 2007. Projecting Heat-Related Mortality Impacts Under a Changing Climate in the New York City Region. *American Journal of Public Health* 97(11), 2028–2034.

Krug, E.G., Kresnow, M., Peddicord, J.P., Dahlberg, L.L., Powell, K.E., Crosby, A.E., & Annest, J.L. 1998. Suicide After Natural Disasters. *New England Journal of Medicine* 338(6), 373–378.

Kulldorff, M., Tango, T., & Park, P.J. 2003. Power Comparisons for Disease Clustering Tests. *Computational Statistics and Data Analysis* 42(4), 665–684.

Lane, K., Charles-Guzman, K., Wheeler, K., Abid, Z., Graber, N., & Matte, T. 2013. Health Effects of Coastal Storms and Flooding in Urban Areas: A Review and Vulnerability Assessment. *Journal of Environmental and Public Health* 2013, 913064.

Leppold, C., Tsubokura, M., Ozaki, A., Nomura, S., Shimada, Y., Morita, T., Ochi, S., Tanimoto, T., Kami, M., Kanazawa, Y., Oikawa, T., & Hill, S. 2016. Sociodemographic Patterning of Long-Term Diabetes Mellitus Control Following Japan's 3.11 Triple Disaster: A Retrospective Cohort Study. *BMJ Open* 6(7).

Lowe, S.R., Sampson, L., Gruebner, O., & Galea, S. 2015. Psychological Resilience After Hurricane Sandy: The Influence of Individual- and Community-Level Factors on Mental Health After a Large-Scale Natural Disaster. *PLoS One* 10(5), e0125761.

Luber, G., & McGeehin, M. 2008. Climate Change and Extreme Heat Events. *American Journal of Preventative Medicine* 35(5), 429–435.

Mason, J.B., White, J.M., Heron, L., Carter, J., Wilkinson, C., & Spiegel, P. 2012. Child Acute Malnutrition and Mortality in Populations Affected by Displacement in the Horn of Africa, 1997–2009. *International Journal of Environ Research Public Health* 9(3), 791–806.

McIver, L., Hashizume, M., Kim, H., Honda, Y., Pretrick, M., Iddings, S., & Pavlin, B. 2015. Assessment of Climate-Sensitive Infectious Diseases in the Federated States of Micronesia. *Tropical Medicine Health* 43(1), 29–40.

McPhate, P. 2017. California Today: The Extreme Heat to Come. *New York Times*, June 22, 2017. Retrieved from www.nytimes.com/2017/06/22/us/california-today-extreme-heat.html?mcubz=0 (accessed 09/2017).

Mills, J.W., & Curtis, A. 2008. Geospatial Approaches for Disease Risk Communication in Marginalized Communities. *Progress Community Health Partnerships* 2(1), 61–72.

Moore, S.M., Monaghan, A., Griffith, K.S., Apangu, T., Mead, P.S., & Eisen, R.J. 2012. Improvement of Disease Prediction and Modeling Through the Use of Meteorological Ensembles: Human Plague in Uganda. *PLoS ONE* 7(9), e44431.

Nakazawa, Y., Williams, R., Peterson, A.T., Mead, P., Staples, E., & Gage, K.L. 2007. Climate Change Effects on Plague and Tularemia in the United States. *Vector Borne Zoonotic Diseases* 7(4), 529–540.

Owrangi, A.M., Lannigan, R., & Simonovic, S.P. 2015. Mapping Climate Change-Caused Health Risk for Integrated City Resilience Modeling. *Natural Hazards* 77(1), 67–88.

Panic, M., & Ford, J.D. 2013. A Review of National-Level Adaptation Planning with Regards to the Risks Posed by Climate Change on Infectious Diseases in 14 OECD Nations. *International Journal of Environmental Research and Public Health* 10(12), 7083–7109.

Paterson, J.A., Ford, J.D., Ford, L.B., Lesnikowski, A., Berry, P., Henderson, J., & Heymann, J. 2012. Adaptation to Climate Change in the Ontario Public Health Sector. *BMC Public Health*, 12, 452.

Patz, J.A., & Hatch, M.J. 2014. Public Health and Global Climate Disruption. *Public Health Reviews* 35(1). Retrieved from https://publichealthreviews.biomedcentral.com/track/pdf/10.1007/BF03391697?site=publichealthreviews.biomedcentral.com (accessed 09/2017).

Platt, J.M., Lowe, S.R., Galea, S., Norris, F.H., & Koenen, K.C. 2016. A Longitudinal Study of the Bidirectional Relationship Between Social Support and Posttraumatic Stress Following a Natural Disaster. *J Trauma Stress* 29(3), 205–213.

Reid, C.E., Brauer, M., Johnston, F.H., Jerrett, M., Balmes J.R., & Elliott, C.T. 2016. Critical Review of Health Impacts of Wildfire Smoke Exposure. *Environmental Health Perspectives* 124(9), 1334–1343.

Rosen, J. 2016. Climate, Environmental Health Vulnerability and Physical Planning: A Review of the Forecasting Literature. *Journal of Planning Literature* 31(1), 3–22.

Schwartz, R.M., Sison, C., Kerath, S.M., Murphy, L., Breil, T., Sikavi, D., & Taioli, E. 2015. The Impact of Hurricane Sandy on the Mental Health of New York Area Residents. *American Journal of Disaster Medicine* 10(4), 339–346.

Seltenrich, N. 2014. Remote-Sensing Applications for Environmental Health Research. *Environmental Health Perspective* 122(10), A268–A275.

Seltenrich, N. 2016. A Satellite View of Pollution on the Ground Long-Term Changes in Global Nitrogen Dioxide. *Environmental Health Perspective* 3(24), A56.

Soneja, S., Jiang, C., Fisher, J., Upperman, C.R., Mitchell, C., & Sapkota, A. 2016. Exposure to Extreme Heat and Precipitation Events Associated with Increased Risk of Hospitalization for Asthma in Maryland, U.S.A. *Environmental Health* 15, 57.

Takaro, T.K., Knowlton, K., & Balmes, J.R. 2013. Climate Change and Respiratory Health: Current Evidence and Knowledge Gaps. *Expert Review of Respiratory Medicine* 7(4), 349–361.

Tanser, F.C., & Le Sueur, D. 2002. The Application of Geographical Information Systems to Important Public Health Problems in Africa. *International Journal of Health Geographics* 1, 4.

Tobias, A., Armstrong, B., Zuza, I., Gasparrini, A., Linares, C., & Diaz, J. 2012. Mortality on Extreme Heat Days Using Official Thresholds in Spain: A Multi-City Time Series Analysis. *BMC Public Health* 12, 133.

Uscher-Pines, L., Vernick, J.S., Curriero, F., Lieberman, R., & Burke, T.A. 2009. Disaster-Related Injuries in the Period of Recovery: The Effect of Prolonged Displacement on Risk of Injury in Older Adults. *Journal of Trauma* 67(4), 834–840.

Waller, L.A., & Gotway, C.A. 2004. *Applied Spatial Statistics for Public Health Data*. Hoboken, NJ: Wiley.

Wang, X.Y., Guo, Y., FitzGerald, G., Aitken, P., Tippett, V., Chen, D., Wang, X., & Tong, S. 2015. The Impacts of Heatwaves on Mortality Differ with Different Study Periods: A Multi-City Time Series Investigation. *PLoS One* 10(7), e0134233.

Waring, S., Zakos-Feliberti, A., Wood, R., Stone, M., Padgett, P., & Arafat, R. 2005. The Utility of Geographic Information Systems (GIS) in Rapid Epidemiological Assessments Following Weather-Related Disasters: Methodological Issues Based on the Tropical Storm Allison Experience. *International Journal Hygiene and Environmental Health* 208(1–2), 109–116.

Weber, C.J. 2010. Update on Global Climate Change. *Urologic Nursing* 30(1), 81–84.

Wegesser, T.C., Pinkerton, K.E., & Last, J.A. 2009. California Wildfires of 2008: Coarse and Fine Particulate Matter Toxicity. *Environmental Health Perspectives* 117(6), 893–897.

Wesolowski, A., Qureshi, T., Boni, M.F., Sundsoy, P.R., Johansson, M.A., Rasheed, S.B., Engo-Monsen, K., & Buckee, C.O. (2015). Impact of Human Mobility on the Emergence of Dengue Epidemics in Pakistan. *Proceedings of National Academy Science USA* 112(38), 11887–11892.

White-Newsome, J.L., McCormick, S., Sampson, N., Buxton, M.A., O'Neill, M.S., Gronlund, C.J., Catalano, L., Conlon, K.C., & Parker, E.A. 2014. Strategies to Reduce the Harmful Effects of Extreme Heat Events: A Four-City Study. *International Journal of Environmental Research and Public Health* 11(2), 1960–1988.

World Health Organization (WHO). 2016. *El Niño May Increase Breeding Grounds for Mosquitoes Spreading Zika Virus, WHO Says*. Geneva, Switzerland: World Health Organization. Retrieved from www.who.int/hac/crises/el-nino/22february2016/en/.

Yao, J., Brauer, M., & Henderson, S.B. 2013. Evaluation of a Wildfire Smoke Forecasting System as a Tool for Public Health Protection. *Environmental Health Perspectives* 121(10), 1142.

Zhang, Z., Assuncao, R., & Kulldorff, M. 2010. Spatial Scan Statistics Adjusted for Multiple Clusters. *Journal of Probability and Statistics* 2010, 642379.

PART 4
Technical Approaches to Formulating Mitigation and Adaptation Strategies

13 Climate Resilience of Urban Systems and Interdependent Infrastructures

Chapter Objectives

This chapter focuses on answering the following questions:

- How do infrastructure interdependencies increase the vulnerability of urban systems to climate change impacts?
- What are some commonly used methodologies and tools to characterize and model infrastructure interdependencies?
- How have advances in knowledge and computing power improved our knowledge of the complex behavior of the interdependent infrastructure systems?

Infrastructure Interdependencies and Urban Systems

Critical infrastructure systems play a pivotal role in supporting essential urban services. Graham and Marvin (2001) coined the term *splintering urbanism* to emphasize the critical role of infrastructure networks in support of key urban innovations in energy and water efficiency, transportation logistics, telecommunications, and urban living. The inventive utilization of "smart" infrastructure supporting economic, social, government, and governance networks further evolved into the broadly defined concept of the "smart city" (Glaeser & Berry, 2006; Hollands, 2008; Komninos, 2009; Paskaleva, 2009). Alongside city governments (e.g., Amsterdam,

Vienna, Southampton, and Yokohama, among others), large business corporations including IBM, Oracle, Siemens, and Invensys are mobilizing resources to support the technological innovations needed to improve the performance of urban infrastructure systems. "Smart cities" seeking to foster sustainability and efficient use of resources, while boosting economic development, are increasingly relying on smart technologies and interdependent infrastructure networks to enable technological transformation and build institutional capacity (Deakin & Allwinkle, 2007; IBM, 2012; Townsend, 2013).

Critical infrastructure systems are highly diverse and interconnected. They incorporate "distributed networks, varied organizational structures and operating models (including multinational ownership), interdependent functions and systems in both the physical space and cyberspace, and governance constructs that involve multi-level authorities, responsibilities, and regulations" (The White House, 2013, online). Among the 16 critical infrastructure sectors identified by the U.S. Department of Homeland Security *Presidential Policy Directive—Critical Infrastructure Security and Resilience* (PPD-21), the energy and communications systems are recognized as "uniquely critical" because they support service delivery across other critical infrastructure sectors (The White House, 2013, online). Other key policy documents including the National Infrastructure Protection Plan (NIPP) 2013—*Partnering for Critical Infrastructure Security and Resilience* (U.S. Department of Homeland Security [DHS], 2013), and PS-Prep™ *The Voluntary Private Sector Preparedness Program* (Federal Emergency Management Agency [FEMA], 2009) further stressed the need for addressing the resiliency of critical infrastructure systems by implementing target criteria and preparedness standards. NIPP 2013 emphasized the importance of developing long-term projections for assessment and management of critical infrastructure interdependencies considering risks, vulnerabilities, sectoral forecasts, and future demand for services (DHS, 2013; Petit et al., 2015).

Across the United States and around the world, increased reliance on technological innovations is layered atop existing, often aging infrastructures (Haberman, 2014), including critical facilities such as power plants, bridges, hospitals, and water and wastewater treatment plants. It has been widely acknowledged that current infrastructure is ill-prepared to meet future challenges (Wilbanks et al., 2014; ASCE, 2017). The increasing complexity of the inter-reliant infrastructure systems has inevitably generated new sources of vulnerability stemming from multi-level cross-sectoral interactions between infrastructure systems, their components, and a larger-scale "system of systems" (Alderson & Doyle, 2010, p. 839). As PPD-21 points out, "just as the physical and cyber elements of critical infrastructure are inextricably linked, so are the vulnerabilities" (The White House, 2013, p. 6). Subsequently, interconnected infrastructures provide multiple "entry" points at which *faults* (i.e., disruptive events) (Stapelberg, 2008, p. 23) may occur as a result of natural and man-made disasters.

Many of the current threats to infrastructure reliability, survivability, and dependability are expected to intensify with climate change, as discussed in Chapter 10. There is mounting evidence of the extent to which engineered systems are becoming vulnerable to unanticipated climate-related disasters including extreme temperatures, low-probability high-impact flood events, tornadoes, droughts, and wildfires.

Furthermore, infrastructure systems could be simultaneously affected by multiple hazards, increasing the risk and amplifying the severity of disruptions (Ayyub, 2014). As Clark Miller (2012) points out in an article for *Future Tense: The Citizen's Guide to the Future* (*Slate*, *New America*, and Arizona State University [ASU]):

> Every building, bridge, levee, road, electricity grid, and power plant in this country has been designed and constructed to a set of engineering standards. Those standards embed static assumptions about the climatic conditions those structures will face. Few of those assumptions remain valid as the climate rapidly shifts away from the normal patterns experienced over the last 100 years. The further climate shifts, year after year, the less accurate those assumptions will be, and the more likely the country will be to face dramatic new systems failures.
>
> (Miller, 2012)

Infrastructure interdependencies often reveal themselves during extreme events in the form of cascading failures throughout multiple infrastructure systems, as illustrated by the devastating impacts of Hurricane Katrina (August 23–31, 2005), Superstorm Sandy (October 22–29, 2012), and more recently, Hurricane Harvey (August 25–September 2, 2017) and Hurricane Irma (September 1–10, 2017). Figures 13.1 depicts the aftermath of Hurricane Irma along the southeast coast of Florida.

A recent report prepared for the National Climate Assessment (NCA) provided a detailed overview of the potential risks and vulnerabilities of regional infrastructure systems to future extreme events, emphasizing the need for "revising engineering standards for buildings and other infrastructures to accommodate projected climate change" (Wilbanks et al., 2014, p. xix). The report predicted a sizeable increase in the demand for infrastructure services over the next 50 years due to a projected increase in population and demand for urban and suburban housing, both expected to nearly double by the year 2100 (Bierwagen et al., 2010, as cited in Wilbanks et al., 2014). The report noted that climate impacts would impose the need for critical infrastructure upgrades and modernization "especially in the regions and cities where much of the current infrastructure is aging and overstressed by demand levels it was not designed to meet" (Wilbanks et al., 2014, p. 16).

Recent reports from the United States have documented deterioration of asphalt pavement as a result of extreme temperatures (Meyer, Amekudzi & O'Hara, 2010; Schwartz et al., 2014). Analyzing the vulnerability of urban infrastructure to multiple hazards, Ayyub (2014) noted that Southern California would be increasingly vulnerable to flooding, drought, heat waves, wildfires, earthquakes, and tsunamis. Zimmerman and Faris (2010) analyzed the impact of climate-related hazards (e.g., heat waves, extreme rainfall events, and sea level rise) on various infrastructure sectors in New York City and drew from the experience of other large cities such as Chicago and London to develop recommendations for climate adaptation strategies.

The exposure of infrastructure systems to climate extremes is expected to increase in many parts of the world. The city of Cape Town in South Africa is expected to be increasingly exposed to storm surge, flooding, drought, fires, and heat waves (Ayyub, 2014). Many coastal cities in Southeast Asia are vulnerable to the simultaneous impacts of typhoons, surge, flooding, and landslides. Similarly, coastal

Figure 13.1 The impact of Hurricane Irma in Palm Beach County, Florida: (a) high winds blow trees over power lines, and (b) tree debris block roads and impede traffic

Source: Diana Mitsova

regions in Central and South America are expected to experience increased exposure to hurricanes, excessive precipitation, and flooding (Ayyub, 2014). Many of the European countries are facing similar challenges (European Environment Agency [EEA], 2017). In 2003 and 2006, record-high temperatures caused tar in road pavement to melt, leading to road closures throughout England (British Broadcasting Corporation [BBC], 2006).

Infrastructure Interdependency Analysis

Infrastructure interdependency analysis incorporates a broad range of perspectives based on both qualitative and quantitative approaches. Qualitative and empirical studies have focused primarily on identifying and defining infrastructure interdependencies at a conceptual level (Rinaldi, Peerenboom & Kelly, 2001; Zimmerman, 2001; Johansson & Jönsson, 2009; Yusta, Correa & Local-Arántegui, 2011; Huang, Liou & Chuang, 2014; Ouyang, 2014). Earlier works, specifically Zimmerman (2001) and Rinaldi, Peerenboom, and Kelly (2001), made major contributions toward understanding and classifying infrastructure interdependencies. Zimmerman (2001) provided a detailed account of functional and spatial interactions between various infrastructures. Functional interdependencies exist where two or more systems and their components are technically and operationally inter-reliant (Zimmerman, 2001; Rinaldi, Peerenboom & Kelly, 2001). Spatial interdependencies arise from co-locating utility lines and other infrastructure network components (Rinaldi, Peerenboom & Kelly, 2001; Zimmerman, 2001). This is often the case along transportation corridors, particularly when distribution networks are laid underground (Zimmerman, 2001). Rinaldi, Peerenboom, and Kelly (2001) expanded the classification by adding two additional interdependency dimensions: cyber (e.g., linkages through information technologies) and logical (i.e., systems and components are linked procedurally even if no other direct dependencies exist). Other scholars have expanded the conceptualization of interdependent infrastructures to include budgetary, socioeconomic, and policy considerations (Dunn, 2005; Kröger, 2008; Tolone et al., 2008; Zhang & Peeta, 2011). More recently, advances in geospatial technologies have offered novel approaches for exploring infrastructure interdependencies within the context of relational, topological, and object-oriented data models.

Infrastructure interdependencies can be classified using the following dimensions:

- **Order and type of interdependencies:** single domain versus multi-domain, cause-effect, generic cascading models, common-mode failure models, etc.
- **Temporal dimensions:** static or dynamic
- **Spatial dimension:** spatially explicit geographic information system (GIS)-based data models versus process-based time-dependent models
- **Modeling techniques:** risk analysis, optimization, graph theory-based models, agent-based models, input-output and computable general equilibrium models, systems dynamics models, hierarchical holographic modeling, dynamic control system models, models based on game theory, operations research models, etc.
- **Type of system represented:** engineering, economic, or social systems

Order and Type of Interdependencies

Drawing upon previous work by Rinaldi, Peerenboom, and Kelly (2001), Petit et al. (2015) developed a comprehensive analytical framework of critical infrastructure dependencies and interdependencies focusing on five dimensions: (i) *operating environment*, (ii) *coupling and response behavior*, (iii) *type of failures*, (iv) *infrastructure characteristics*, and (v) *state of operations*. In the proposed conceptual framework, the *operating environment* refers to local and regional contexts of service delivery including socioeconomic factors, regulatory and public policy considerations, health, safety and security issues, and political factors (Petit et al., 2015). The *coupling and response behavior* addresses the flexibility and complexity of response mechanisms after interdependent infrastructure systems are disrupted (Petit et al., 2015). The *type of failures* includes initial disruption caused by an extreme event: *cascading* effects (e.g., inoperability of one infrastructure leads to inoperability of another infrastructure) and *escalating* effects (e.g., inoperability of one infrastructure leads to higher levels of inoperability in one or multiple infrastructures) (Petit et al., 2015, p. 11). Furthermore, *infrastructure characteristics* are dependent upon decision-making structures, preparedness and mitigation planning, and task schedule dependencies of restoration and recovery. The final element included in this conceptual framework is the *state of operations* defined as normal, disrupted, or in repair (Petit et al., 2015, p. 11). While mostly qualitative, these conceptual studies have framed important research questions and have inspired a growing body of research literature on analyzing and quantifying infrastructure interdependencies (Yusta, Correa & Local-Arántegui, 2011; Ouyang, 2014; Roe & Schulman, 2015).

There has been much discussion in the literature of first-order infrastructure failures (resulting from the direct impacts of a disruptive event) and higher-order (second- and third-order) failures resulting from the indirect effects of initial disruptions (Chang, McDaniels, Mikawoz & Peterson, 2007; McDaniels, Chang, Peterson, Mikawoz & Reed, 2007; Johansson & Jönsson, 2009; Miles, Jagielo & Gallagher, 2016). Conceptually, these effects are defined as disruptions resulting from relationships (functional, logistic, geographic, or logical) between different types of infrastructure systems and infrastructure components (Rinaldi, Peerenboom & Kelly, 2001). For example, if two systems or components (*a* and *b*) are connected through one or more of the interdependencies described by Rinaldi, Peerenboom, and Kelly (2001), and system *c* relies on system *b* for service delivery to retain its functionality, then there is a second-order (indirect) dependency between *a* and *c* (Johansson & Hassel, 2010). Higher-order indirect impacts of disruptions of interdependent infrastructures cannot be fully captured without using the predictive capabilities of simulation models (Johansson & Hassel, 2010). Robert (2004) developed a three-tier approach to cascading infrastructure failures which includes: (1) an initial assessment of a wide range of potential failures of interrelated networks; (2) analysis of the possible cascading effects; and (3) creating a pool of transferred effects to determine key infrastructure vulnerabilities.

Representing Spatio-Temporal Dynamics in GIS

Knowledge of hazard exposure, potential damage, and recovery timelines are critical for the protection of critical infrastructures and the services they provide (Tolone et al., 2008). Traditional data models in GIS are well suited for spatial queries based on location or feature attributes but are constrained in their representations of time-dependent event specifications (Peuquet, 2002). Lo and Yeung (2002) defined data models as "a description or view of the real world" at a specific level of data abstraction that "tends to be tailored to a specific application or problem context"

(Lo & Yeung, 2002, p. 79). GIS-based data models vary in classification (how a collection of objects is created) and encapsulation (the way attributes and operations are bundled) (Lo & Yeung, 2002).

In the GIS environment, changes over time are modeled using *discrete-state spatial models* (Brown, Riolo, Robinson, North & Rand, 2005). Modeling of continuous processes with GIS requires discretization of space and time where the attributes of the spatial objects are updated at each time-step (Brown, Riolo, Robinson, North & Rand, 2005). Discrete time-steps are typically represented by *snapshots* or *time-stamped* map layers (Cromley & McLafferty, 2012, p. 248). Mapping the infrastructure restoration process can be visualized using a sequence of maps that depict the progression of the recovery efforts. Visualization techniques are based on the layering-relational concept where the data models are associated with a series of spatial attributes managed through a relational database management system (RDBMS). Changes in geographic objects occurring over space and time can be analyzed using the *time-stamped entries* that appear as separate records in the attribute table (e.g., 2000, 2005, 2010, etc.) (Cromley & McLafferty, 2012). In *map animations*, time-stamped maps are displayed as a dynamic sequence. Map animations can be adjusted to reflect chronological order and rate of change (Cromley & McLafferty, 2012).

Recent developments in geographic information science and technology offered new perspectives in spatiotemporal data modeling. Peuquet and Duan (1995) proposed an Event-based Spatio-Temporal Data Model (ESTDM) in which temporal information was associated with individual features and their topological relationships. A temporal vector in the form of "an event list" represented feature changes over space. An "ordered progression through time" was used to record changes in values of an attribute (property or a state) appended to the observed spatial feature (Peuquet & Duan, 1995, p. 12). Erwig, Gueting, Schneider, and Vazirgiannis (1999) discussed the fundamentals of designing database queries for moving objects. The authors described a collection of abstract spatiotemporal data types and outlined a number of applications discussing the strengths and weaknesses of the proposed approach (Erwig, Gueting, Schneider & Vazirgiannis, 1999). Other approaches to studying spatiotemporal data models incorporated event-based processes and change queries capturing relationships between events and causal relationships between events and the spatial object's state (Chen & Jiang, 2000). Tolone et al. (2004) identified three potential contributions of the evolving field of geographic information science (GIScience) to modeling critical infrastructure systems and their interactions: (i) visualization and spatial overlays, (ii) spatial data analysis, and (iii) rule-based approaches including agent-based modeling.

Approaches to Modeling Infrastructure Interdependencies

Modeling of infrastructure interdependencies focuses on identification of potential threats and vulnerabilities as well as on predicting the behavior of critical "lifelines" during extreme events. This section briefly reviews three commonly used approaches to modeling infrastructure interdependencies: (1) risk analysis; (2) economic input-output models; and (3) simulation models.

Risk Analysis

Risk analysis perspectives are particularly useful for understanding the vulnerability of critical infrastructure systems. The *Supplement* to the *2013 National Infrastructure Protection Plan* outlined a framework for risk management of critical infrastructure assets to help identify threats and vulnerabilities, risk reduction strategies, and potential courses of action (DHS, 2013). The importance of considering risk in climate-related assessments has been widely recognized by decision-makers, scientists, and practitioners. The *Summary for Policymakers* of the Intergovernmental Panel on Climate Change (IPCC) Fifth Assessment Report (AR5) specifically stressed that "iterative risk management is a useful framework for decision-making in complex situations characterized by large potential consequences, persistent uncertainties, long timeframes, potential for learning, and multiple climatic and non-climatic influences changing over time" (IPCC, 2014, p. 9).

Risk analysis theory decomposes risk into four broad analytical categories including risk characterization, risk assessment, risk management, and risk communication. By definition, risk is a function of the likelihood of occurrence of an adverse event and its severity (Haimes, Lambert, Li, Schooff & Tulsiani, 1995; Haimes, Kaplan & Lambert, 2002). A risk-based methodology in the context of infrastructure resilience explicitly considers scenarios, assumptions, metrics, and potential processes and outcomes that meet decision-making criteria. *Hierarchical Holographic Modeling* (HHM) represents a set of qualitative and quantitative approaches to address questions regarding the nature of disruptive events, probability of occurrence, and expected consequences (Haimes, Lambert, Li, Schooff & Tulsiani, 1995). The *Risk Filtering, Ranking, and Management* (RFRM) approach offers a comprehensive framework conducted in eight steps involving identification, assessment, prioritization, and management of risks (Haimes, Kaplan & Lambert, 2002; Lambert, Haimes, Schoof & Tulsiani, 2001). RFRM enables "filtering" or selecting a critical subset of risk categories from options available at various levels of decision-making and assists the development of multiple risk scenarios. RFRM involves the creation of a risk matrix that facilitates alignment of risk categories with defensive capabilities grouped in three classes: redundancy, robustness, and resilience (Haimes, Kaplan & Lambert, 2002).

Economic Input-Output Models

There is a growing body of scientific literature applying input-output and macro-level computable general equilibrium models to assess economic consequences of disasters (Li, Crawford-Brown, Syddall & Guan, 2013; Santos, Yu, Pagsuyoin & Tan, 2014; Xu, Wang, Hong, He & Chen, 2015; Kelly, 2015; Wing, Rose & Wein, 2015). Resurreccion and Santos (2013) introduced a dynamic inoperability input-output risk assessment model based on empirical distributions derived from the inventory-to-sales ratio (ISR) of the manufacturing and trade sectors from the Bureau of Economic Analysis database. The model employed a multi-objective optimization framework to minimize economic losses and sector inoperability in critically disrupted systems. Apostolakis and Lemon (2005) applied multi-attribute utility theory to three types of interdependent infrastructure systems (e.g., electrical power, water, and natural gas) to identify strategies that would minimize potential ripple effects leading to systemic vulnerabilities. Santos, Yu, Pagsuyoin, and Tan (2014) developed a dynamic input-output model using event-tree analysis to investigate cascading effects across interdependent economic sectors. The proposed inoperability index ranging from 0 and 1 is intended to measure loss of output due to production disruption and could be used to adjust inoperability parameters within the recovery timeline to accelerate the returm to the normal level of business operations (Santos, Yu, Pagsuyoin & Tan, 2014).

Simulation Models

In graph theory, large networks are modeled as directed graph topologies in which interdependencies are represented as *causal relationships*. In a causal network, a failure in one node or component affects the performance of other nodes or components unless a buffering strategy is in place (Bloomfield, Popov, Salako & Wright, 2007). Hernandez-Fajardo and Dueñas-Osorio (2013) proposed a graph model of interdependent networks and simulated damage propagation in power and water systems based on interdependencies, fragilities, and cascading failures. The findings indicated that the magnitude of the initial component damage affected the magnitude of the subsequent propagation of failures in interdependent networks (Hernandez-Fajardo & Dueñas-Osorio, 2013). Holden, Val, Burkhard, and Nodwell (2013) developed a flow network model to simulate the operation of interdependent infrastructure systems at a community level under normal conditions and during disruptions. The proposed model considers the uncertainties associated with the damaged assets performance by treating relevant parameters as random variables (Holden, Val, Burkhard and Nodwell, 2013).

Ouyang and Dueñas-Osorio (2011) introduced a simulation methodology for representing generalized interdependent effects of infrastructure disruptions. Pre- and post-failure efficiencies served as a basis for predicting the interdependent effects. The method was illustrated using artificial power (*p*p) and gas (*g*g) networks. Karamlou and Bocchini (2017) developed a methodology to examine damage uncertainty and to predict the functionality of individual components using the concept of "functional-fragilities." Bocchini and Frangopol (2011) evaluated the propagation of functionality loss at the network level and the associated cascading failures using graph theory and combinatorial optimization. Figure 13.2 displays a graphical representation of the interdependencies between electrical power, transportation, and communication networks.

Analysis of Infrastructure Services, Failures, and Interdependencies Using GIS

GIS technologies have been extensively used to map hazards, exposure risk, and damage to critical infrastructures. Map overlays of hazard zones with infrastructure networks can provide an initial assessment of the existing risk factors and facilitate decision-making in preparedness, mitigation, response, and recovery. The *Transportation Routing Analysis Geographic Information System* (TRAGIS) is a client-server

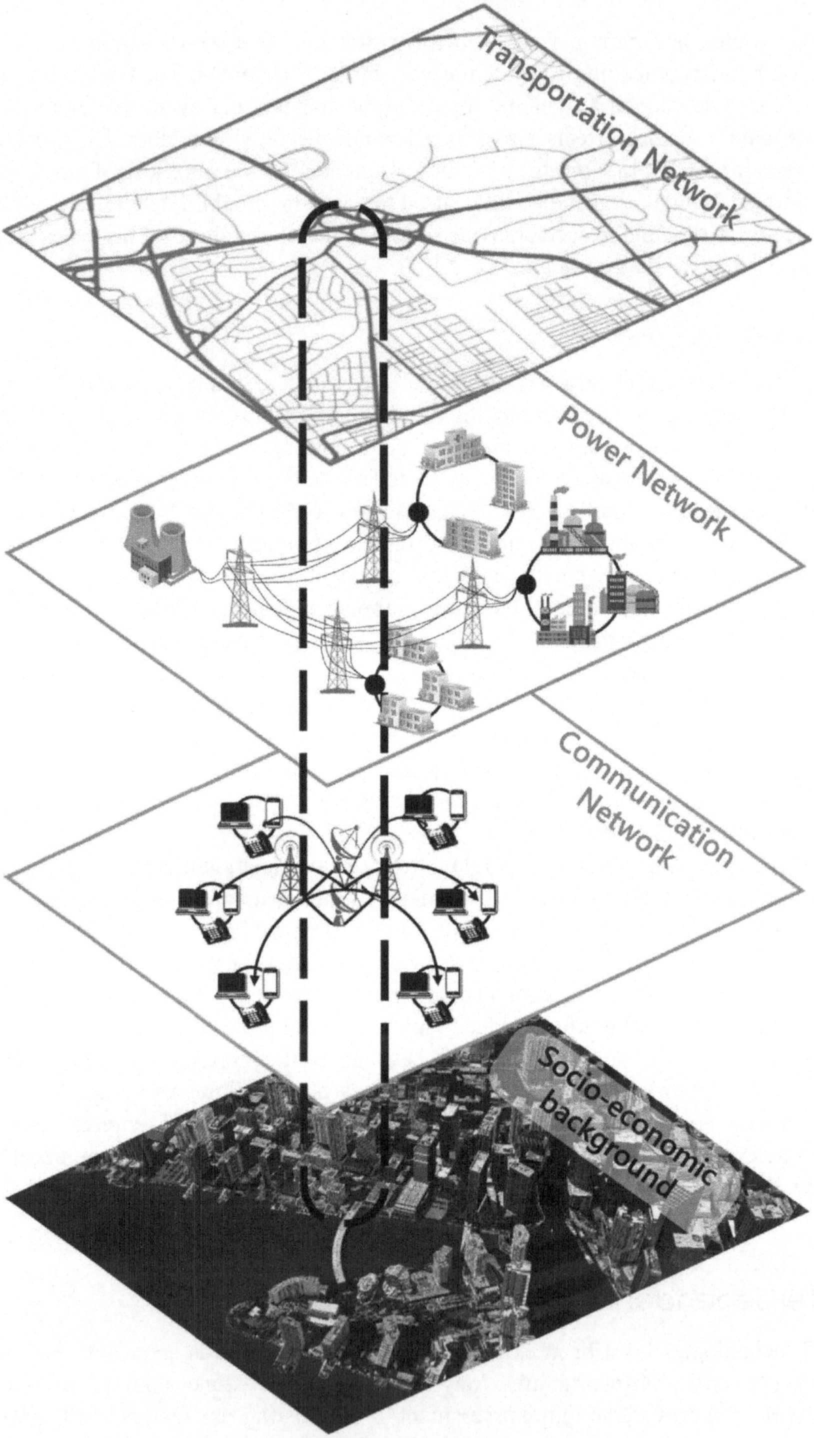

Figure 13.2 Schematic representation of infrastructure interdependencies

Source: Bocchini et al. (2015)

application developed by the Oak Ridge National Laboratory. Released in 2012, the web-based version of the application, WebTRAGIS, provides users with a database of highway, rail, and waterway routes in the United States. WebTRAGIS performs routing routines using network flow models (Pederson, Dudenhoeffer, Hartley & Permann, 2006). WebTRAGIS also provides improved population density measures per link, daytime and nighttime population estimates, and information on proximity of critical facilities (e.g., hospitals, schools, fire stations, etc.). Users can consider several criteria for selecting a routing option and can choose route modifiers such as way-point routing, network maintenance, graphic route blocking, graphic selection of origins and destinations, and metadata display (Tuttle, 2014). The developers consider adding additional critical infrastructures and intermodal implementations (Tuttle, 2014).

Co-location of infrastructure assets and critical facilities can increase risk and vulnerability. Johansson and Hassel (2010) proposed a methodology to examine the types of functional dependencies of an electrified railway system that arise from critical geographic proximities. The analysis of the infrastructure interdependencies is conducted using square grid cells to model the criticality of the geographic location under two assumptions of cell size. The study found that highly dispersed networks with few nodes such as railway systems were quite different from high-density infrastructure networks such as power lines and water distribution systems. As the authors further noted, "although it may be theoretically possible to model interdependent infrastructure systems of basically any size, substantial practical difficulties exist regarding systems mapping and modeling" (Johansson & Hassel, 2010, p. 1343).

Lambert and Patterson (2002) developed a framework of critical schedule dependencies that can affect the speed of recovery of a disrupted transportation network in the aftermath of a major hurricane using critical path analysis. In order to evaluate the effects of combined failures of multiple infrastructures due to co-location, Ouyang, Tian, Wang, Hong, and Mao (2017) proposed the concept of spatially localized failures (SLFs) of interdependent infrastructures using as a testbed **China**'s railway system. The authors modeled node-centered, district-based, and circle-shaped modes of co-located infrastructure failures and proposed a search engine that could identify the most critical locations. The *Location-Based Critical Infrastructure Interdependency* (LBCII) tool developed by Abdalla and Niall (2010) employs extensive geospatial analytical capabilities and web-based services to support information sharing and decision-making in earthquake-prone areas.

Reed, Kapur, and Christie (2009) examined the interdependencies between electrical systems and telecommunications using a networked infrastructure model, an input-output inoperability model, and a set of resilience measures within the context of post-disaster recovery after Hurricane Katrina in four states: **Louisiana, Florida, Mississippi, and Alabama**. A high-resolution fragility measure of the Louisiana coastline was derived overlaying georeferenced data of the percent of customers without power with a spatially explicit wind speed factor calculated as the ratio of maximum wind speed to the equivalent design wind speed established by the American Society of Civil Engineers (ASCE). Reed, Wang, Kapur, and Zheng (2015) developed a model to predict power and telecommunications systems performance during multi-hazard events, taking into consideration peak wind speed, storm surge elevation, and maximum precipitation. Tonn, Guikema, Ferreira, and Quiring (2016) developed a power outage prediction model using data for Hurricane Isaac, which made landfall in the coastal areas of the U.S. state of **Louisiana** in August 2012. Using random forest and quantile regression forest models, the study examined the relationship between power outages and the spatial and temporal distribution of wind speed, storm surge, and rainfall at a zip code level (Tonn, Guikema, Ferreira & Quiring, 2016).

Espada, Apan, and McDougall (2015) developed a two-stage GIS-based modeling approach to characterize infrastructure vulnerability and formulate climate adaptation strategies for **Queensland, Australia**, in the aftermath of the record-breaking 2010/2011 floods. The study examined the vulnerability of transportation networks, power distribution, and sewerage systems under a range of climate-related parameters. A set of connectivity rules using the *Utility Network* tool in ArcGIS was established for each network feature class. The results from the exposure analysis of the critical infrastructure nodes were incorporated in a multi-level interdependency model. Nodes with significant potential for synchronous failures due to geographic proximity to a flood-prone area were incorporated in a model of indirect or higher-order effects (Espada, Apan & McDougall, 2015). The study formulated a hierarchy of seven climate adaptation strategies depending on the severity of the anticipated cascading infrastructure failures. Protection, better management, and modification and augmentation of assets were identified as the most effective short-term strategies. Long-term strategies considered substitution, isolation, and even elimination at particularly vulnerable locations (Espada, Apan & McDougall, 2015).

Baloye and Palamuleni (2016) developed the *Abeokuta Critical Infrastructure Information System* (ACIIS) to enable tracking of critical infrastructure performance during extreme events. The city of **Abeokuta, Nigeria**, was selected as a case study for the research project. ACIIS employed a client-server architecture which allowed users to access the application through a browser and perform analyses using data processing services (Baloye & Palamuleni, 2016). The geospatial database implemented in PostGIS[1] contained information on seven infrastructure classes including roads, power lines, water distribution, hospitals, emergency services, ATMs, and gas stations. Web services provided by the Environmental Systems Research Institute (ESRI) and the application program interface (API) of Google Maps were used to enable data retrieval and access to data processing algorithms (Baloye & Palamuleni, 2016).

Yang, Hu, and Jaeger (2016) examined the impact of tropical cyclones on the highway network in **China**. The study evaluated the applicability of the fuzzy set theory in risk assessment and uncertainty analysis and employed spatial factor analysis to map functional damage to a number of highway network segments as a result of weather-related hazards (Yang, Hu & Jaeger, 2016). Ramachandran, Shoberg, Long, Corns, and Carlo (2015) used high-resolution orthoimagery to develop a database of critical infrastructure networks such as electrical and communication transmission lines. The investigators used GIS, combinatorial graph theory, and critical path analysis to construct a priority matrix of activities for optimizing restoration timelines.

Considering various aspects of the recovery process, Orabi, El-Rayes, Senouci, and Al-Derham (2009) developed a multi-objective optimization model using a genetic algorithm to prioritize recovery projects related to disruptions of the transportation network following a natural disaster. The study compared seven hypothetical projects based on the location of the closed links of the transportation network, the volume of traffic affected, and the number of restoration/recovery activities. Two optimization objectives were formulated: (i) maximizing the performance of the transportation network, and (ii) minimizing the recovery costs (Orabi, El-Rayes, Senouci & Al-Derham, 2009). The study found that prioritizing the repair of bridges was among the most desirable solutions as it minimized performance loss and the duration of the transportation network reconstruction (Orabi, El-Rayes, Senouci & Al-Derham, 2009).

Agent-Based Models of Infrastructure Interdependencies

Agent-based models (ABMs) are discrete-event simulation models designed to mimic the complex behavior and interaction among multiple autonomous agents (Brown, Riolo, Robinson, North & Rand, 2005; Pederson, Dudenhoeffer, Hartley & Permann, 2006; Ouyang, 2014). An agent-based model is an evolutionary framework where urban processes are time dependent. ABMs derive global interactions from the behavior of individual agents facing a range of discrete choice possibilities (Zeigler, Praehofer & Kim, 2000). The interactions between agents occur at a micro-level in conformity with a predetermined set of rules (Brown, Riolo, Robinson, North & Rand, 2005; Pederson, Dudenhoeffer, Hartley & Permann, 2006; Ouyang, 2014). Agent-based models have been adopted by several disciplines, including ecology, economics, urban planning, and engineering, among others, to simulate complex phenomena.

ABMs are well suited for conducting simulations of the interactions between critical interdependent infrastructures. Pederson, Dudenhoeffer, Hartley, and Permann (2006) provide detailed information on recently developed agent-based models for infrastructure analysis. Although the majority of the existing ABMs have been developed outside the GIS environment, many of them present opportunities for integration and visualization with GIS. The U.S. National Laboratories have created a variety of agent-based tools to model critical infrastructure interdependencies (a detailed overview of these models can be found in Pederson, Dudenhoeffer, Hartley & Permann, 2006, and Ouyang, 2014, among others). The focus of this section is on ABM tools with a graphical interface that supports GIS-based visualization and analysis.

The *Critical Infrastructure Modeling System* (CIMS) is a simulation framework developed by Idaho National Laboratory (INL). CIMS offers 3D visualization capabilities. All objects in the simulation framework are georeferenced by latitude and longitude, enabling spatiotemporal analysis (Pederson, Dudenhoeffer, Hartley & Permann, 2006). CIMS has been designed to represent the behavior of systems and system components based on topologies within a single infrastructure or between various infrastructures (Dudenhoeffer, Permann & Sussman, 2002). Within the framework, each type of infrastructure interdependency (e.g., physical, informational, geospatial, policy, and societal) is defined using logical formalism (Dudenhoeffer, Permann & Sussman, 2002). CIMS is specifically designed to simulate cascading effects in interconnected infrastructures. The problem space contains an initial set of cascading failures, a set of events that can cause such failures, and a set of observed outcomes (Dudenhoeffer, Permann & Sussman, 2002). The simulation framework has been further enhanced by incorporating genetic algorithms for probabilistic analysis (Permann, 2007).

Sandia National Laboratory (SNL) created a series of tools for infrastructure analysis with a variety of applications including interactions between economic agents using Monte Carlo simulation (*ASPEN*) (Basu, Pryor & Quint, 1998), interactions between economic actors and electricity markets (*ASPEN-EE* [Electricity Enhancement]) (Barton, Eidson, Schoenwald, Stamber & Reinert, 2000), and explicit modeling of telecommunication networks (*CommASPEN*) (Barton, Edison, Schoenwald, Cox & Reinert, 2004).

The *Fast Analysis Infrastructure Tool* (FAIT) developed by the National Infrastructure Simulation and Analysis Center (NISAC) (operated jointly by Sandia National Laboratory and Los Alamos National Laboratory) is one of the first modeling frameworks containing a comprehensive spatial database to enable rapid analysis capabilities and mapping of object-oriented infrastructure interdependencies (Kelic, Warren & Phillips, 2008). Encoded in Java Expert System Schell (JESS), which is a rule-based expert systems engine for Java API, FAIT provides capabilities for object-oriented modeling of the interconnections between infrastructure assets based on service boundaries, asset attributes, and economic impacts (Pederson, Dudenhoeffer, Hartley & Permann, 2006). FAIT's users have access to a web-based interactive mapping interface via Google Earth and other advanced visualization capabilities (Kelic, Warren & Phillips, 2008). Users are also provided with opportunities to customize the rules and use customary data to improve and expand the infrastructure analysis (Kelic, Warren & Phillips, 2008). FAIT facilitates regional impact infrastructure analysis following a disaster by extending the rules applicable to individual assets to classes of demand infrastructure (e.g., emergency services and medical centers) and categories of service delivery (e.g., electricity and telecommunications) (Pederson, Dudenhoeffer, Hartley & Permann, 2006).

The *Urban Infrastructure Suite* (UIS) developed by Los Alamos National Laboratory (LANL) integrates several interoperable infrastructure simulation and population models linked through a standard interface to support integrated regional analyses of critical infrastructures and potential impacts of disruptions (Barrett et al., 2001; Barrett, Eubank & Marathe, 2005; Atkins et al., 2008). The overall objective is to develop an integrated modeling environment that can simulate the complex interactions of sociotechnical systems (Atkins et al., 2008). UIS models include the following systems: transportation, water, energy, telecommunications, commodity flows, and public health (Pederson, Dudenhoeffer, Hartley & Permann,

2006; Atkins, 2008). As part of UIS, the *Transportation Analysis Simulation System* (TRANSIMS) is an agent-based microsimulation modeling system intended for use in transportation planning (Paleti, Pendyala, Bhat & Konduri, 2011). The simulation framework is designed for integrated transportation system modeling, including land use planning, travel behavior, and modeling of traffic flows and route-destination choices (Paleti, Pendyala, Bhat & Konduri, 2011). The simulation platform also provides tools for energy consumption and air quality modeling. The *Water Infrastructure Simulation Environment* and the *Interdependent Energy Infrastructure Simulation System* (IEISS) provide extensive capabilities for damage assessments and analysis of water infrastructure.

Fort Future, developed by the U.S. Army Corps of Engineers in collaboration with Argonne National Laboratory, is a comprehensive web-based simulation platform with extensive capabilities for modeling a broad range of adversarial agents. The platform is fully integrated within an interactive GIS environment and offers capabilities to create dynamic plans using the *Virtual Installation* tool, which enables users to test ideas and develop decision trees to evaluate new installations of the U.S. Department of Defense (DoD) (Pederson, Dudenhoeffer, Hartley & Permann, 2006).

Adaptation Strategies for Critical Interdependent Infrastructures

Climate adaptation strategies for interdependent infrastructures "should be robust, . . . flexible and adjustable; include no-regret (valuable even without climate change) and co-benefits (valuable to multiple sectors) actions" (Wilbanks et al., 2014, p. 76). Short-term climate adaptation strategies usually seek to establish improved standards for new construction or respond to already existing, identifiable threats (Wilbanks et al., 2014). Long-term strategies take into account risk, uncertainty, and potential "trigger points" indicating when future actions should be initiated (Wilbanks et al., 2014, p. 76). Summarizing the recommendations of the National Infrastructure Advisory Council (NIAC), Berkeley III and Wallace (2010, p. 18) propose the formulation of long-term resilience goals and strategies for critical infrastructures that incorporate "high-impact scenarios." Furthermore, robust adaptation strategies for infrastructure systems requires better integration of existing modeling frameworks with climate projections. Consideration of climate extremes with appropriate margins of uncertainty should become a key component of critical infrastructure protection plans and future large-scale investments in infrastructure upgrades (Wilbanks et al., 2014). It is acknowledged, however, that data sharing and interoperability remain a key challenge (Johansson & Hassel, 2010; Wilbanks et al., 2014; Chang, McDaniels, Fox, Dhariwal & Longstaff, 2014).

In 2013, the U.S. Government Accountability Office (GAO) prepared a report to Congress providing an assessment of the "vulnerabilities, threats, and potential consequences" of the interactions among regional infrastructure systems, anticipated cascading failures, resiliency measures, and potential gaps in knowledge

(GAO, 2013, p. 2). In order to enhance regional efforts to implement infrastructure protection measures and improve resilience strategies, the DHS has taken steps to expand risk assessment and protective activities "within groups of related infrastructures, regions, and systems to place great emphasis on resilience" (GAO, 2013, p. 2). As part of these efforts, DHS developed the Regional Resiliency Assessment Program (RRAP) through which priority is given to assessments of specific geographic areas and regions (GAO, 2013). RRAP underscores regional contexts as the basis of interdependency analysis.

Despite the advancements in infrastructure interdependency analysis, further research is needed to (a) identify comprehensive resilience measures that will reflect the needs of social, economic, and technical systems; (b) draw contributions from diverse stakeholders involved in designing optimal strategies for responding to current and future threats; and (c) address the degree of complexity characterizing the response of large-scale sociotechnical systems to climate change. Further integration of engineering models and simulation platforms with contributions from social and environmental sciences provides emerging opportunities to strengthen community resilience, accounting for broader societal impacts beyond physical infrastructure.

Note

1 PostGIS (postgis.net) is an extension to PostgreSQL (postgresql.org) that allows for the development of a spatial object-relational database to support queries of the locational characteristics of geographic objects in SQL (postgis.net).

References

Abdalla, R.M., & Niall, K.K. 2010. *Location-Based Infrastructure Interdependency: New Concept, New Modeling Approach*. DRDC TR 2009-130. Toronto, Canada: Defense Research & Development. Retrieved from www.dtic.mil/get-tr-doc/pdf?AD=ada526442.

Alderson, D.L., & Doyle, J.C. 2010. Contrasting Views of Complexity and Their Implications for Network-Centric Infrastructures. IEEE *Transactions on Systems, Man, and Cybernetics-Part A: Systems and Humans* 40(4), 839–852.

American Society of Civil Engineers (ASCE). 2017. *2017 Report Card of America's Infrastructure*. Retrieved from https://www.infrastructurereportcard.org/.

Apostolakis, G.E., and Lemon, D.A. 2005. A Screening Methodology for the Identification and Ranking of Infrastructure Vulnerabilities Due to Terrorism. *Risk Analysis* 25, 361–376.

Atkins, K., Barrett, C.L., Beckman, R., Bisset, K., Chen, J., Eubank, S., Feng, A., Feng, X., Harris, S.D., Lewis, B., Kumar,V.S., Marathe, M.V., Marathe, A., Mortveit, H., & Stretz, P. 2008. An Interaction Based Composable Architecture for Building Scalable Models of Large Social, Biological, Information and Technical Systems. *CTWatch Quarterly: Cyberinfrastructure Technology Watch* 4(1), 46–53.

Ayyub, B.M. 2014. Systems Resilience for Multihazard Environments: Definition, Metrics, and Valuation for Decision Making. *Risk Analysis* 34(20), 340–355.

Bagheri, E., Baghi, H., Ghorbani, A.A., & Yari, A. 2007. An Agent-Based Service-Oriented Simulation Suite for Critical Infrastructure Behavior Analysis. *International Journal of Business Process Integration and Management* 2(4), 312–326.

Baloye, D.O., & Palamuleni, L.G. 2016. Modelling a Critical Infrastructure-Driven Spatial Database for Proactive Disaster Management: A Developing Country Context. *Jàmbá: Journal of Disaster Risk Studies* 8(1), 1–14.

Barrett, C., Beckman, R., Berkbigler, K., Bisset, K., Bush, B., Campbell, K., Eubank, S., Henson, K., Hurford, J., Kubicek, D., Marathe, M., Romero, P., Smith, J., Smith, L., Speckman, P., Stretz, P., Thayer, G., Eeckhout, E., & Williams, M.D. 2001. *TRANSIMS: Transportation Analysis Simulation System*. Technical Report LA-UR-00-1725. Los Alamos National Laboratory Unclassified Report, 2001.

Barrett, C., Eubank, S., & Marathe, M. 2005. Modeling and Simulation of Large Biological, Information and Socio-Technical Systems: An Interaction Based Approach. In Goldin, G., Smolka, S., & Wegner, P. (Eds.) *Interactive Computation: The New Paradigm*, 356–487. New York, NY: Springer-Verlag.

Barton, D.C., Eidson, E.D., Schoenwald, D.A., Stamber, K.L., & Reinert, R.K. 2000. *Aspen EE: An Agent-Based Model of Infrastructure Interdependency*. Sandia report, SAND2000-2925. Retrieved from http://prod.sandia.gov/techlib/access-control.cgi/2000/002925.pdf.

Barton, D.C., Edison, E.C., Schoenwald, D.A., Cox, R.G., & Reinert, R.K. 2004. *Simulating Economic Effects of Disruptions in the Telecommunications Infrastructure*. Sandia report, SAND2004-0101. Retrieved from http://prod.sandia.gov/techlib/access-control.cgi/2004/040101.pdf.

Basu, N., Pryor, R.J., & Quint, T. 1998. ASPEN: A Microsimulation Model of the Economy. *Computational Economics* 12(December), 223–241.

Berkeley III, A.R., & Wallace, M. 2010. *A Framework for Establishing Critical Infrastructure Resilience Goals*. Final Report and Recommendations by the Council, National Infrastructure Advisory Council. Retrieved from www.dhs.gov/xlibrary/assets/niac/niac-a-framework-for-establishing-critical-infrastructure-resilience-goals-2010-10-19.pdf (accessed 09/16/17).

Bierwagen, B.G., Theobald, D.M., Pyke, C.R., Choate, A., Groth, P., Thomas, J.V., & Morefield, P. 2010. National Housing and Impervious Surface Scenarios for Integrated Climate Impact Assessments. *Proceedings of the National Academy of Sciences of the United States of America* 107(49), 20887–20892.

Bloomfield, R., Popov, P., Salako, K., & Wright, D. 2007. *Deliverable D2.2.4 Report on Service Oriented Interdependency Analysis*. Integrated Risk Reduction of Information-Based Infrastructure Systems (IRRIIS), IST Project 027568. Retrieved from www.irriis.org/File7864.pdf?lang=2&oiid=9247&pid=572 (accessed 03/02/2017).

Bocchini, P., Davison, B., Esnard, A.-M., Lamadrid, A., Mitsova, D., Sapat, A., Sause, R., & Snyder, L. 2015. *Probabilistic Resilience Assessment of Interdependent Systems (PRAISys), funded by the National Science Foundation (CRISP2)*. Presented at the National Workshop on Resilience Research, October 22–23, Washington, DC. Retrieved from www.prasys.org.

Bocchini, P., & Frangopol, D.M. 2011. A Stochastic Computational Framework for the Joint Transportation Network Fragility Analysis and Traffic Flow Distribution Under Extreme Events. *Probabilistic Engineering Mechanics* 26(2), 182–193.

Bocchini, P., Frangopol, D.M., Ummenhofer, T., & Zinke T. 2014. Resilience and Sustainability of Civil Infrastructure: Toward a Unified Approach. *Journal of Infrastructure Systems* 20(2). doi: 04014004-1-04014004-16.

British Broadcasting Corporation (BBC). 2006, July 18. *Roads Melt as Temperatures Soar*. Retrieved from http://news.bbc.co.uk/2/hi/5189734.stm (accessed 06/12/2017).

Brown, D.G., Riolo, R., Robinson, D.T., North, M., & Rand, W. 2005. Spatial Process and Data Models: Toward an Integration of Agent-Based Models and GIS. *Journal of Geographical Systems* 7, 25–47.

Chang, S.E., McDaniels, T.L., Mikawoz, J., & Peterson, K. 2007. Infrastructure Failure Interdependencies in Extreme Events: Power Outage Consequences in the 1998 Ice Storm. *Natural Hazards* 41(2), 337–358.

Chang, S.E., McDaniels, T., Fox, J., Dhariwal, R., & Longstaff, H. 2014. Toward Disaster-Resilient Cities: Characterizing Resilience of Infrastructure Systems with Expert Judgments. *Risk Analysis* 34(3), 416–434.

Chen, J., & Jiang, J. 2000. An Event-Based Approach to Spatiotemporal Data Modeling in Land Subdivision System for Spatiotemporal Process of Land Subdivision. *GeoInformatica* 4(4), 387–402.

Cromley, E.K., & McLafferty, S.L. 2012. *GIS and Public Health* (2nd edition). New York, NY: Guilford Press.

Deakin, M., & Allwinkle, S. 2007. Urban Regeneration and Sustainable Communities: The Role Networks, Innovation, and Creativity in Building Successful Partnerships. *Journal of Urban Technology* 14(1), 77–91.

Dudenhoeffer, D.D., Permann, M.R., & Sussman, E.M. 2002. A Parallel Simulation Framework for Infrastructure Modeling and Analysis. In Yucesan, E., Chen, C.H., Snowdon, J.L., & Charnes, J.M. (Eds.) *Proceedings of the 2002 Winter Simulation Conference of the Institute of Electrical and Electronics Engineers*, pp. 1971–1977. Piscataway, NJ.

Dunn, M. 2005. The Socio-Political Dimensions of Critical Information Infrastructure Protection (CIIP). *International Journal of Critical Infrastructures* 1(2/3), 258–268.

Erwig, M., Gueting, R.H., Schneider, M., & Vazirgiannis, M. 1999. Spatio-Temporal Data Types: An Approach to Modeling and Querying Moving Objects in Databases. *GeoInformatica* 3(3), 269–296.

Espada, R.J.R., Apan, A., & McDougall, K. 2015. Vulnerability Assessment and Interdependency Analysis of Critical Infrastructures for Climate Adaptation and Flood Mitigation. *International Journal of Disaster Resilience in the Built Environment* 6(3), 313–346.

European Environment Agency (EEA). 2017. *Climate Change, Impacts and Vulnerability in Europe 2016.* Retrieved from www.eea.europa.eu/publications/climate-change-impacts-and-vulnerability-2016/#parent-fieldname-title (accessed 08/28/2017).

Federal Emergency Management Agency (FEMA). 2009. Voluntary Private Sector Accreditation and Certification Preparedness Program. *Federal Register* 74(199), 53286, Friday, October 16, 2009, Notices, Department of Homeland Security. Retrieved from www.gpo.gov/fdsys/pkg/FR-2009-10-16/pdf/E9-24968.pdf

Glaeser, E.L., & Berry, C.R. 2006. Why Are Smart Places Getting Smarter? *Taubman Center Policy Brief.* Cambridge, MA: Taubman Centre, 2006–2.

Graham, S., & Marvin, S. 2001. *Splintering Urbanism: Networked Infrastructures, Technological Mobilities and the Urban Condition.* London, UK: Routledge.

Haberman, C. 2014. A Disaster Brought Awareness but Little Action on Infrastructure. *The New York Times*, March 3, 2014. Retrieved from www.nytimes.com/2014/03/03/us/a-disaster-brings-awareness-but-little-action-on-infrastructure.html?hp&_r=0.

Haimes, Y.Y. 2009. *Risk Modeling, Assessment, and Management* (3rd edition). Hoboken, NJ: John Wiley & Sons, Inc.

Haimes, Y.Y., & Jiang, P. 2001. Leontief-Based Model of Risk in Complex Interconnected Infrastructures. *Journal of Infrastructure Systems* 7(1), 1–12.

Haimes, Y.Y., Kaplan, S., & Lambert, J.H. 2002. Risk Filtering, Ranking, and Management Framework Using Hierarchical Holographic Modeling. *Risk Analysis* 22(2), 383–397.

Haimes, Y.Y., Lambert, J., Li, D., Schooff, R., & Tulsiani, V. 1995. Hierarchical Holographic Modeling for Risk Identification in Complex Systems, 1027–1032. In *Systems, Man and Cybernetics*, IEEE International Conference on Intelligent Systems for the 21st Century, October 22–25, Vancouver, BC, Canada.

Hernandez-Fajardo, I., & Dueñas-Osorio, L. 2013. Probabilistic Study of Cascading Failures in Complex Interdependent Lifeline Systems. *Reliability Engineering & System Safety*. doi: 10.1016/j.ress.2012.10.012, 260–272.

Holden, R., Val, D.V., Burkhard, R., & Nodwell, S. 2013. A Network Flow Model for Interdependent Infrastructures at the Local Scale. *Safety Science* 53, 51–60.

Hollands, R.G. 2008. Will the Real Smart City Please Stand Up? *City* 12(3), 303–320.

Huang, C.-N., Liou, J.H., & Chuang, Y.-C. 2014. A Method for Exploring the Interdependencies and Importance of Critical Infrastructures. *Knowledge-Based Systems* 55, 66–74.

IBM. 2012. *Smarter, More Competitive Cities*. White Paper, PUB03003USEN-00. Somers, NY: IBM.

Intergovernmental Panel on Climate Change (IPCC). 2014. Summary for Policymakers. In Field, C.B., Barros, V.R., Dokken, D.J., Mach, K.J., Mastrandrea, M.D., Bilir, T.E., Chatterjee, M., Ebi, K.L., Estrada, Y.O., Genova, R.C., Girma, B., Kissel, E.S., Levy, A.N., MacCracken, S., Mastrandrea, P.R., & White, L.L. (Eds.) *Climate Change 2014: Impacts, Adaptation, and Vulnerability. Part A: Global and Sectoral Aspects. Contribution of Working Group II to the Fifth Assessment Report of the Intergovernmental Panel on Climate Change*, 1–32. Cambridge, UK/New York, NY: Cambridge University Press.

Johansson, J., & Hassel, H. 2010. An Approach for Modeling Interdependent Infrastructures in the Context of Vulnerability Analysis. *Reliability Engineering & System Safety* 95(12), 1335–1344.

Johansson, J., & Jönsson, H. 2009. A Model for Vulnerability Analysis of Interdependent Infrastructure Networks. In *Safety, Reliability and Risk Analysis: Theory, Methods and Applications*, Proceedings of ESREL 2008 and 17th

SRA-Europe Conference, Valencia, September 2008, Taylor & Francis Group, London, 2491–2499.

Karamlou, A., & Bocchini, P. 2017. From Component Damage to System-Level Probabilistic Restoration Functions for a Damaged Bridge. *Journal of Infrastructure Systems* 23(3). doi: https://doi.org/10.1061/(ASCE)IS.1943-555X.0000342

Kelic, A., Warren, D.E., & Phillips, L.R. 2008. *Cyber and Physical Infrastructure Interdependencies*. Sandia Report, SAND2008-6192. Retrieved from http://prod.sandia.gov/techlib/access-control.cgi/2008/086192.pdf.

Kelly, S. 2015. Estimating Economic Loss from Cascading Infrastructure Failure: A Perspective on Modeling Interdependency. *Infrastructure Complexity* 2(7), 1–13. doi: 10.1186/s40551-015-0010-y.

Komninos, N. 2009. Intelligent Cities: Towards Interactive and Global Innovation Environments. *International Journal of Innovation and Regional Development* (Inderscience Publishers) 1(4), 337–355.

Kröger, W. 2008. Critical Infrastructures at Risk: A Need for a New Conceptual Approach and Extended Analytical Tools. *Reliability Engineering & System Safety* 93(12), 1781–1787.

Lambert, J.H., Haimes, Y.Y., Schoof, R., & Tulsiani, V. 2001. Identification, Ranking, and Management of Risks in a Major System Acquisition. *Reliability, Engineering, and System Safety* 72(3), 315–325.

Lambert, J.H., & Patterson, C.E. 2002. Prioritization of Schedule Dependencies in Hurricane Recovery of a Transportation Agency. *Journal of Infrastructure Systems* 8(3), 103–111.

Li, J., Crawford-Brown, D., Syddall, M., & Guan, D. 2013. Modeling Imbalanced Economic Recovery Following a Natural Disaster Using Input-Output Analysis. *Risk Analysis* 33(10), 1908–1923.

Lo, C.P., & Yeung, A.K. 2002. *Concepts and Techniques of Geographic Information Systems*. Upper Saddle River, NJ: Prentice-Hall, Inc.

McDaniels, T., Chang, S., Peterson, K., Mikawoz, J., & Reed, D.M. 2007. An Empirical Framework for Characterizing Infrastructure Failure Interdependencies. *Journal of Infrastructure Systems* 13(3), 175–184.

Meyer, M.D., Amekudzi, A., & O'Hara, J.P. 2010. Transportation Asset Management Systems and Climate Change. *Transportation Research Record: Journal of the Transportation Research Board* 2160, 12–20. doi: 10.3141/2160-02.

Miles, S.B., Jagielo, N., & Gallagher, H. 2016. Hurricane Isaac Power Outage Impacts and Restoration. *Journal of Infrastructure Systems* 22(1), 0501500, 1–9.

Miller, C.A. 2012, November 5. Will Politicians Actually Heed the Lessons of Sandy? The Superstorm Demonstrated that Our Infrastructure Is Not Ready for Climate Change. *Future Tense: The Citizen's Guide to the Future. Slate, New America, and Arizona State University (ASU) [Online]*. Retrieved from www.slate.com/articles/technology/future_tense/2012/11/hurricane_sandy_demonstrated_that_our_infrastructure_is_not_ready_for_climate.html (accessed 08/15/17).

Orabi, W., El-Rayes, K., Senouci, A.B., & Al-Derham, H. 2009. Optimizing Postdisaster Reconstruction Planning for Damaged Transportation Networks. *Journal of Construction Engineering and Management* 135(10), 1039–1048. doi: 10.1061/(ASCE)CO.1943-7862.0000070.

Ouyang, M. 2014. Review on Modeling and Simulation of Interdependent Critical Infrastructure Systems. *Reliability Engineering & System Safety* 121, 43–60.

Ouyang, M., & Dueñas-Osorio, L. 2011. Efficient Approach to Compute Generalized Interdependent Effects Between Infrastructure Systems. *Journal of Computing in Civil Engineering* 394–406. doi: 10.1061/(ASCE)CP.1943-5487.0000103.

Ouyang, M., Tian, H., Wang, Z., Hong, L., & Mao, Z. 2017. Critical Infrastructure Vulnerability to Spatially Localized Failures with Applications to Chinese Railway System. *Risk Analysis*.

Paleti, R., Pendyala, R.M., Bhat, C.R., & Konduri, K.C. 2011. *A Joint Tour-Based Model of Tour Complexity, Passenger Accompaniment, Vehicle Type Choice, and Tour Length*. Retrieved from JointTourAttributesVehicleTypeTourLength_KonduriPaletiPendyalaBhat_UnAbridgedVersion_July31.pdf.

Paskaleva, K. 2009. Enabling the Smart City: The Progress of E-city Governance in Europe. *International Journal of Innovation and Regional Development* 1(4), 405–422.

Pederson, P., Dudenhoeffer, D., Hartley, S., & Permann M. 2006. *Critical Infrastructure Interdependency Modeling: A Survey of U.S. and International Research*. Technical Report INL/EXT-06-11464, Idaho National Laboratory Idaho Falls, Idaho, Technical Support Working Group, Washington, DC. Retrieved from http://cip.management.dal.ca/publications/Critical%20Infrastructure%20Interdependency%20Modeling.pdf.

Permann, M.R. 2007. *Genetic Algorithms for Agent-Based Infrastructure Interdependency Modeling and Analysis*. SpringSim 2007, INL/CON-07-12317, Idaho National Laboratory, Idaho Falls, ID. Retrieved from https://pdfs.semanticscholar.org/a9f4/d6e2e595aefa866fed1f4ecf72fe98bcb2b0.pdf (accessed 06/12/2017).

Petit, F., Verner, D., Brannegan, D., Buehring, W., Dickinson, D., Guziel, K., Haffenden, R., Phillips, J., & Peerenboom, J. 2015. Analysis of Critical Infrastructure Dependencies and Interdependencies, ANL/GSS-15/4, Risk and Infrastructure

Science Center, Global Security Sciences Division, Argonne National Laboratory. Retrieved from www.ipd.anl.gov/anlpubs/2015/06/111906.pdf.

Peuquet, D., & Duan, N. 1995. An Event-Based Spatiotemporal Data Model (ESTDM) for Temporal Analysis of Geographical Data. *International Journal of Geographical Information Systems* 9(1), 7–24.

Peuquet, D.J. 2002. *Representations of Space and Time*. New York, NY: Guilford Press.

Ramachandran, V., Shoberg, T., Long, S. K., Corns, S., & Carlo, H. 2015. Identifying Geographical Interdependency in Critical Infrastructure Systems Using Open Source Geospatial Data in Order to Model Restoration Strategies in the Aftermath of Large-Scale Disasters. *International Journal of Geospatial and Environmental Research* 2(1), Article 4. Retrieved from http://dc.uwm.edu/ijger/vol2/iss1/4.

Reed, D.A., Kapur, K.C., & Christie, R.D. 2009. Methodology for Assessing the Resilience of Networked Infrastructure. *IEEE Systems Journal* 3(2), 174–180.

Reed, D., Wang, S., Kapur, K., & Zheng, C. 2015. Systems-Based Approach to Interdependent Electric Power Delivery and Telecommunications Infrastructure Resilience Subject to Weather-Related Hazards. *Journal of Structural Engineering* 142(8). doi: https://doi.org/10.1061/(ASCE)ST.1943-541X.0001395.

Resurreccion, J., & Santos, J. 2013. Uncertainty Modeling of Hurricane-Based Disruptions to Interdependent Economic and Infrastructure Systems. *Natural Hazards* 69(3), 1497–1518.

Rinaldi, M., Peerenboom, J.P., & Kelly, T.K. 2001. Identifying, Understanding, and Analyzing Critical Infrastructure Interdependencies. *IEEE Control System Magazine* 12/2001, 11–25. doi: 0272–1708/01/$10.00©2001IEEE. Retrieved from www.ce.cmu.edu/~hsm/im2004/readings/CII-Rinaldi.pdf (accessed 05/07/16).

Robert, B. 2004. A Method for the Study of Cascading Effects Within Lifeline Networks. *International Journal of Critical Infrastructures* 1(1), 86–99.

Roe, E., & Schulman, P.R. 2015. Comparing Emergency Response Infrastructure to Other Critical Infrastructures in the California Bay-Delta of the United States: A Research Note on Inter-Infrastructural Differences in Reliability Management. *Journal of Contingencies and Crisis Management* 23(4), 193–200.

Santos, J., Yu, K., Pagsuyoin, S., & Tan, R. 2014. Time-Varying Disaster Recovery Model for Interdependent Economic Systems Using Hybrid Input-Output and Event-Tree Analysis. *Economic Systems Research* 26(1), 60–80.

Schwartz, H.G., Meyer, M., Burbank, C.J., Kuby, M., Oster, C., Posey, J., Russo, E.J., & Rypinski, A. 2014. Transportation. In Melillo, J.M., Terese (T.C.) Richmond, & Yohe, G.W. (Eds.) *Climate Change Impacts in the United States: The Third National Climate Assessment*, 130–149. U.S. Global Change Research Program, chapter 5. doi: 10.7930/J06Q1V53. Retrieved from http://nca2014.globalchange.gov/report/sectors/transportation.

Stapelberg, R.F. 2008. Infrastructure Systems Interdependencies and Risk-Informed Decision Making (RIDM): Impact Scenario Analysis of Infrastructure Risks Induced by Natural, Technological and Intentional Hazards. *Systemics, Cybernetics and Informatics* 6(5), 21–27.

Stergiopoulos, G., Kotzanikolaou, P., Theocharidou, M., Lykou, G., & Gritzalis, D. 2016. Time-Based Critical Infrastructure Dependency Analysis for Large-Scale and Cross-Sectoral Failures. *International Journal of Critical Infrastructure Protection* 12, 46–60.

Tolone, W.J., Wilson, D., Raja, A., Xiang, W.-N., Hao, H., Phelps, S., & Johnson, E.W. 2004. Critical Infrastructure Integration Modeling and Simulation. In Chen, H., Moore, R., Zeng, D.D., & Leavitt, J. (Eds.) *Intelligence and Security Informatics. ISI 2004. Lecture Notes in Computer Science*, Vol. 3073. Berlin, Heidelberg: Springer.

Tolone, W., Lee, S.-W., Xiang, W.-N., Blackwell, J., Yeager, C., Schumpert, A., & Johnson, W. 2008. An Integrated Methodology for Critical Infrastructure Modeling and Simulation. In Papa, M., & Shenoi, S. (Eds.) *Critical Infrastructure Protection II, Vol. 290, International Federation for Information Processing (IFIP),* Chapter 19, 257–268. Boston: Springer.

Tonn, G.L., Guikema, S.D., Ferreira, C.M., & Quiring, S.M. 2016. Hurricane Isaac: A Longitudinal Analysis of Storm Characteristics and Power Outage Risk. *Risk Analysis* 36(10), 1936–1947. doi: 10.1111/risa.12552.

Townsend, A.M. 2013. *Smart Cities: Big Data, Civic Hackers, and the Quest for a New Utopia*. New York, NY: W.W. Norton & Co., Inc.

Trusted Information Sharing Network (TISN) for Critical Infrastructure Resilience. 2015. *Critical Infrastructure Resilience Strategy: Plan*. Attorney General's Department (AGD), Commonwealth of Australia. Retrieved from www.tisn.gov.au/Documents/CriticalInfrastructureResilienceStrategyPlan.PDF.

Tuttle, M. 2014. *WebTRAGIS* [PowerPoint Slides]. Retrieved from https://energy.gov/sites/prod/files/2014/06/f16/2%20WebTRAGIS.pdf.

United States Department of Homeland Security (DHS). 2013. National Infrastructure Protection Plan (NIPP). 2013. *Partnering for Critical Infrastructure*

Security and Resilience. Retrieved from www.dhs.gov/publication/nipp-2013-partnering-critical-infrastructure-security-and-resilience.

United States Government Accountability Office (GAO). 2013. *Critical Infrastructure Protection: DHS Could Strengthen the Management of the Regional Resiliency*, GAO-13-616 U, Assessment Program Report to the Chairman, Subcommittee on Cybersecurity, Infrastructure Protection, and Security Technologies, House of Representatives, Committee on Homeland Security. Retrieved from www.gao.gov/assets/660/656344.pdf.

Ungerer, M.J., & Goodchild, M.F. 2010. Integrating Spatial Data Analysis and GIS: A New Implementation Using the Component Object Model (COM). *International Journal of Geographical Information Science* 16(1), 41–53.

The White House. 2013. *Presidential Policy Directive—Critical Infrastructure Security and Resilience (PPD-21)*. Retrieved from https://obamawhitehouse.archives.gov/the-press-office/2013/02/12/presidential-policy-directive-critical-infrastructure-security-and-resil. Also available at https://fas.org/irp/offdocs/ppd/ppd-21.pdf.

Wilbanks, T., Fernandez, S., Backus, G., Garcia, P., Jonietz, K., Kirshen, P., Savonis, M., Solecki, W., & Toole, L. 2014. *Climate Change and Infrastructure, Urban Systems, and Vulnerabilities*. Technical Report to the U.S. Department of Energy in Support of the National Climate Assessment, 119 pp., Washington, DC: Island Press. Retrieved from www.eenews.net/assets/2014/03/06/document_cw_01.pdf.

Wing, I., Rose, A., & Wein, A. 2015. Economic Consequence Analysis of the ARkStorm Scenario. *Natural Hazards Review* 17(4). https://doi.org/10.1061/(ASCE)NH.1527-6996.0000173.

Xu, W., Wang, Z., Hong, L., He, L., & Chen, X. 2015. The Uncertainty Recovery Analysis for Interdependent Infrastructure Systems Using the Dynamic Inoperability Input-Output Model. *International Journal of Systems Science* 46(7), 1299–1306.

Yang, S., Hu, F., & Jaeger, C. 2016. Impact Factors and Risk Analysis of Tropical Cyclones on a Highway Network. *Risk Analysis* 36(2), 262–278.

Yusta, J.M., Correa, G.J., & Local-Arántegui, R. 2011. Methodologies and Applications for Critical Infrastructure Protection: State-of-the-Art. *Energy Policy* 39(10), 6100–6119.

Zeigler, B.P., Praehofer, H., & Kim, T.G. 2000. *Theory of Modeling and Simulation: Integrating Discrete Event and Continuous Complex Dynamic Systems*. Cambridge, MA: Academic Press Books Elsevier.

Zhang, P., & Peeta, S. 2011. A Generalized Modeling Framework to Analyze Interdependencies Among Infrastructure Systems. *Transportation Research Part B: Methodological* 45(3), 553–579.

Zimmerman, R. 2001. Social Implications of Infrastructure Network Interactions. *Journal of Urban Technology* 8(3), 97–119.

Zimmerman, R., & Faris, C. 2010. Infrastructure Impacts and Adaptation Challenges. *Annals of the New York Academy of Sciences* 1196, 63–86. doi: 10.1111/j.1749-6632.2009.05318.x.

14 Urban Growth Modeling and Decision Support Systems

Chapter Objectives

This chapter provides a brief overview of the evolution of urban growth modeling (UGM) and its various applications. It seeks to answer the following questions:

- What is the role of urban simulation models in environmental impact assessments?
- What is the contribution of UGM in modeling the complex interactions between urban development and climate change?
- What are some applications of UGM in flood analysis, green infrastructure planning, and evaluation of policy measures?

Introduction

Urban growth modeling offers a broad spectrum of tools to project future patterns of development and evaluate their potential impacts on urban morphology and environmental resources (Balbi & Giupponi, 2009; Fisher-Vanden, Wing, Lanzi & Popp, 2013; Nay, Abkowitz, Chu, Gallagher & Wright, 2014; Brown, Alexander, Holzhauer & Rounsevell, 2017). This chapter provides a brief overview of several urban growth modeling and planning support systems applications. Planning support systems (PSSs) are nested structures of techniques used to provide a holistic approach to decision-making. Hopkins, Kaza, and Pallathucheri (2005, p. 599) define planning support systems "as tools and techniques to enhance the effectiveness of planning through information technologies." Geertman and Stillwell (2004, p. 293) emphasize that "it is not so much the technology, with its capabilities and restrictions that dictates the support function performed by the PSS, but the specific needs of the planning context in which the PSS is designated to operate."

Urban simulation models have been used to estimate the environmental impacts of changes in urban morphology and structure, derive physical urban landscape parameters to facilitate integration with other models and projections (e.g., atmospheric, hydrology/water quality, ecological, economic, etc.), and develop applications relevant to planning and decision-making contexts (Balbi & Giupponi, 2009; Nay, Abkowitz, Chu, Gallagher & Wright, 2014; Masson et al., 2014; Choi & Lee, 2016; Mahmoud, Duker, Conrad, Thiel & Ahmad, 2016; Deal & Pan, 2017; Brown, Alexander, Holzhauer & Rounsevell, 2017). Some of the studies cited here do not refer explicitly to climate adaptation. Nevertheless, they provide valuable insights as they examine important, closely related issues of sustainability, flood protection, and environmental quality.

Modeling the Complex Interactions between Urban Development and Climate Change

One of the first regional growth models integrating the output of a macro-scale economic model with a micro-scale cellular automata model was developed by White and Engelen (1997) to address long-term global climate change impacts on the island nations of the Caribbean. The project was created as a demonstration tool in support of climate adaption actions endorsed by the United Nations Environment Programme (UNEP). The dynamic modeling framework was applied to the **Caribbean island of St. Lucia** to examine the impacts of sea level rise and temperature increase on land development and the economy of the coastal communities (White & Engelen, 1997). Population and employment projections developed at the macro-level were used to inform the micro-scale spatial cellular automata model integrating land demand with suitability and accessibility factors operating on finer spatiotemporal scales. The model was loosely coupled with the geographic information system (GIS) environment. Land use, suitability, and accessibility inputs to the constraint automata were created in GIS. The output of the integrated model determined the demand for land in various economic sectors such as tourism, agriculture, exports, shopping, and manufacturing over a 40-year period (White & Engelen, 1997). The framework allowed for the simulation of different scenarios with an emphasis on the impacts of climate change on coastal resources.

Masson et al. (2014) proposed a comprehensive modeling approach integrating five components: (1) a large-scale economic model estimating per-capita energy use; (2) an urban growth model simulating spatial dynamic at a regional scale; (3) an urban morphology model at the neighborhood/street block scale; (4) a local climate zones model; and (5) a model accounting for building energy use. The Non-Equilibrium Dynamical Urban Model (NEDUM) projected future urban development based on household decisions and economic activities in 10-year increments. The **cities of Paris and Toulouse, France**, were selected for the simulations. The spatial allocation of new development was estimated by coupling NEDUM with a modified version of the SLEUTH (Slope, Land use, Exclusion, Urban extent, Transportation, Hillshade) urban growth model (Masson et al., 2014). The SLEUTH model (Clarke & Gaydos, 1998) output was adjusted to generate future growth scenarios (Aguejdad, Hidalgo, Doukari, Masson & Houet, 2012). Urban form analysis at the block level was conducted using a newly developed model GENUS (GENerator of interactive Urban blockS) (Masson et al., 2014). The local climate zone approach developed by Stewart and Oke (2012) was combined with GENUS to characterize the urban heat island effect with higher spatial accuracy using over 60 variables. The urban climate effects were further modeled using building characteristics, including building envelope and energy performance. The findings highlight the complex choices that cities face in order to reduce climate change impacts. For example, the study compared the "business-as-usual" scenario to a "green belt" scenario in which future development was constraint by an urban growth boundary. The results indicated that in the short run, the "green belt" scenario would reduce the cost of new development but if population growth continues the restrictions would lead to a considerable increase in property values, which might eventually push new development outward (Masson et al., 2014).

Until recently, agent-based models (AMBs) had limited applications in addressing the complex policy choices related to climate change (Balbi & Giupponi, 2009; Gerst et al., 2013). Previous studies have demonstrated that ABMs are well suited to represent spatial planning processes related to land use and water resources (Couclelis, 2002; Torrens, 2003; Parker, Manson, Janssen, Hoffmann & Deadman, 2003; Batty, 2005; Barthel et al., 2008). More recently, ABMs have been used to simulate market decisions of firms and individual agents involved in the process of adoption of innovations and diffusion of new technologies. Eppstein, Grover, Marshall, and Rizzo (2011), for example, developed a spatially explicit agent-based model to illustrate how fuel cost, accessibility to charging stations, financial incentives, and households' socioeconomic and demographic characteristics would affect consumer decisions to purchase hybrid and electrical vehicles. Multi-agent modeling approaches have also been employed to simulate interactions between coupled human-natural systems (Monticino, Acevedo, Callicott, Cogdill & Lindquist, 2007) and the impact of climate change on local economies (Berman, Nicolson, Kofinas, Tetlichi & Martin, 2004). Further, ABMs have been implemented to represent electricity markets as complex adaptive systems in the context of identifying potential mechanisms to reduce greenhouse gas emissions (Batten & Grovez, 2006; Beckenbach & Briegel, 2010). At a higher level of abstraction, ABMs have been used as system-of-systems approaches integrating global actions on climate change with domestic policies (Gerst et al., 2013) and macroeconomic decisions with the consumer behavior of households (Hasselmann, 2008).

Wenkel et al. (2013) developed the *LandCaRe 2020* (Land, Climate, and Resources) decision support system, which provided an interdisciplinary framework to integrate climate science, ecology, agronomy, economics, and computer science to support climate adaptation efforts in agriculture. The proposed platform incorporated three climate data analysis models, land use change model, several ecological models with varying spatiotemporal resolutions, and two econometric models.

For the purposes of the study, the platform performed detailed impact assessments related to plant phenology, soil processes, the degree of irrigation demand, primary net productivity, carbon sequestration, and crop yield (Wenkel et al., 2013). The platform is intended to serve a large number of local and regional stakeholders in their assessments of potential agricultural adaptation strategies at various scales (Wenkel et al., 2013).

Integration of Urban Simulation Models with Environmental Impact Assessment Models

Urban growth modeling plays a key role in assessing the long-term viability of proposed and emerging patterns of development (Brown, Alexander, Holzhauer & Rounsevell, 2017). Solecki and Oliveri (2004) conducted a study of future land use/land cover change in the **New York metropolitan region**. The study employed emissions scenarios for 2020 and 2050 using the Special Report on Emissions Scenarios (SRES) developed by the Intergovernmental Panel on Climate Change (IPCC) and included in the Third (TAR) and Fourth (AR4) Assessment Reports. The authors developed land use/land cover projections aligned with emissions scenarios A2 and B2 using Clarke's Urban Growth Model, Land Cover Deltratron Model, and SLEUTH.[1] The landscape characteristics derived from the urban growth simulations were used to calculate albedo, percent vegetative cover, and soil moisture (Solecki & Oliveri, 2004). The derived parameters were used in regional climate models to determine regional climate profiles over a large metropolitan area encompassing 31 counties. That study was commissioned by New York Climate and Health Project and sought to identify potential future changes in regional air quality to evaluate risks of amplified health impacts (Solecki & Oliveri, 2004).

Arthur-Hartranft, Carlson, and Clarke (2003) used imagery from Landsat Thematic Mapper (TM) to estimate the Normalized Difference Vegetation Index (NDVI) and radiant surface temperature as a function of land cover and surface moisture conditions. Landsat data were also used to generate land cover maps using the maximum likelihood supervised classification method. The land cover data provided inputs for generating urban growth projections for **southeastern Pennsylvania** using SLEUTH (Clarke & Gaydos, 1998). Changes in microclimate and hydrology were evaluated through a set of urban growth scenarios taking into account various rates of land conversion and implementation of measures for green space conservation. The study found that the most significant changes in the urban heat island effect measured as changes in radiant surface temperature and evapotranspiration occurred when development proceeded without environmental constraints. The study emphasized that increased pace of development and land conversion could result in steady increases in radiant surface temperature and a decrease in soil moisture (Arthur-Hartranft, Carlson & Clarke, 2003). Urban growth projections were also used to predict representative runoff coefficient for each land cover class using a runoff to rainfall ratio (Arthur-Hartranft, Carlson & Clarke, 2003, p. 385). The coefficient was estimated via a linear regression model accounting for percent developed, forested, and agricultural land; elevation change; and downstream area. The research showed that urban growth could result in predictable changes in hydrological and microclimate regimes (Arthur-Hartranft, Carlson & Clarke, 2003). The findings have implications for policy development to mitigate the impacts of future climate conditions through sustainable land development practices.

Clemonds and Liu (2004) used the SLEUTH model to examine land cover change in the **Houston metropolitan area** and to analyze the propagation of the heat island effect. Cogan, Davis, and Clarke (2001) integrated the SLEUTH model with a wildlife habitat model to predict the impact of urban development on biodiversity. Hester and Feller (2002) explored the effect of the land cover change in the Middle Rio Grande Basin of **New Mexico** on groundwater quantity and quality. Platt (2006) used a cellular automata model of urban development in conjunction with a conceptual model for selecting primary areas for forest thinning with the objective to reduce damage from wildfires in the fire-prone areas of **Boulder County, Colorado**. The combined WHAMED (Wildfire Hazard Mitigation and Exurban Development) model was implemented in the SELES (Spatially Explicit Landscape Event Simulator) modeling environment. The output of the cellular automata model was used to generate "priority areas" for wildfire protection according to criteria outlined by the community protection zone (CPZ) and the wildland-urban interface (WUI) guidelines (Platt, 2006). The results of the study were useful for understanding patterns of future urbanization, how they might affect the fire-prone areas in the study region, where the mitigation efforts should be targeted, and where financial resources should be allocated.

Mitsova (2014) integrated downscaled climate projections of seven general circulation models (GCMs) with the BASINS Climate Assessment Tool (CAT) developed by the United States Environmental Protection Agency (Imhoff, Kittle, Gray & Johnson, 2007) to examine changes in watershed hydrology based on future land cover projections in a large watershed in southwestern **Ohio** (Figure 14.1). Land use/land cover changes were simulated using the cellular automata and Markov chain analysis (*CA-Markov*) module in IDRISI (Eastman, 2012). The estimates of future rainfall variability were based on the assumptions of two IPCC greenhouse gas emissions scenarios. Seasonal variability in mean streamflow discharge, the 100-year flood, and 7Q10 low-flow were evaluated. The uncertainty in the modeling results was assessed using a Monte Carlo simulation approach. Probability density functions were generated to estimate the probability of exceedance of the observed average seasonal hydrologic responses to changing climatic conditions. The Storm Water Management Model (SWMM) was used to conduct sensitivity analysis and evaluate a number of stormwater best management practices. The study found that disconnection of impervious surfaces, rain harvesting, rain gardens, street planters, infiltration basins, and porous pavement could mitigate anticipated increases in the frequency of 100-year flood events and maintain adequate soil moisture conditions during summer dry spells and low flows (Mitsova, 2014). Figure 14.1 displays existing land use/land cover for the year 2010 as well as projections of future patterns of development for 2020 and 2030 using the IDRISI cellular automata and Markov chain analysis (CA-Markov) module (Eastman, 2012).

Sun, Deal, and Pallathucherill (2009) introduced the Land-use Evolution and Impact Assessment Model (LEAM), a comprehensive urban planning support system incorporating a CA-based urban growth model. The proposed framework

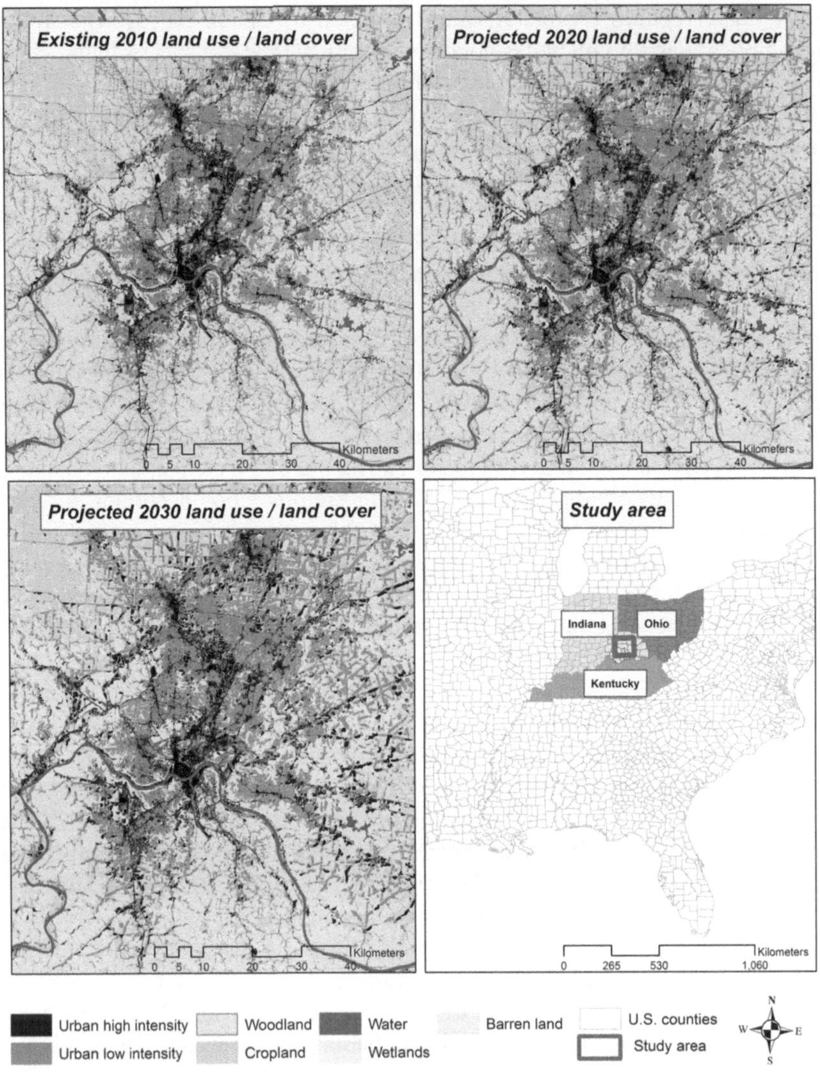

Figure 14.1 Existing land use/land cover in the Cincinnati–Middletown, OH–KY–IN Metropolitan Statistical Area; 2020 and 2030 projected land use/land cover changes in the study area using the IDRISI CA-Markov module (Eastman, 2012)

Source: Map created by the authors

consists of three components: (1) a land use change model; (2) an impact assessment tool; and (3) a feedback component. A strength of the modeling framework is its ability to incorporate different assumptions and policies as part of the inputs to the land use dynamics model. The impact assessment tool consists of several submodels that provide estimates of future air quality, water quality, travel demand, and housing values. Planners and decision-makers could use the feedback component to evaluate alternatives and help adjust policies and plans according to desired outcomes (Sun, Deal & Pallathucherill, 2009).

Planning for Flood Protection

Hoymann and Goetzke (2016) used the GIS-based Land Use Scanner Model (LSM) to simulate patterns of urban development in **Germany** by the year 2030. The simulation was intended to identify measures that could reduce climate change impacts and facilitate future adaptation planning. The study considered a range of adaptation and mitigation measures related to land management practices. Each projected land use scenario was evaluated using a number of indicators composed of multiple sub-measures. Each indicator was formulated as an index score using as a baseline the 2009 national average (Hoymann & Goetzke, 2016). Flood protection measures included in the evaluation of the land use change scenarios aimed at reducing the impact of floods under projected climatic variability. The authors suggested categorizing the areas prone to a 100-year flood event as *priority* or *conservation* areas. In the proposed framework, the designation of the flood-prone areas depended on the amount of existing infrastructure and public services. The proposed conservation areas aimed at increasing carbon sequestration while the priority areas required protective measures for existing infrastructure and housing (Hoymann & Goetzke, 2016).

Voskamp and van den Ven (2015) developed the *Adaptation Support Tool* (AST) to implement "green" and "blue" climate adaptation measures aimed at reducing vulnerability to high-impact low-probability flood events and increase the provision of ecosystem services. The authors emphasized the importance of involving professional planners and other local stakeholders in designing *green-blue* infrastructure measures and other components of the planning support system. The framework was applied to selected neighborhoods in the cities of **Rotterdam** and **Delft** in the **Netherlands**. The proposed green-blue measures incorporated the core concepts of sustainable stormwater management and best practices combined with the principles of green infrastructure design and green space conservation (Voskamp & van den Ven, 2015). The framework consists of over 30 measures intended to improve rainwater harvesting and storage, facilitate runoff attenuation, and contribute to mitigating the urban heat island effect by harnessing the cooling power of water bodies and vegetation (Voskamp & van den Ven, 2015).

Pathirana, Denekew, Veerbeek, Zevenbergen, and Banda (2014) examined the probability of increased flood risk in four tropical cities, taking into consideration urban expansion and higher frequency of extreme rainfall events. Numerical experiments were developed through the integration of a land use change model and a mesoscale atmospheric model. For example, future urban development of the **city of Mumbai, India**, was simulated using Dinamica-EGO, a landscape dynamics model. The results of the urban simulation were used as inputs to the mesoscale atmospheric model. Using historical rainfall records and future land use, the model simulated hourly 10-year and 50-year rainfall events. The results indicated that these events would be equivalent to a future frequency of 3-year and 22-year flood events, respectively (Pathirana, Denekew, Veerbeek, Zevenbergen & Banda, 2014). The findings suggested a significant increase in extreme rainfall in three of the four study areas. Increased flood risk was found to be highly correlated with the increased pace of urbanization and land conversion to urban uses (Pathirana, Denekew, Veerbeek, Zevenbergen & Banda, 2014).

Choi and Lee (2016) adapted the "What-if" planning support system (Klosterman, 2001a, 2001b) to simulate future patterns of development and develop a spatial plan for the **Seoul metropolitan area** and **Gyeonggi Province, South Korea**. Future land use scenarios incorporated as driving forces future population growth, economic activities, environmental concerns and objectives, and other relevant policies (Choi & Lee, 2016). Several environmental indicators were developed to quantify the impact of the proposed land use change scenarios on natural ecosystems, microclimate, air quality, and vulnerability to natural hazards. The study identified several areas of concern where future planning initiatives could help reduce vulnerability and increase resilience to climatic and other environmental stressors (Choi & Lee, 2016).

Mahmoud, Duker, Conrad, Thiel, and Ahmad (2016) conducted a comprehensive analysis of the urban expansion in **Abuja, Nigeria**, one of West Africa's fastest growing cities. Urban settlement growth was projected using the Land Change Modeler (LCM) in IDRISI GIS. The study found an increased development pressure in the floodplains and projected a considerable loss of arable land and vegetation. The findings indicated that the current patterns of development could amplify the likelihood of catastrophic floods. Rapid land conversion and increasing temperatures would likely facilitate the formation of urban heat islands. The authors suggested mitigating these impacts through green urban design (Mahmoud, Duker, Conrad, Thiel & Ahmad, 2016).

Green Infrastructure Planning

Green infrastructure can provide multiple ecosystem services to mitigate some of the undesirable impacts of urban development including mitigation of the urban heat island effect, stormwater runoff attenuation, and filtering of pollutants. Yang and Lo (2003) linked the SLEUTH urban growth model (Clarke & Gaydos, 1998) to a landscape change model to identify the best possible scenario for the future development of **Atlanta, Georgia**. Three scenarios were proposed and investigated. The first scenario extrapolated current trends of "rampant" development at the urban fringe into the future (Yang & Lo, 2003). This type of development, as shown by the modeling results, would cause a substantial decrease in woodland and open space in the study area. The other two scenarios focused on measures to expand the existing green infrastructure. The low-impact scenario envisioned a slight increase in riparian setbacks, protection of wetlands, and limited development in the floodplains. The outcomes of this scenario showed an improvement of open space conservation (Yang & Lo, 2003). The high-impact scenario envisioned a more aggressive growth management approach resulting in reducing the amount of newly developed land, increasing the green space areas by almost 30%, and enforcing stronger protection measures for the ecologically sensitive areas (Yang & Lo, 2003).

Mitsova, Shuster, and Wang (2011) developed a Markov chain–based CA model with multi-criteria evaluation (MCE) using IDRISI GIS. The study compared two scenarios of future urban development—a "business as usual" scenario and a green infrastructure (GI) conservation scenario protecting environmentally sensitive areas at a regional scale. Some of the policy measures incorporated in the GI conservation scenario included preservation of urban open space, restriction of urban development on steep slopes and in floodplains, protection of wetlands, and the establishment of vegetated buffers along impaired streams (Mitsova, Shuster & Wang, 2011). The two scenarios were compared using a set of landscape-level metrics to quantify the possible environmental impacts. The study found an overall reduction of the spatial extent of urban development under the GI conservation scenario and decreased fragmentation of natural areas. The study highlighted the importance of considering policy measures to protect impaired streams across jurisdictional boundaries and provided useful assessments that could inform watershed managers and municipalities seeking to reduce pollutant loadings to local streams (Mitsova, Shuster & Wang, 2011).

Jat, Choudhary, and Saxena (2017) used the SLEUTH model to project future urban development in **Ajmer City, India**. The city has experienced exponential population growth in recent years and has been selected as one of the **Rajasthan** province's pilot smart growth cities. Multispectral satellite data were processed and used as inputs in the urban growth model. The results from the urban growth simulation showed that there was a high potential for expansion of the urbanized areas. The study also modeled vertical growth in the form of multi-story development as a means to mitigate the environmental impacts of land conversion and preserve arable land and open space (Jat, Choudhary & Saxena, 2017).

Exploring the Effects of Policy Measures

Research has shown that urban simulation models can represent complex social processes and changes in the natural and built environment. Despite the advancements, more research is needed into the dynamics of urban systems and climate change adaptation. Future studies could benefit from stakeholder participation and integration of "real-world" policies and plans into the modeling frameworks. Several studies reviewed in this chapter showed the feasibility of aligning urban growth projections with the output of climate models to improve long-term environmental assessments, flood forecast modeling, and assessment of future urban heat island effects (Nay, Abkowitz, Chu, Gallagher & Wright, 2014; Mitsova, 2014; Choi & Lee, 2016; Deal & Pan, 2017). The coupled urban growth–climate modeling systems can also inform policy scenarios and urban design alternatives to mitigate anticipated adverse impacts (Drewniak et al., 2013).

The study by Hoymann and Goetzke (2016) revealed the possibility of conflicting mitigation and adaptation objectives. For example, inner-city redevelopment was found to contribute to climate mitigation while simultaneously exacerbating the urban heat island effect by concentrating urban activities and reducing the amount of land available for green space (Hoymann & Goetzke, 2016). The study results demonstrated that a scenario-based urban growth modeling approach with a set of indicators could optimize the outcomes of the decision process aimed at achieving multiple policy objectives.

Masson et al. (2014) suggest that climate impact studies often focus on specific parameters while omitting important considerations and variables that could enable more comprehensive approaches to climate adaptation planning. The "systemic modeling approach" implemented by Masson et al. (2014) allows for a holistic evaluation of various urban climate futures against the output of integrated multidisciplinary platforms and is capable of generating dynamic urban growth scenarios that could be used to evaluate adaptation actions. Recent studies have explored the possibility of coupling cellular automata models with statistical techniques including linear, logistic, and multi-level regression models and artificial neural network to enhance their predictive and explanatory power (Duncan & Jones, 2000; Couclelis, 2002; Li & Yeh, 2002; Paez & Scott, 2004; Lee & Lathrop, 2006; Mahajan & Venkatachalam, 2009; Fisher-Vanden, Wing, Lanzi & Popp, 2013).

Note

1 For detailed descriptions of these models, see Clarke (2008).

References

Aguejdad, R., Hidalgo, J., Doukari, O., Masson, V., & Houet, T. 2012. *Assessing the Influence of Long-Term Urban Growth on Urban Climate*. 6th International Congress on Environmental Modeling and Software, July 1–5, 2012, UFZ. Leipzig, Germany.

Arthur-Hartranft, T., Carlson, T.N., & Clarke, K.C. 2003. Satellite and Ground-Based Microclimate and Hydrologic Analyses Coupled with a Regional Urban Growth Model. *Remote Sensing of Environment* 86, 385–400.

Balbi, S., & Giupponi, C. 2009. *Reviewing Agent-Based Modelling of Socio-Ecosystems: A Methodology for the Analysis of Climate Change Adaptation and Sustainability*. Working Papers, University of Venice.

Barthel, R., Janisch, S., Schwarz, N., Trifkovic, A., Nickel, D., Schultz, C., & Mauser, W. 2008. An Integrated Modelling Framework for Simulating Regional-Scale Actor Responses to Global Change in the Water Domain. *Environmental Modelling and Software* 23(9), 1095–1121.

Batten, D., & Grovez, G. 2006. NEMSIM: Finding Ways to Reduce Greenhouse Gas Emissions Using Multi-Agent Electricity Modelling. In Perez, P. (Ed.)

Complex Science for a Complex World: Exploring Human Ecosystems with Agents Canberra, Australia: Australian National University Press.

Batty, M. 2005. Approaches to Modeling in GIS: Spatial Representation and Temporal Dynamics. In Mcguire, D., Batty, M., & Goodchild, M.F. (Eds.) *GIS, Spatial Analysis and Modeling*, 41–62. Redlands, CA: ESRI Press.

Beckenbach, F., & Briegel, R. 2010. Multi-Agent Modeling of Economic Innovation Dynamics and Its Implications for Analyzing Emission Impacts. *International Economics and Economic Policy* 7, 317–341.

Berman, M., Nicolson, C., Kofinas, G., Tetlichi, J., & Martin, S. 2004. Adaptation and Sustainability in a Small Arctic Community: Results of an Agent-Based Simulation Model. *Arctic* 57, 401–414.

Brown, C., Alexander, P., Holzhauer, S., & Rounsevell, M.D.A. 2017. Behavioral Models of Climate Change Adaptation and Mitigation in Land-based Sectors. *Wiley Interdisciplinary Reviews: Climate Change* 8(2), e448.

Choi, H.-S., & Lee, G.-S. 2016. Planning Support Systems (PSS)-Based Spatial Plan Alternatives and Environmental Assessment. *Sustainability* 8, 286.

Clarke, K.C. 2008. A Decade of Cellular Urban Modeling with SLEUTH: Unresolved Issues and Problems. In Brail, R.K. (Ed.) *Planning Support Systems for Cities and Regions*. Cambridge, MA: Lincoln Institute of Land Policy.

Clarke, K.C., & Gaydos, L. 1998. Loose-Coupling a Cellular Automaton Model and GIS: Long-Term Urban Growth Prediction for San Francisco and Washington-Baltimore. *International Journal of Geographic Information Science* 12, 699–714.

Clemonds, M., & Liu, H. 2004. *Exploring Urban Heat Island effects in the Houston Metropolitan Area using Satellite Remote Sensing Data*. Proceedings of the 100th Annual Meeting of the American Association of Geographers, Philadelphia, PA.

Cogan, C.B., Davis, F.W., & Clarke, K.C. 2001. *Application of Urban Growth Models and Wildlife Habitat Models to Assess Biodiversity Losses*. Santa Barbara, CA: The University of California–Santa Barbara Institute for Computational Earth System Science. U.S. Department of the Interior, U.S. Geological Survey—Biological Resources Division. Gap Analysis Program.

Couclelis, H. 2002. Modeling Frameworks, Paradigms and Approaches. In Clarke, K., Parks, B., & Crane, M. (Eds.) *Geographic Information Systems and Environmental Modeling*. Upper Saddle River, NJ: Prentice Hall Series in Geographic Information Science.

Deal, B., & Pan, H. 2017. Discerning and Addressing Environmental Failures in Policy Scenarios Using Planning Support System (PSS) Technologies. *Sustainability* 9, 13.

Drewniak, B., Kotamarthi, R., Jacob, R., Chen, F., Catlett, C., Ching, J., & Wu, W. 2013. Urban: Landscape and Climate Change, Workshop Summary, August 28–29, 2013, Argonne National Laboratory, Lemont, Illinois.

Duncan, C., & Jones, K. 2000. Using Multilevel Models to Model Heterogeneity: Potential and Pitfalls. *Geographical Analysis* 32(4), 279–305.

Eastman, J.R. 2009. *IDRISI Taiga Guide to GIS and Image Processing*. Worcester, MA: Clark Labs, Clark University.

Eastman, J.R. 2012. *IDRISI Selva Tutorial, Manual Version 17.0*, Clark University, January 2012. Retrieved from http://uhulag.mendelu.cz/files/pagesdata/eng/gis/idrisi_selva_tutorial.pdf (accessed 11/09/2015).

Eppstein, M.J., Grover, D.K., Marshall, J.S., & Rizzo, D.M. 2011. An Agent-Based Model to Study Market Penetration of Plug-in Hybrid Electric Vehicles. *Energy Policy* 39(6), 3789–3802.

Fisher-Vanden, K., Wing, I., Lanzi, E., & Popp, D. 2013. Modeling Climate Change Feedbacks and Adaptation Responses: Recent Approaches and Shortcomings. *Climatic Change* 117(3), 481–495.

Geertman, S., & Stillwell, J. 2004. Planning Support Systems: An Inventory of Current Practice. *Computers, Environment and Urban Systems* 28, 291–310.

Gerst, M.D., Wanga, P., Roventini, A., Fagiolo, G., Dosi, G., Howarth, R.B., & Borsuk, M.E. 2013. Agent-Based Modeling of Climate Policy: An Introduction to the ENGAGE Multi-Level Model Framework. *Environmental Modelling & Software* 44, 62–75.

Hasselmann, K. 2008. Simulating Human Behavior in Macroeconomic Models Applied to Climate Change. *Proceedings of the Heraeus seminar on Energy and Climate* 2.

Hester, D.J., & Feller, M.R. 2002. Landscape Change Modeling: Groundwater Resources of the Middle Rio Grande Basin, New Mexico. In Bartolino, J.R., & Cole, J.C. (Eds.) *U.S. Geological Survey Circular* 1222, 20–21. Reston, VA: U.S. Geological Survey.

Hopkins, L.D., Kaza, H., & Pallathucheril, V.G. 2005. Representing Urban Development Plans and Regulations as Data: A Planning Data Model. *Environment and Planning B: Planning and Design* 32, 597–615.

Hoymann, J., & Goetzke, R. 2016. Simulation and Evaluation of Urban Growth for Germany Including Climate Change Mitigation and Adaptation Measures. *ISPRS International Journal of Geo-Information* 5, 101–123.

Imhoff, J.C., Kittle, J.L., Gray, M.R., & Johnson, T.E. 2007. Using the Climate Assessment Tool (CAT) in US EPA BASINS Integrated Modeling System to Assess Watershed Vulnerability to Climate Change. *Water Science and Technology* 56, 49–56.

Jat, M.K., Choudhary, M., & Saxena, A. 2017. Urban Growth Assessment and Prediction Using RS, GIS and SLEUTH Model for a Heterogeneous Urban Fringe. *The Egyptian Journal of Remote Sensing and Space Sciences*. Retrieved from https://doi.org/10.1016/j.ejrs.2017.02.002.

Klosterman, R. 2001a. Planning Support Systems: A New Perspective on Computer-Aided Planning. In Brail, R.K., & Klosterman, R.E. (Eds.) *Planning Support Systems*. Redlands, CA: ESRI Press.

Klosterman, R. 2001b. The What If? Planning Support System. In Brail, R.K., & Klosterman, R.E. (Eds.) *Planning Support Systems*. Redlands, CA: ESRI Press.

Lee, S., & Lathrop, R.G. 2006. Subpixel Analysis of Landsat ETM+ Using Self-Organizing Map (SOM) Neural Networks for Urban Land Cover Characterization. *IEEE Transactions on Geoscience and Remote Sensing* 44(6), 1642–1654.

Li, X., & Yeh, G.O. 2002. Neural-Network-Based Cellular Automata for Simulating Multiple Land Use Changes Using GIS. *International Journal of Geographical Information Science* 16(4), 323–343.

Mahajan, Y., & Venkatachalam, P. 2009. Neural Network Based Cellular Automata Model for Dynamic Spatial Modeling in GIS. In Gervasi, O., Taniar, D., Murgante, B., Lagana, A., & Mun, Y. (Eds.) *Computational Science and Its Applications—ICCSA 2009*. Berlin, Heidelberg: Springer.

Mahmoud, M.I., Duker, A., Conrad, A., Thiel, M., & Ahmad, H.S. 2016. Analysis of Settlement Expansion and Urban Growth Modelling Using Geo-Information for Assessing Potential Impacts of Urbanization on Climate in Abuja City, Nigeria. *Remote Sensing* 8, 220–244.

Masson, V., Marchadier, C., Adolphe, L., Aguejdad, R., Avner, P., Bonhomme, M., Bretagne, G., Briottet, X., Bueno, B., de Munck, C., Doukari, O., Hallegatte, S., Hidalgo, J., Houet, T., Le Bras, J., Lemonsu, A., Long, N., Moine, M.-P., Morel, T., Nolorgues, L., Pigeon, G., Salagnac, J.-L., Viguié, V., & Zibouche, K. 2014. Adapting Cities to Climate Change: A Systemic Modelling Approach. *Urban Climate* 10(2), 407–429.

Mitsova, D. 2014. Coupling Land Use Change Modeling with Climate Projections to Estimate Seasonal Variability in Runoff from an Urbanizing Catchment Near Cincinnati, Ohio. *ISPRS, International Journal of Geo-Information* 3, 1256–1277.

Mitsova, D., Shuster, W., & Wang, X. 2011. A Cellular Automata Model of Land Cover Change to Integrate Urban Growth with Open Space Conservation. *Landscape and Urban Planning* 99(2), 141–153.

Monticino, M., Acevedo, M., Callicott, B., Cogdill, T., & Lindquist, C. 2007. Coupled Human and Natural Systems: A Multi-Agent-Based Approach. *Environmental Modelling & Software* 22(5), 656–663.

Nay, J.J., Abkowitz, M., Chu, E., Gallagher, D., & Wright, H. 2014. A Review of Decision-Support Models for Adaptation to Climate Change in the Context of Development. *Climate and Development* 6(4), 357–367.

Paez, A., & Scott, D. 2004. Spatial Statistics for Urban Analysis: A Review of Techniques with Examples. *GeoJournal* 61(1), 53–67. doi: 10.1007/s10708-004-0877-x.

Parker, D.C., Manson, S.M., Janssen, M.A., Hoffmann, M.J., & Deadman, P. 2003. Multi-Agent Systems for the Simulation of Land-Use and Land-Cover Change: A Review. *Annals of the Association of American Geographers* 93(2), 314–337.

Pathirana, A., Denekew, H.B., Veerbeek, W., Zevenbergen, C., & Banda, I.T. 2014. Impact of Urban Growth-Driven Land Use Change on Microclimate and Extreme Precipitation: A Sensitivity Study. *Atmospheric Research* 138, 59–72.

Platt, R. 2006. A Model of Exurban Land-Use Change and Wildfire Mitigation. *Environment and Planning B: Planning and Design* 33, 749–765.

Solecki, W.D., & Oliveri, C. 2004. Downscaling Climate Change Scenarios in an Urban Land Use Change Model. *Journal of Environmental Management* 72, 105–115.

Stewart, I.D., & Oke, T.R. 2012. Local Climate Zones for Urban Temperature Studies. *Bulletin of American Meteorological Society* 93, 1879–1900.

Sun, Z., Deal, B., & Pallathucherill, V.G. 2009. The Land-Use Evolution and Impact Assessment Model: A Comprehensive Urban Planning Support System. *URISA Journal* 21(1), 55–62.

Torrens, P. 2003. Automata-Based Models of Urban Systems. In Longley, P.A., & Batty, M. (Eds.) *Advanced Spatial Analysis: The CASA Book of GIS*, 61–81. Redlands, CA: ESRI Press.

Voskamp, L.M., & van den Ven, F.H.M. 2015. Planning Support System for Climate Adaptation: Composing Effective Sets of Blue-Green Measures to Reduce Urban Vulnerability to Extreme Weather Events. *Building and Environment* 83, 159–167.

Wenkel, K.-O., Berg, M., Mirschel, W., Wieland, R., Nendel, C., & Kostner, B. 2013. LandCaRE—an Interactive Decision Support System for Climate Change Impact Assessment and the Analysis of Potential Agricultural Land use Adaptation Strategies. *Journal of Environmental Management* 127, S168–S183.

White, R., & Engelen, G. 1997. Cellular Automata as the Basis of Integrated Dynamic Regional Modelling. *Environment and Planning B: Planning and Design* 24, 235–246.

Yang, X., & Lo, C.P. 2003. Modeling Urban Growth and Landscape Changes in the Atlanta Metropolitan Area. *International Journal of Geographical Information Science* 17(5), 463–488.

15 Internet-Based GIS Applications to Facilitate the Adaptation of the Built Environment to Climate Change

Chapter Objectives

This chapter catalogs several widely used web-based geographic information system (GIS) applications seeking to answer the following questions:

- What capabilities do these tools, platforms, and application provide to assist climate mitigation and adaptation planning?
- Which of these tools support interactive mapping?
- Which tools are particularly useful for conducting advanced local and regional analyses?
- Which tools support data download and decision support?

Introduction

While climate change policies require action at a national and global level, most of the impacts, as discussed in the previous chapters and elsewhere, are felt locally (Torresan, Critto, Rizzi, & Marcomini, 2012). For many local governments and their constituencies, planning for climate mitigation and adaptation is unlike any of the well-established planning practices. While a typical planning project may encompass 3 to 5 years from the first design charrette to the actual implementation, planning to address climate change requires a level of forethought that spans over decades. Planning horizons that extend thus far into the future are naturally fraught with uncertainties (Rozum & Carr, 2013). Furthermore, climate adaptation planning requires scientific knowledge from the natural and social sciences and the ability to process vast amounts of data and modeling outputs presented in a format that can be easily shared with the end user (Grant, Baldwin, Lieske, & Martin, 2015). The sheer number of unknowns and the difficulties in communicating issues that many consider as belonging to a distant future have hindered some of the earlier efforts to discuss climate change with stakeholders and the general public (Barnett, Waters, Pendercast, & Puleston, 2013). Additional challenges arise from lack of a common knowledge base about climate issues. Research, policy-making, and planning practices indicate that building a knowledge base as a mutually agreed-upon reference frame is instrumental in creating common ground for coordination of plans and actions across jurisdictions and agencies (MacEachren & Brewer, 2004). Last but not least, many local governments simply do not possess the resources and skills needed to address these challenges.

The realization that the success of climate adaptation planning partly depends on the availability of data and visualization tools has led to rapid development of Internet-based GIS tools, platforms, and applications supported by federal agencies, universities, research centers, and non-profit organizations (Ernst & Blaha, 2015). Visualization is a communicative process in which a message "selected from a set of possible messages" is encoded and transmitted from an information source to a destination (Shannon, 1948, p. 379). *Geovisualization* (geographic visualization), more specifically, is a broad term used for a set of tools and techniques such as cartography, image analysis, information analysis, exploratory data analysis, and GIS to support exploration, understanding, and communication of spatial phenomena (Nöllenburg, 2007). Human-centric geovisualization approaches communicate geospatial information in ways that, when combined with human understanding, allow for data exploration, knowledge construction, and decision-making processes (Nöllenburg, 2007; Esnard, 2012). Sheppard (2012) provides a comprehensive guide to how visual imagery, media, 3D and 4D visualizations, and newer tools like video games can be used by planners and other experts who engage the public, build awareness, and promote mitigation and adaptation solutions to climate change. Overall, a range of static (e.g., graphics and maps), interactive (e.g., websites and participatory GIS), and dynamic (e.g., 3D simulation models and web-GIS) tools can contribute to intelligence gathering and knowledge creation for climate adaptation planning processes. Visualization tools and geospatial technologies can reveal spatial trends and patterns that might not be detected when paper maps are manually reviewed or if geographic information is presented in a spreadsheet.

In order to facilitate the use of a wide variety of tools, several classifications and taxonomies have been developed. Carr (2014) organized the decision support tools reported by the participants in the Ecosystem-Based Management (EBM) Network survey in several categories starting with the most commonly used, followed by tools with regional and national coverage, location-specific tools potentially applicable to other regions, and projects using these tools. Ernst and Blaha (2015) categorized the most commonly used climate adaptation tools into three categories: (a) web-based decision support tools, (b) decision support software, and (c) applications developed by individual cities and regions. Rozum and Carr (2013) selected 10 of the most widely used ecosystem-based climate planning tools and classified their functionalities in terms of visualization, modeling, and decision support. Condon, Cavens, and Miller (2009) reviewed 12 planning and urban design tools offering analytical capabilities in support of climate mitigation strategies for the built environment. The authors evaluated these tools according to their scope, methods, scale, and decision support capabilities. Among these tools, *INDEX-Cool Spots* and *I-PLACE's* offer the most extensive array of functionalities in support of climate mitigation efforts (Condon, Cavens & Miller, 2009).

The approach adopted in this chapter is two-fold. First, in order to be consistent with the organization of the book, we categorize the web-based applications according to their topical area (e.g., climate data portals, sea level rise visualization, flood analysis and mapping, heat effects mapping, climate change vulnerability mapping, ecosystem-based approaches, and greenhouse gas [GHG] emissions inventories and portals). Next, we identify tools that offer different capabilities in each topical area. In doing so, we categorize the available tools in each thematic area as viewers, digital data portals, software platforms with modeling capabilities, and software platforms with decision support capabilities. Viewers are web-based browsers that support interactive mapping. Most digital data portals are specifically designed as data catalogs, and some are equipped with a viewer. Several freely available software platforms offer modeling capabilities that can be used to quantify climate variability and change and their potential impacts on the natural and built environments. Finally, there is a growing pool of tools that are designed explicitly as decision support applications with options to generate and assess scenarios.

Viewers and Digital Data Portals: High-Resolution Data for Climate Assessments in the United States

National Aeronautics and Space Administration (NASA) Climate Data Services (CDS)

https://nex.nasa.gov/nex

NASA's CDS website provides users, including researchers and citizen scientists, with a broad set of tools to conduct climate assessments and visualize climate data (Carriere, 2015). Historical records (1950–2005) of monthly average maximum temperature, minimum temperature, precipitation as well as climate models' outputs are available through the NASA Earth Exchange (NEX) Downscaled Climate Projections (NEXDCP30) (NASA, 2017, https://nex.nasa.gov/nex/projects/1356). The *NASA Earth Exchange* provides access to the *U.S. Downscaled Climate Projections (NEXDCP30)* and the *Global Daily Downscaled Projections (NEX-GDDP)* datasets, which contain the output of 30 climate models for the period from 2006–2099 (Meinshausen et al., 2011). The dataset takes into account all four Representative Concentration Pathways (RCPs) emission scenarios based on the *Coupled*

Model Intercomparison Project Phase 5 (CMIP5) developed for the Intergovernmental Panel on Climate Change (IPCC) Fifth Assessment Report (AR5) (Taylor, Stouffer, & Meehl, 2012).

The United States Geological Survey (USGS) National Climate Change Viewer (NCCV)

www2.usgs.gov/climate_landuse/clu_rd/nccv.asp

NCCV is a clearinghouse and viewer for the *NEX-US-DCP30* climate data for two of the RCP scenarios (USGS, 2016). *NEX-US-DCP30* provides downscaled climate data with a resolution of 800 meters (NASA, 2017, https://cds.nccs.nasa.gov/nex). The *NEX-US-DCP30* parameter values are averaged for four climatology periods (1981–2010, 2025–2049, 2050–2074, and 2075–2099) to facilitate data access and use (USGS, 2016).

National Oceanic and Atmospheric Administration (NOAA)'s Weather and Climate Toolkit (WCT)

www.ncdc.noaa.gov/wct

The *Weather and Climate Toolkit (WCT)* available through NOAA's National Centers for Environmental Information is an open-source software platform for downloading climate data and conducting data format conversions (NOAA-NCEI, 2016). Scientific communities can benefit from WCT's export features and interoperability with common software packages such as ArcGIS, Google Earth, QGIS, and R, among others. Standard formats for data export and conversion include KMZ, shapefile, GeoTIFF, Environmental Systems Research Institute (ESRI) grid, and gridded NetCDF (NOAA-NCEI, 2016).

National Center for Atmospheric Research (NCAR) Climate Data Guide and GIS Program

https://climatedataguide.ucar.edu/climate-data

https://gis.ucar.edu

The NCAR *Climate Data Guide*, according to the website, provides observational datasets, tools, and methods used for weather monitoring, analyses, and earth system model evaluation (Schneider, Deser, Fasullo, & Trenberth, 2013). The data guide is described as "a living repository" and "a community-authored resource" and provides users, expert contributors, and dataset developers with information on data limitations to facilitate appropriate use and application (NCAR, 2017a).

NCAR's GIS Program is of particular interest to the geoscience community, given the availability of a suite of tools (e.g., *GIS Climate Change Scenarios* portal, *Climate Inspector*, *Extreme Heat Climate Inspector*) for atmospheric and climate change analyses and research (NCAR, 2017b). The *GIS Climate Change Scenarios* portal allows users to access global climate model data in standard GIS data formats (NCAR, 2017b). *Climate Inspector*, an interactive web application, enables visualization of temperature and precipitation changes throughout the 21st century (NCAR, 2017b). The *Extreme Heat Climate Inspector* is another interactive web

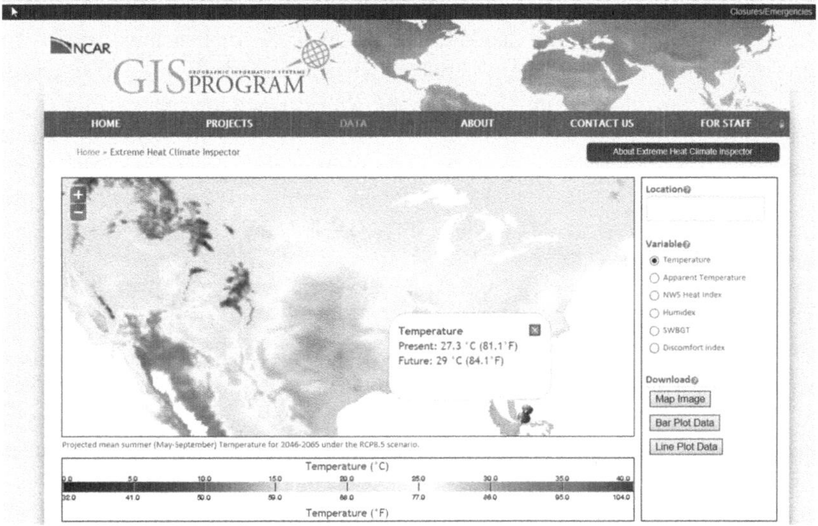

Figure 15.1 NCAR's GIS Program: Extreme Heat Climate Inspector

Source: National Center for Atmospheric Research, https://gis.ucar.edu/heatinspector, used with permission

application that allows users to visualize projected increase in temperatures (NCAR, 2017b). Figure 15.1 shows the home page of the NCAR's GIS Program.

World Resources Institute (WRI) Climate Analysis Indicators Tool (CAIT)

http://cait.wri.org

WRI's *Climate Analysis Indicators Tool* provides free and open-source climate and emissions data. CAIT offers nine tools: Emissions, Paris Contributions, Pre-2020 Pledges Map, Emissions Projections, Indonesia Climate Data, Equity, Google Public Data, Business Emissions and Targets, and Global Forest Watch (WRI, 2009). Data are at the country level, and the mapping functions allow for the display of a variety of past, current, and projected climate data. An example is the Indonesian Climate Data Explorer (PINDAI), which presents climate data and information at the provincial level (WRI, 2009).

Decision Support Tools and Scenario Planning

The Climate Wizard developed by The Nature Conservancy (TNC)

www.climatewizard.org

Climate Wizard is a web application that allows access to historical and future climate change data (temperature and precipitation) for geographic locations in the United States (TNC, 2009). Historical data cover the years 1951 to 2006, while future projections cover two time periods: 2040–2069 and 2070–2100. Temperature and precipitation data can be displayed as the average for that location and time period or as a magnitude of change between two time periods (TNC, 2009). The data output can be downloaded in ASCII format (World Geodetic System [WGS] 84), making it usable in many GIS applications. Past projections are based on three sets of data with 4-kilometer, 12-kilometer, and 50-kilometer resolutions (TNC, 2009).

NOAA National Weather Service (NWS) Climate Prediction Center (CPC)

www.cpc.ncep.noaa.gov/products/forecasts

The Climate Prediction Center provides the official NOAA 3-month temperature and precipitation forecast for the continental United States and Alaska (NOAA-NWS-CPC, 2015). The NOAA long-range prediction projects changes in temperature and precipitation for 3-month time periods over the upcoming year. The forecast indicates expected above- and below-normal conditions. The Probability of Exceedance maps provide a 3-month outlook for a temperature variation from the baseline (average of years 1981 to 2010) (NOAA-NWS-CPC, 2015). Figure 15.2 displays seasonal temperature and precipitation outlook for selected months in 2018.

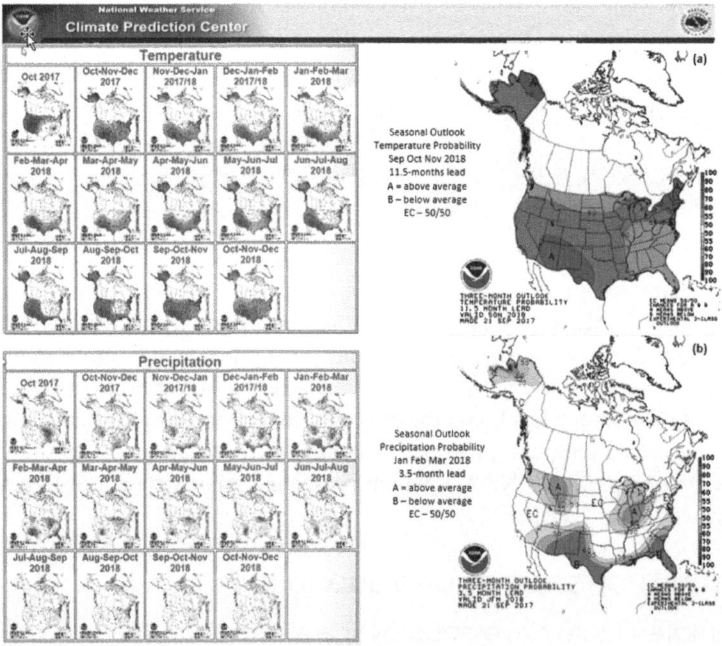

Figure 15.2 Two-class monthly and seasonal climate forecasts: (a) temperature probability outlook for Sep–Oct–Nov 2018, 11.5-month lead; (b) precipitation probability outlook for Jan–Feb–Mar 2018, 3.5-month lead

Source: NOAA-NWS-CPC, www.cpc.ncep.noaa.gov/products/predictions/long_range/two_class.php, used with permission

Web-Based Tools for Sea Level Rise and Flood Mapping

NOAA Sea Level Rise Viewer

https://coast.noaa.gov/digitalcoast/tools/slr

Sea Level Rise Viewer allows for the visualization of the effect of sea level rise on coastal areas in the United States, excluding Alaska (NOAA Office of Coastal Management [OCM], 2017). The effect on the land of sea level rise of 0 to 6 feet can be observed, in both depth and extent of the inland intrusion. Flooding is calculated using coastal elevation as well as some hydrological modeling, but erosion and other future conditions are not modeled (NOAA-OCM, 2017). Certain locations provide rising water superimposed on photographs for additional visualization. The *Local Scenarios* tab allows users to display five local sea level rise scenarios by year or by individual scenario (NOAA-OCM, 2017). In addition, flood frequency, marsh migration, and socioeconomic vulnerability can be displayed where socioeconomic vulnerability is represented using the Social Vulnerability Index (SoVI) developed by Cutter, Boruf, and Shirley (2003). Figure 15.3 shows projected sea level rise in Savannah, Georgia.

Climate Change Adaptation Modeler (CCAM) in Clark Labs' TerrSet

https://clarklabs.org/terrset/climate-change-adaptation-modeler

According to the website, CCAM, which is fully integrated with Clark Labs's TerrSet system, is a suite of tools for modeling climate scenarios and assessing impacts of global warming and sea level rise (Clark Labs, 2015). CCAM supports tools for downloading downscaled climate projections, obtaining a range of bioclimatic variables for species modeling, and developing crop suitability scenarios (Clark Labs, 2015).

TNC Coastal Resilience Mapping Portal

www.coastalresilience.org

Since its inception in 2008, *CoastalResilience.org* has rapidly evolved into a hub of online mapping decision support tools and apps to guide local communities in their planning efforts to reduce vulnerability to coastal hazards (TNC, 2016; Ferdaña, 2015). Developed and supported by The Nature Conservancy and its partners, CoastalResilience.org currently provides technical assistance and communicates the key role of natural habitats in increasing coastal resilience both nationally and globally (TNC, 2016; Ferdaña, 2015). In 2014, CoastalResilience.org became the winner of the *Global Disaster Resilience App Challenge* in the *Best Professional and Scientific App* category (United Nations International Strategy for Disaster Risk Reduction [UNISDR], 2014). The award was announced in New York by ESRI and the United Nations International Strategy for Disaster Risk Reduction during the 2014 *Climate Week* (UNISDR, 2014; Ferdaña, 2015).

The CoastalResilience.org platform highlights ecological and socioeconomic vulnerabilities and supports the development of apps and data layers to facilitate

Figure 15.3 Screenshots of NOAA Sea Level Rise Viewer displaying expected levels of inundation for Savannah, Georgia

Source: Image courtesy of NOAA's Office of Coastal Management, https://coast.noaa.gov/digitalcoast/tools/slr, used with permission

decision-making related to coastal flood risk and climate adaptation. The Coastal Resilience mapping portal currently includes data and visualization tools for 17 U.S. coastal states: Alabama, California, Connecticut, Florida, Georgia, Hawaii, Louisiana, Maine, Maryland, Mississippi, New Jersey, New York, North Carolina, South Carolina, Texas, Virginia, and Washington), the Caribbean (Grenada, St. Vincent and the Grenadines, and the U.S Virgin Islands), and Central America (Belize, Guatemala, Honduras, and Mexico) (Ferdaña, 2015). Data include sea level rise (SLR), storm surge, social variables and economic assets, community vulnerability, and natural resources. Additionally, a variety of variables can be mapped including habitats (e.g., oyster reefs and salt marshes), threatened and endangered species, bathymetry, salinity, and coastal management practices (TNC, 2016). Figure 15.4 shows the relative risk to communities in Galveston Bay and Jefferson County, Texas, to a storm surge similar to 2008's Hurricane Ike plus 1 meter of SLR (for the year 2100).

USGS Coastal Change Hazards Portal version 1.1.58

https://marine.usgs.gov/coastalchangehazardsportal

This USGS *Coastal Hazards Change Portal (v.1.1.58)* provides a suite of methods and tools for supporting the management of coastal infrastructure and resources (USGS, 2017). The site is organized into three main tabs:

(i) *Extreme storms*: (1) assessment of vulnerability to severe storms and related processes such as dune erosion and inundation, and (2) capabilities for generating real-time storm-induced coastal change and scenario-based analyses (USGS, 2017)

(ii) *Shoreline change*: assessment of shoreline change, including short-term (< 30 years) and long-term (78 + years) rates of shoreline change (USGS, 2017)

(iii) *Sea level rise*: assessment of vulnerability to sea level rise based on three distinct methods; one of these methods is the Coastal Vulnerability Index, which shows where and to what extent physical coastal change is likely (low, moderate, high, very high) based on a combination of variables—geomorphology, regional coastal slope, tide range, wave height, relative sea level rise, and shoreline erosion and accretion rates (USGS, 2017)

Figure 15.5 displays the level of coastal vulnerability for the Atlantic seaboard and the Gulf of Mexico.

Surging Seas—Climate Central

http://sealevel.climatecentral.org

Surging Seas provides a searchable mapping application and tools to explore the effects of sea level rise on coastal regions in the United States (Climate Central, 2016). The *Risk Finder* application provides graphical information about the risks of sea level rise. The *Risk Zone Map* provides the same information in a GIS format, displaying changes in inundation extent based on user-selected water levels

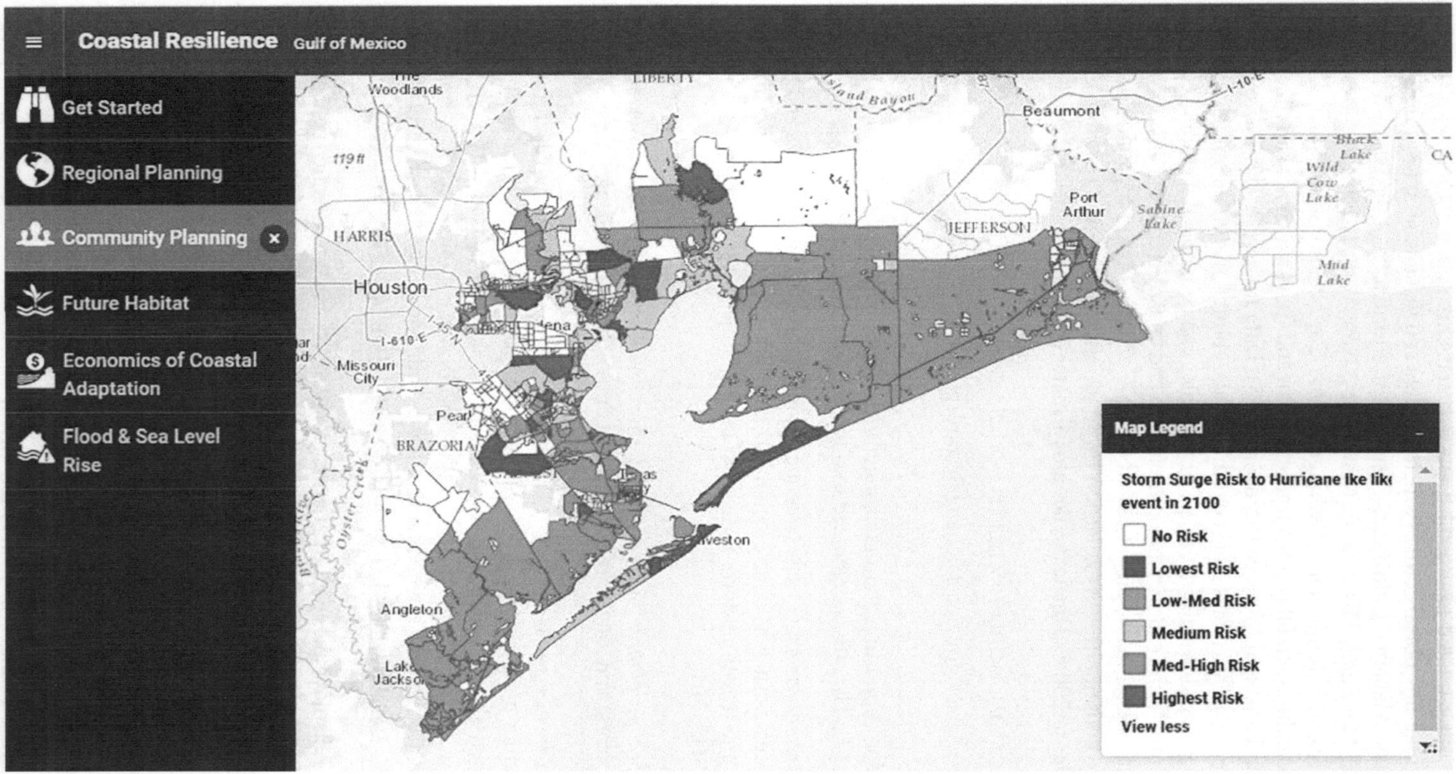

Figure 15.4 A screenshot from the Gulf of Mexico Coastal Resilience mapping portal showing relative risk to communities in Galveston Bay and Jefferson County, Texas, to a storm surge similar to 2008's Hurricane Ike plus 1 meter of SLR (for the year 2100)

Source: © The Nature Conservancy, http://maps.coastalresilience.org/gulfmex/#, used with permission

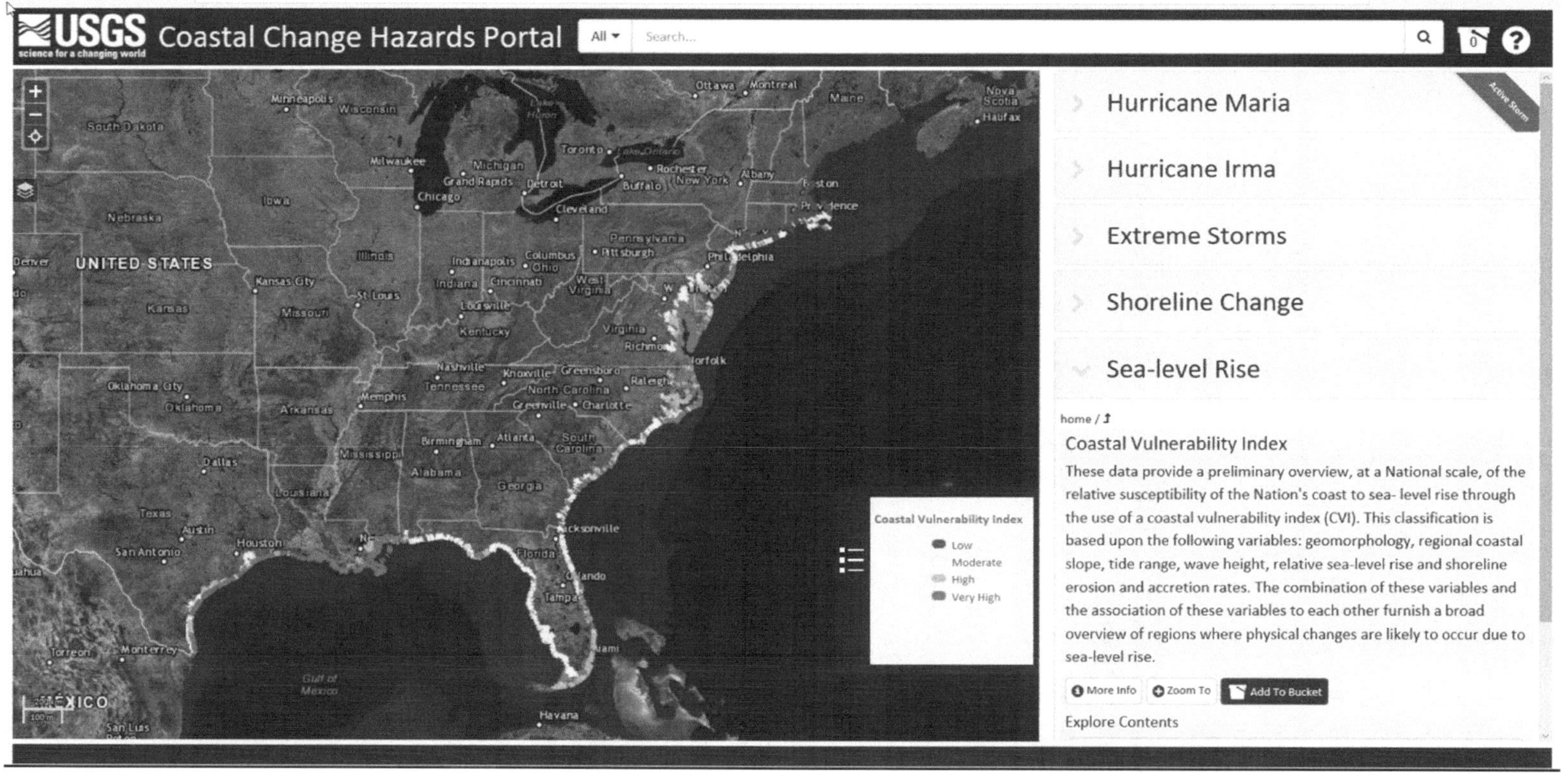

Figure 15.5 USGS Coastal Vulnerability Index available through the USGS Coastal Change Hazards Portal

Source: USGS, https://marine.usgs.gov/coastalchangehazardsportal, used with permission

(Climate Central, 2016). The risk zone mapping is based on updated NOAA LiDAR (Light Detection and Ranging) elevation data. The application also displays population density, social vulnerability, ethnicity, income, poverty, and landmarks. The *Mapping Choices* tool allows for the comparison of the water levels at a particular place given different temperature projections (Climate Central, 2016). The *Seeing Choices* application produces an animated map of sea level change (Climate Central, 2016). The web platform also hosts additional tools to provide decision support for coastal cities, including capabilities to conduct assessments of energy infrastructure at risk for inundation over time and links to other mapping resources (Climate Central, 2016).

Web-Based Tools for Mapping Temperature and Urban Heat Islands

CoolClimate Maps: Average Annual Household Carbon Footprint by Zip Code

http://coolclimate.berkeley.edu/maps

CoolClimate tools provide information for climate action decisions and plans aimed at reducing greenhouse gas emissions (University of California Berkeley CoolClimate Network, 2013). One of the tools is the interactive carbon footprint map. Figure 15.6 displays the interface of the CoolClimate tool and depicts the carbon footprint (metric tons of CO_2) of U.S. households by zip code. By clicking on a point on the map, users can obtain the zip code, the name of the municipality, and the value of the annual average carbon footprint per household. The carbon footprint is based on econometric models of demand for energy, transportation, food, goods,

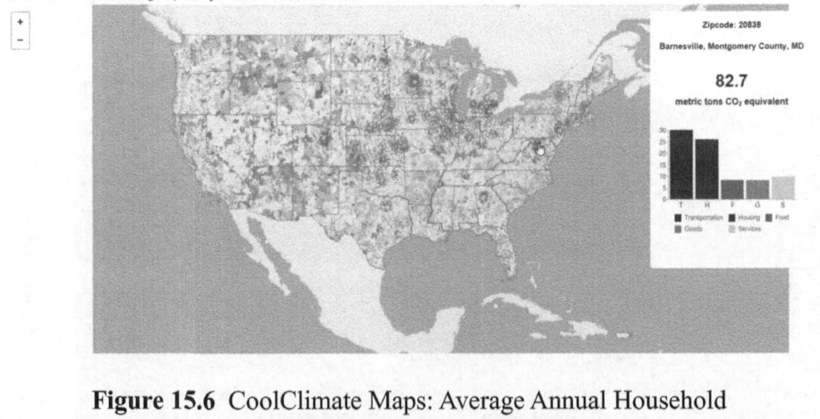

Figure 15.6 CoolClimate Maps: Average Annual Household Carbon Footprint by Zip Code

Source: CoolClimate Network, University of California Berkeley, http://coolclimate.berkeley.edu/maps, used with permission

and services consumed. The baseline data are derived from household surveys (UC Berkeley CoolClimate Network, 2013). Comparative analysis of carbon footprint across municipalities and regions could support the development of local climate action plans.

NASA's Socioeconomic Data and Applications Center (SEDAC)

http://sedac.ciesin.columbia.edu/data/set/sdei-global-uhi-2013

NASA's *Socioeconomic Data and Applications Center* (SEDAC) supports an online platform that grants access to the *Global Urban Heat Island Dataset* (NASA-SEDAC, 2013). The dataset provides a satellite-derived surface temperature on a global scale. The data represent the average summer (40 days, July–August) daytime maximum and nighttime minimum temperatures in degrees Celsius (NASA-SEDAC, 2013). The mapping tool also shows the difference in temperature between urban areas and a 10-kilometer buffer. Urban areas are identified with the use of SEDAC's Global Rural-Urban Mapping Project, Version 1 (GRUMPv1). The surface temperatures were obtained from SEDAC's Global Summer Land Surface Temperature (LST) Grids (NASA-SEDAC, 2013). Figure 15.7 shows the summer daytime urban-rural temperature difference in the eastern part of the United States.

Trust for Public Land (TPL) Urban Heat Risk Explorer App

http://tplgis.org/UrbanHeatRiskApp

TPL's *Urban Heat Risk Explorer App* shows the heat risk in six locations: Chicago, Illinois; Chattanooga, Tennessee; Denver, Colorado; Los Angeles, California; New York, New York; and Tucson, Arizona (TPL, 2014; Law, 2016). The application is developed as a prototype for the 2014 *ESRI Climate Resilience App Challenge* and displays the daytime urban heat island, nighttime urban heat island, impervious surface, urban tree canopy, and cooling centers (Law, 2016).

Urban Planning Tools for Climate Adaptation

Criterion Planning Support Tools

http://crit.com/our-portfolio

Criterion, founded in 1980, focuses on sustainable community development by seeking innovative and resilient solutions to community development challenges. *INDEX*, *Climate Neighbor*, and *RegGIS* are among the planning support tools featured on the Criterion website. INDEX has been widely used for building planning scenarios since 1994 (Allen, 2008). More recently, INDEX has become part of SPARC (Scholarly Publishing and Academic Resources Coalition), which supports open-source cloud-based services and provides access to a wide range of GIS data from multiple collaborating agencies. Participating organizations and other end users can visualize future land use and transportation planning scenarios, evaluate their environmental impact, and rank the outcomes based on a set of sustainability indicators (Criterion Planners, 2014a).

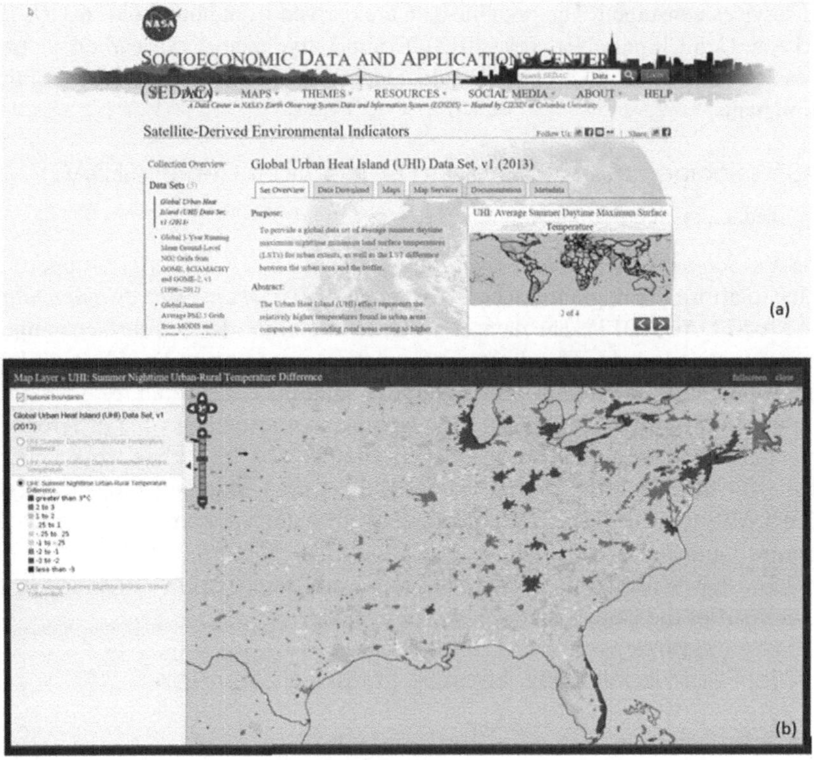

Figure 15.7 Interactive mapping of the urban heat island effect: (a) access to the mapping portal through the SEDAC website; (b) Map Layer UHI: Summer Daytime Urban-Rural Temperature Difference

Source: Center for International Earth Science Information Network (CIESIN), Columbia University. 2016. Global Urban Heat Island (UHI) Data Set, 2013: Summer Daytime Urban-Rural Temperature Difference *[Map]. Palisades, NY: NASA SEDAC, http://dx.doi.org/10.7927/H4H70CRF, used with permission*

Climate Neighbor is a pilot energy efficiency assessment tool implemented in **Chula Vista, California**. The tool, developed in a spreadsheet format, takes into consideration the Leadership in Energy and Environmental Design (LEED) for Neighborhood Development standards and transforms them into estimates of energy savings and CO_2 emissions reduction. The tool is particularly useful for comparing a baseline development scenario to proposed alternatives (Criterion Planners, 2014b). Criterion Planners (2014c) offers a range of other tools, including the *LEED-ND Community Audit Checklist*, technical assistance in evaluating *EcoDistricts*, and *Transformative Tools*, a clearinghouse of global sustainability rating tools. *RegGIS Regional Priority Tool* is an online GIS-based tool to conduct regional assessments and assign LEED-based priority points to projects that address key local concerns (Criterion Planners, 2014d).

UrbanSim

www.urbansim.com/platform

UrbanSim (Waddell, Liu, & Wang, 2008) is an open-source urban simulation and visualization platform that allows users to create scenarios, simulate short (year to year) and long-term (up to 30 years) planning actions, and evaluate the impacts of urban development. The platform allows end users to select different levels of geography, including U.S. metropolitan areas, census blocks, or parcels. According to the website, the *UrbanSim* cloud platform is intended to assist planning agencies and decision-makers in evaluating "the effects of infrastructure and policy choices on community outcomes such as motorized and non-motorized accessibility, housing affordability, greenhouse gas emissions, and the protection of open space and environmentally sensitive habitats" (UrbanSim, www.urbansim.com/platform). UrbanSim seeks to expand its global audience by launching a new cloud-based platform in 2017.

Calthorpe Analytics: UrbanFootprint and RapidFire Tools

http://calthorpeanalytics.com

UrbanFootprint (an open-source web-based platform) and *RapidFire* (a spreadsheet-based tool) are land use, policy, and resource planning tools developed by Calthorpe Analytics in collaboration with partnering institutions (Calthorpe Analytics. 2016). These platforms facilitate scenario development for alternative land use patterns and futures. The tools support evaluation of the impacts of land use decisions on carbon emissions, energy use, water consumption, and other relevant metrics (Calthorpe Analytics, 2016).

GRaBS Adaptation Action Planning Toolkit

www.ppgis.manchester.ac.uk/grabs

The *Green and Blue Space Adaptation for Urban Areas and Eco-Towns* (GRaBS) mapping toolkit provide access to spatial information and analysis for a host of European cities. The tool has been used to conduct climate adaptation assessments of the cities of Amsterdam in the Netherlands; Bratislava, Slovakia; Catania in Sicily, Italy; Genoa, Italy; Kalamaria, Greece; Klaipeda, Lithuania; Malmo, Sweden; and Styria, Austria (University of Manchester School of Environment and Development [SED], 2017). Climate adaptation assessments have been also conducted for northwest England, Southampton, and Sutton using the newer *Assessment Tool* (University of Manchester-SED, 2017). The tool provides access to past climatological data, climate projections, socioeconomic data, and critical infrastructure assets. GRaBS allows end users to examine climate change scenarios and supports analyses of climate adaptation decisions. The *EcoCities Spatial Portal* (www.ppgis.manchester.ac.uk/ecocities) allows for a participatory planning process for climate mitigation planning in Manchester, UK. The tool supports the exploration of climatological data, hazards, infrastructure assets, population growth, and land use change (University of Manchester-SED, 2017).

Tools in Support of Ecosystem-Based Approaches

InVEST (Integrated Valuation of Ecosystem Services and Trade-offs)

www.naturalcaptialproject.org/invest

InVEST, a suite of free, open-source software models, has been widely used to inform natural resource management decisions (Sharp et al., 2015). The models characterize a number of ecological functions and predict changes based on alterations of baseline conditions (Sharp et al., 2015). The software package enables the analysis of a range of ecosystem services from habitat quality to climate regulation, water purification, and aquaculture (Sharp et al., 2015). The assessment tools have a variety of ecosystem restoration planning applications.

NatureServe Vista

www.natureserve.org/conservation-tools/natureserve-vista

NatureServe Vista is a free ArcMap extension for landscape and seascape scenario-based assessments (NatureServe, 2013). The software supports modeling of ecological processes and can be integrated with other tools for ecosystem services valuation, suitability analysis, and decision support (NatureServe, 2013). The software generates reports and maps with broad applicability in resource management, energy and green infrastructure planning, and climate change adaptation (NatureServe, 2013).

SLAMM (Sea Level Affecting Marshes Model)

https://toolkit.climate.gov/tool/sea-level-affecting-marshes-model-slamm

SLAMM, first developed in the mid-1980s, is open-source software used to simulate potential impacts of long-term sea level rise on wetlands and natural shorelines. According to the website, the tool supports the computation of relative sea level rise for time sequences of 5 to 25 years (United States Climate Resilience Toolkit, 2017). The tool enables the assessment of various wetland inundation scenarios. The results can be summarized in tabular and graphical formats for display and further analysis with GIS (United States Climate Resilience Toolkit, 2017).

i-Tree

www.itreetools.org/about.php

i-Tree is a free software package for urban and rural forestry analysis developed by the U.S. Forest Service (i-Tree Streets User's Manual v5.x., n.d.; Willis & Petrokofsky, 2017). The tools available through the software package provide estimates of tree coverage and tree canopy and support the analysis of related ecosystem services including water quality, shading, and temperature moderation (i-Tree Streets User's Manual v5.x., n.d.; Willis & Petrokofsky, 2017).

Coastal Resilience

http://maps.coastalresilience.org/seflorida

In 2016, The Nature Conservancy, Florida Atlantic University, and the Southeast Florida Regional Climate Change Compact's Shoreline Resilience Working Group collaborated to develop an analytical framework to assess the suitability and feasibility of nature-based shoreline stabilization options to protect the shores and

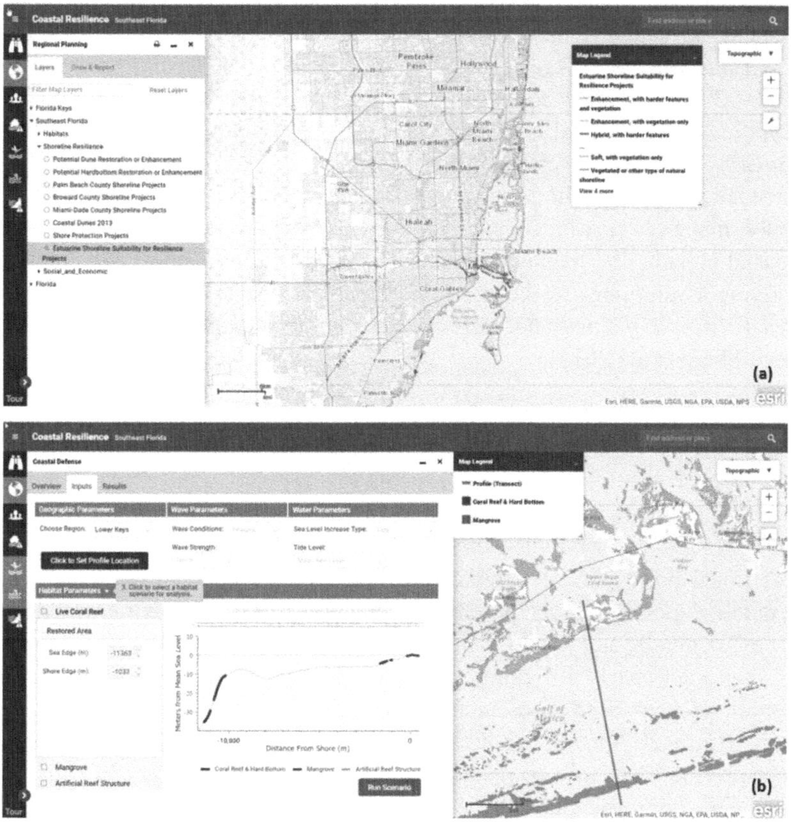

Figure 15.8 Tools available through the Coastal Resilience.org website: (a) suitability mapping for the implementation of living shorelines in Southeast Florida; (b) applying the award-winning Coastal Defense App to quantify the protection afforded by coral reefs and mangroves in the Lower Keys, Florida

Source: © The Nature Conservancy, the Natural Capital Project, the University of California at Santa Cruz, University of Southern Mississippi, NOAA, and the United States Geological Survey, http://maps.coastalresilience.org/seflorida/#, used with permission

respond to coastal hazards in Southeast Florida. The pilot study, titled *Suitability Analysis for Living Shorelines Development in Southeast Florida's Estuarine Systems*, seeks to assess the feasibility of various erosion control and flood abatement projects using a set of criteria such as wave/wake environment, water depth, existing shoreline condition, and presence/absence of habitat (Mitsova et al., 2018). Weights for the decision parameters are derived from expert judgment and local institutional knowledge. The weighted parameters are mathematically aggregated to identify locations suitable for living shorelines (Mitsova et al., 2018). Despite Florida's history of disturbance, the results reveal multiple opportunities to enhance shoreline protection using soft and hybrid techniques. Online maps of the project results are hosted on The Nature Conservancy's CoastalResilience.org website (Figure 15.8a).

The *Coastal Defense App* (Figure 15.8b) launched by *Coastal Resilience.org* in February 2015 is another collaborative project developed by The Nature Conservancy, the Natural Capital Project, NOAA, USGS, the University of California at Santa Cruz, and the University of Southern Mississippi (Ferdaña, 2015). The app is intended to quantify the protection afforded by coastal habitats including tidal marshes, mangroves, seagrasses, beach dunes, and coral reefs by reducing the risk of coastal erosion and inundation under the storm and non-storm conditions. The underlying assumptions of the one-dimensional model are derived from the *Integrated Valuation of Environmental Services and Trade-offs* of the Natural Capital Project (Sharp et al., 2015).

Guides to Selecting the Appropriate Tool

Several organizations including the Lincoln Institute of Land Policy (2009), ESRI (2015), NatureServe (Rozum & Carr, 2013), the Trust for Public Land (Ernst & Blaha, 2015), the EBM Tools Network (Carr, 2014), and the International Council for Local Environmental Initiatives (ICLEI) have devoted resources to prepare detailed guides to assist practitioners in navigating through the web of tools, "how to" manuals, and data portals. A survey conducted by the EBM Tools Network identified nearly 130 applications and written methodologies designed to carry out science-based assessments of climate change impacts on marine ecosystems (Carr, 2014). Among the most widely used tools cited in these publications are the TNC's *Coastal Resilience* (www.coastalresilience.org), NOAA's *Digital Coast* and the Coastal Services Center *Coastal Inundation Toolkit*, the *Climate Wizard* (another tool developed by TNC), and the *Climate-Smart Cities Decision Support Tool* (DST) (sponsored by the Trust for Public Land) (Carr, 2014; Ernst & Blaha, 2015).

The *Sea Level Rise and Coastal Flood Web Tools Comparison Matrix* (Climate Central, 2015) provides a detailed overview of dozens of tools with national, regional, and local coverage and scope. The matrix, developed collaboratively by Climate Central, The Nature Conservancy, and NOAA's Office for Coastal Management, is available on the Climate Central website. Potential users seeking a web-based decision support tool can refer to the matrix to gain an insight into the analytical capabilities and required skill level. The matrix consists of six primary searchable databases including introductory information, an overview of tools that can facilitate sea level rise and flood risk assessments under various scenarios, comparisons of shoreline analysis tools, technical information, and other miscellaneous details (Rizza, Rozum & Ferdaña, 2015). The matrix also provides information on each tool's geographic scope, advantages, limitations, availability of data for download, primary outputs, target audiences, links, and points of contact (Climate Central, 2015; Rizza, Rozum & Ferdaña, 2015).

References

Allen, E. 2008. Clicking Toward Better Outcomes: Experience with INDEX, 1994–2006. In: Brail, R.K. (Ed.) *Planning Support Systems for Cities and Regions*, 139–166, Cambridge, MA: Lincoln Institute of Land Policy.

Barnett, J., Waters, E., Pendercast, S., & Puleston, A. 2013. Barriers to Adaptation to Sea Level Rise: The Legal, Institutional and Cultural Barriers to Adaptation to Sea Level Rise in Australia. *NCCARF Publication 35/13*, National Climate Change Adaptation Research Facility, Gold Coast.

Calthorpe Analytics. 2016. *Urban Footprint* and *RapidFire Tools*. Retrieved from http://calthorpeanalytics.com/.

Carr, S. 2014. Results Announced from Global Survey of Tools and Resources for Addressing Climate Change Impacts on Marine Ecosystems, a Report for OpenChannels—a Forum for Ocean Planning and Management, available at www.openchannels.org/ (accessed 07/08/2017).

Carriere, L. 2015. Advancing Research and Applications with NASA Climate Model Data, NASA Center for Climate Simulation (NCCS). NASA Goddard Space Flight Center, Retrieved from www.nas.nasa.gov/SC13/demos/demo20.html.

Center for International Earth Science Information Network (CIESIN), Columbia University. 2016. *Global Urban Heat Island (UHI) Data Set, 2013*. Palisades, NY: NASA Socioeconomic Data and Applications Center (SEDAC). Retrieved from http://dx.doi.org/10.7927/H4H70CRF.

Clark Labs. 2015. *IDRISI TerrSet—Climate Change Adaptation Modeler (CCAM)*. Worcester, MA: Clark University. Retrieved from https://clarklabs.org/terrset/climate-change-adaptation-modeler/.

Climate Central. 2015. *Sea Level Rise and Coastal Flood Web Tools Comparison Matrix—National*. The Nature Conservancy, NOAA's Office for Coastal Management, Climate Central. Retrieved from http://sealevel.climatecentral.org/matrix/national.html?v=1).

Climate Central. 2016. *Surging Seas—Climate Central*. Retrieved from http://sealevel.climatecentral.org/.

Condon, P.M., Cavens, D., & Miller, N. 2009. *Urban Planning Tools for Climate Change Mitigation*. Policy Focus Report. Cambridge, MA: Lincoln Institute of Land Policy.

Criterion Planners. 2014a. *SPARC with INDEX*. Retrieved from http://crit.com/portfolio/sparc-with-index/.

Criterion Planners. 2014b. *Climate Neighbor*. Retrieved from http://crit.com/portfolio/climate-neighbor/.

Criterion Planners. 2014c. *Sustainability Assessment*. Retrieved from http://crit.com/our-portfolio/sustainability-assessment/.

Criterion Planners. 2014d. *RegGIS Regional Priority Tool*. Retrieved from http://crit.com/portfolio/reggis-regional-priority-tool/.

Cutter, S.L., Boruf, B.J., & Shirley, W.L. 2003. Social Vulnerability to Environmental Hazards. *Social Science Quarterly* 84(2), 242–261.

Environmental Systems Research Institute (ESRI). 2015. *Mapping and Modeling Weather and Climate with GIS*. Redlands, CA: ESRI Press.

Ernst, C., & Blaha, K. 2015. *Decision Support Tools for Climate Change Planning*. Produced under a grant from the Macarthur Foundation for The Trust For Public Land's Climate-Smart Cities program. Retrieved from www.tpl.org/sites/default/files/files_upload/ClimateSmart%20DecisionSupport.pdf (accessed 07/02/17).

Esnard, A-M. 2012. Visualizing Information. In: Weber, R., & Crane R. (Eds.) *Handbook of Urban Planning*. New York: Oxford University Press Inc.

Ferdaña, Z. 2015. Coastal App Wins Global Disaster Resilience App Challenge, The Nature Conservancy. *Silver Jackets Newsletter*. Retrieved from https://silverjackets.nfrmp.us/Resources/Silver-Jackets-Newsletter/The-Buzz-January-2015/Coastal-Resilience-Wins-Challenge.

Grant, B., Baldwin, C., Lieske, S.N., & Martin, K. 2015. Using Participatory Visual Methods for Information Exchange About Climate Risk in Canal Estate Communities. *Australian Journal of Maritime & Ocean Affairs* 7(1), 23–37. DOI: 10.1080/18366503.2015.1014012.

i-Tree Streets User's Manual v5.x. (n.d.). Retrieved from www.itreetools.org/resources/man . . . ual_v5.pdf (accessed 09/01/2017).

Law, D. 2016. Urban Heat Risk Explorer App. *ArcGIS Online*. Retrieved from www.arcgis.com/home/item.html?id=b61f07e61cd340e1a4bbc58c953de1e1.

MacEachren, A.M., & Brewer, I. 2004. Developing a Conceptual Framework for Visually-Enabled Geo-Collaboration. *International Journal of Geographical Information Science* 18, 1–34.

Meinshausen, M., Smith, S.J., Calvin, K., Daniel, J.S., Kainuma, M.L.T., Lamarque, J-F., Matsumoto, K., Montzka, S.A., Raper, S.C.B., Riahi, K., Thomson, A., Velders, G.J.M., & van Vuuren, D.P.P. 2011. The RCP Greenhouse Gas Concentrations and Their Extensions from 1765 to 2300. *Climatic Change* 109, 213–241.

Mitsova, D., Bergh, C., Guannel, G., Lustic, C., Renda, M., Byrne, J., Graves, A., Reed, S., Alhawiti, R., Cresswell, K., & Goldberg, A. 2018. Spatial Decision Analysis of Nature-Based Shoreline Stabilization Options in South Florida's Estuarine Environments. *Journal of Environmental Planning and Management*. DOI: 10.1080/09640568.2017.1398637

National Aeronautics and Space Administration (NASA). 2017. *NASA Earth Exchange (NEX) Downscaled Climate Projections (NEX-DCP30)*. https://cds.nccs.nasa.gov/nex/.

National Aeronautics and Space Administration Socioeconomic Data and Applications Center (NASA-SEDAC). 2013. Retrieved from http://sedac.ciesin.columbia.edu/data/set/sdei-global-uhi-2013.

National Center for Atmospheric Research (NCAR). 2017a. *Climate Data Guide*. Retrieved from https://climatedataguide.ucar.edu/climate-data.

National Center for Atmospheric Research (NCAR). 2017b. *GIS Program*. Retrieved from https://gis.ucar.edu/.

National Oceanic and Atmospheric Administration National Centers for Environmental Information (NOAA-NCEI). 2016. *NOAA's Weather and Climate Toolkit (WCT)*. Retrieved from www.ncdc.noaa.gov/wct/.

National Oceanic and Atmospheric Administration Office for Coastal Management (NOAA-OCM). 2017. NOAA Sea Level Rise Viewer. *Digital Coast*. Retrieved from https://coast.noaa.gov/digitalcoast/tools/slr.

National Oceanic and Atmospheric Administration National Weather Service Climate Prediction Center (NOAA-NWS-CPC). 2015. *Three-Month Outlooks—Official Forecasts*. Retrieved from www.cpc.ncep.noaa.gov/products/predictions/long_range/seasonal.php?lead=1.

The Nature Conservancy (TNC). 2009. *Climate Wizard*. Retrieved from www.climatewizard.org/.

The Nature Conservancy (TNC). 2016. *Coastal Resilience*. Retrieved from http://coastalresilience.org/.

NatureServe. 2013. *NatureServe Vista 3.0*. URL: www.natureserve.org/conservation-tools/natureserve-vista.

Nöllenburg, M. 2007. Geographic Visualization. In: Kerren A., Ebert A., & Meyer J. (Eds.). *Human-Centered Visualization Environments*. Dagstuhl Castle, Germany: GI-Dagstuhl Research Seminar, LNCS 4417, Springer-Verlag.

Rizza, D., Rozum, J., & Ferdaña, Z. 2015. *Online Matrix Guides Planners to Available Sea Level Rise and Coastal Flood Risk Web Tools*. Retrieved from http://blogs.planning.org/postdisaster/2015/06/23/online-matrix-guides-planners-to-available-sea-level-rise-and-coastal-flood-risk-web-tools/.

Rozum, J.S., & Carr, S.D. 2013. *Tools for Coastal Climate Adaptation Planning: A Guide for Selecting Tools to Assist with Ecosystem-Based Climate Planning*. Coastal-Marine Ecosystem-Based Management Tools Network, NatureServe, Arlington, VA.

Schneider, D.P., Deser, C., Fasullo, J., & Trenberth, K.E. 2013. Climate Data Guide Spurs Discovery and Understanding. *Eos Transactions AGU* 94(13), 121–122.

Shannon, C.E. 1948. A Mathematical Theory of Communication. *The Bell System Technical Journal* 27, 379–423.

Sharp, R., Tallis, H.T., Ricketts, T., Guerry, A.D., Wood, S.A., Chaplin-Kramer, R., Nelson, E., Ennaanay, D., Wolny, S., Olwero, N., Vigerstol, K., Pennington, D., Mendoza, G., Aukema, J., Foster, J., Forrest, J., Cameron, D., Arkema, K., Lonsdorf, E., Kennedy, C., Verutes, G., Kim, C.K., Guannel, G., Papenfus, M., Toft, J., Marsik, M., Bernhardt, J., Griffin, R., Glowinski, K., Chaumont, N., Perelman, A., Lacayo, M., Mandle, L., Hamel, P., Vogl, A.L., Rogers, L., & Bierbower, W. 2015. *InVEST +VERSION+ User's Guide*. The Natural Capital Project, Stanford University, University of Minnesota, The Nature Conservancy, and World Wildlife Fund. Retrieved from http://data.naturalcapitalproject.org/nightly-build/release_default/rele . . . (accessed 10/04/17).

Sheppard, S.R.J. 2012. *Visualizing Climate Change: A Guide to Visual Communication of Climate Change and Developing Local Solutions*. New York: Routledge.

Taylor, K.E., Stouffer, R.J., & Meehl, G.A. 2012. An Overview of CMIP5 and the Experiment Design. Bull. *American Meteor Society* 93, 485–498.

Torresan, S., Critto, A., Rizzi, J., & Marcomini, A. 2012. Assessment of Coastal Vulnerability to Climate Change Hazards at the Regional Scale: The Case Study of the North Adriatic Sea. *Natural Hazards and Earth System Sciences* 12, 2347–2368.

Trust for Public Land (TPL). 2014. *Urban Heat Risk Exposure*. Retrieved from http://tplgis.org/UrbanHeatRiskApp/.

United Nations International Strategy for Disaster Risk Reduction (UNISDR). 2014. *UN Recognition for Disaster Resilience Apps*. Retrieved from www.unisdr.org/archive/39500.

United States Climate Resilience Toolkit. 2017. *SLAMM (Sea Level Affecting Marshes Model)*. Retrieved from https://toolkit.climate.gov/tool/sea-level-affecting-marshes-model-slamm.

United States Geological Survey (USGS). 2016. *National Climate Change Viewer (NCCV) Home*. Retrieved from www2.usgs.gov/climate_landuse/clu_rd/nccv.asp.

United States Geological Survey (USGS). 2017. *USGS Coastal Change Hazards Portal, v.1.1.58*. Retrieved from https://marine.usgs.gov/coastalchangehazards portal/.

University of California Berkeley (UC Berkeley) CoolClimate Network. 2013. *CoolClimate Maps: Average Annual Household Carbon Footprint by Zip Code*. Retrieved from http://coolclimate.berkeley.edu/maps.

University of Manchester School of Environment and Development (SED). 2017. *GRaBS Adaptation Action Planning Toolkit*. Retrieved from www.ppgis.manchester.ac.uk/grabs/.

UrbanSim. 2017. Available at www.urbansim.com/platform.

Waddell, P., Liu, X., & Wang, L. 2008. UrbanSim: An Evolving Planning Support System for Evolving Communities. In: Brail, R.K. (Ed.) *Planning Support Systems for Cities and Regions*, 103–138. Cambridge, MA: Lincoln Institute of Land Policy.

Willis, K.J. & Petrokofsky, G. 2017. The Natural Capital of City Trees. *Science* 356(6336), 374–376.

World Resources Institute (WRI). 2009. *Climate Analysis Indicators Tool (CAIT)*, version 7.0. Washington, D.C.: World Resources Institute. Retrieved from http://cait.wri.org.

Index

Note: Page numbers in **bold** indicate tables and page numbers in *italic* indicate figures and boxes on the corresponding page.

A

B

C

D

E

F

G

H

N

V

W

Y

Z